The Ten Dimensions of Kabb

A Scientist's Guide to Entering the Invisi

The Ten Dimensions of Kabbalistic Biology: A Scientist's Guide to Entering the Invisible Worlds of Life

Copyright © 2025 by Michael Kosoy

Library of Congress Cataloging-in-Publication Data

The Ten Dimensions of Kabbalistic Biology / by Michael Kosoy

Illustrations by Jenia Kosoy

978-1-965117-16-3 Paperback

978-1-965117-18-7 Hardcover

Revised edition; originally published by KTAV Publishing House in 2024.

1. New Science. 2. Biology. 3. Kabbalah.

KB One Health LLC, Fort Collins, Colorado

www.kbonehealth.com

kosoymichael@gmail.com

Printed in the USA

New Edition

The Ten Dimensions of Kabbalistic Biology

A Scientist's Guide to Entering the
Invisible Worlds of Life

Michael Kosoy, *Ph.D.*

Illustrations by Jenia Kosoy

KB One Health, Fort Collins, Colorado

To Jenia:

The book wouldn't be the same without you.

Contents

The List of Illustrations:

1. Knowledge interaction models
2. How a 2D-subject sees a 3D cube
3. The carrot's journey through the Flatland
4. A carrot and a beet travel through the Flatland
5. Two carrots travel together
6. Pheasants as a new dimension for foxes
7. Life at the edge of order and chaos
8 - Butterfly effect
9. Punctuated evolution of horses
10. The Cosmic Accordion
11. A half-full glass
12. Is zebra vertically striped?
13. The DIKW pyramid
14. The streetlight effect
15. Inhabitants of the DIKW worlds
16. Perceived levels in the description of diseases
17. The Petri dish
18. A dot-to-dot rabbit
19. Terra Incognita
20. Dogs, not "dog-ness"
21. Duck or rabbit
22. Ecological successions
23. Junk DNA
24. Spider net
25. Network between viruses and animals
26. Funnels
27. Crystal structure
28. Circular model
29. Two or three pillars
30. Three triads
31. Five *Partzufim*

Introduction
Welcome! (and a Caution): This Book is not for Everyone

If you picked up this book, you were most likely intrigued by the title. After all, the words "Kabbalah" and "Biology" are rarely seen together in a book aimed at serious readers. Consider this a first warning! This book will *not* reference religion, mysticism, spirituality, rituals, or the New Age movement. These topics are acceptable in other books, but they are not the focus of this one. However, before I explain why I have used the word "kabbalistic" in the title of this book, I would like to issue a few more warnings.

This book is indeed about biology as the "Science of Life." However, it is unlikely that all readers will apply the same definitions to the words "Science" and "Life." Therefore, Part One will clarify exactly how we interpret the meaning of these words.

As a biologist, it has been my lifelong passion to better understand the world of Nature and living things. From early childhood, I was drawn to the study of life on this planet. On my 13th birthday, I received a book titled *The Naturalist on the River Amazon,* written by the British naturalist Henry Bates. I was captivated by the author's adventures, including illustrations of animals I had never seen before and insights into the lives of the native Brazilians and wild animals. The more I read, the more compelled I became to learn all I could about wildlife in exotic places. Later, as a student, I was intrigued by the ability of life forms to change, survive, and thrive under a wide variety of conditions.

Today, I look back on 45-plus years of active biological research and recognize that, despite occasional "breakthroughs" in science, we have yet to truly access the mysteries of biological life. As our existence—and that of our planet--is now more vulnerable than ever, we are facing a vast information gap that must be filled. This can only be done by learning to receive information from previously untapped sources. Even more important, we must be ready to convert this information into knowledge that can make a difference in specific situations.

For 25 years, I worked as a research biologist at the Centers for Disease Control and Prevention, investigating infectious diseases that originate in nature and are transmitted to human beings. I have written around two hundred scientific articles, which earned many thousands of citations in scientific literature. The discoveries about diseases in nature have enabled me to work with subjects ranging from viruses to elephants and at levels of research from gene sequencing to ecosystem analysis.

In the course of my explorations, I have participated in many international collaborations in more than 30 countries in both Americas, Asia, Africa, and Europe. I am listing those activities to confirm that I am a biologist who has not just absorbed information from biology books and articles. I have also actively participated in biological research under various conditions, including working with colleagues in America and abroad and leading scientific projects and programs.

While this book is indeed about biology, it is time for a third warning. The contents of this book are unlike any biology course my colleagues and I have taken or taught, whether as neophyte scientists or academically advanced ones. The reason? Because such a course does not yet exist! The primary reason I decided to write this book is to make the knowledge within it available to those who seek to pursue unseen and unknown truths.

We professional biologists have accumulated mountains of data about biological organisms—and some proposed scientific concepts can help us comprehend biological systems. However, the point I make in Part Two is that we cannot restrict our knowledge of biological life and its manifestations to "visible" data. We must also include the "invisible" reality that forms more than 90% of any biological object or process! Unfortunately, most biologists have not learned how to investigate the invisible aspects. In all fairness, the reason is simple: no one else has learned to do it, either. And without a teacher, there are no students.

The drama surrounding the COVID-19 pandemic highlighted one important reality: *the biological world not only surrounds us, but it also includes and forms us.* It was also clear that the present level of scientific information could not be transformed into enough knowledge to reorient an unprepared world facing the epidemic.

The pandemic was not the first indication of an impending health disaster. There were others, including the accelerated appearance of other emergent infectious diseases, confusion in discriminating between health and disease, escapes of genetically modified microorganisms from laboratories, rapidly diminished biodiversity across the globe, and many other ecological disasters. All of the above have shaken the foundations of biological science.

As modern science has developed, particularly in the sphere of molecular biology, an unprecedented amount of information about biological organisms, their structure, and their functionality has also emerged. Unexpectedly, the massive increase in biological information has been accompanied by increasing difficulty for observers to understand biological systems above the molecular level. The situation is quite paradoxical, as scientists expected that new information would lead to a better understanding and more accurate predictions of biological systems.

The thesis that "information is *not* knowledge" will be one of the main revelations of this book. Unfortunately, the real challenge lies in our ability to think outside the box rather than just accumulate information.

I have long pondered the question: Can we acquire biological knowledge without introducing a kabbalistic perspective? For several objective reasons, I now believe it cannot. In 1994, I began my career in research at the Center for Disease Control and simultaneously enrolled in the first Kabbalah school in Atlanta. Very quickly, I recognized that Kabbalah held opportunities to guide me in defining and resolving some biological theories and practice problems. Both areas of exploration loomed over researchers during my professional history as a scientist.

Over many centuries, the tradition rooted in the old books of Kabbalah developed ways for us to perceive intangible worlds. I discovered that access to kabbalistic teachings that had remained secret for millennia was becoming available during a crisis in scientific thinking. Welcome to the new world of quantum theory, quickly proclaimed a "new science." Unfortunately, the tectonic changes that occurred in physics were not accompanied by a handy switch by which we could receive new information about biological worlds.

In Part Three, I will introduce selected directions developed in Kabbalah that have given us information that was hidden--either intentionally or unintentionally--beneath many layers of deviations from the truth. The idea of invoking Kabbalah as a new means of knowledge remains very challenging. It requires us to conjoin the integrity and professionalism of biological science with the authenticity of the kabbalistic tradition. One must also be aware of the obvious danger of an inaccurate presentation of kabbalistic treasures. My Kabbalah teacher, Samuel Avital, often reminded me often that it is imperative not to utter a word until you are confident that the word is 1) true, 2) necessary, and 3) kind. A written word may present a risk of unforeseen consequences!

I consider Part Four the most important section of this book. There, I will share concrete principles from the kabbalistic tradition intended to guide readers to the hidden worlds of biological life. These additional orientation points can be used under critical circumstances—situations that become increasingly obvious with the development of theoretical and applied biology.

Those who have questions must determine for themselves the ultimate value of new potential tools for understanding. Some studies of biological worlds may not require conceptual input from Kabbalah. I do not discriminate against any existing concept in biology, nor do I harbor doubts about any facts obtained during biological investigations. Even if the reader is not a professional biologist, the following reality affects us all: every one of us lives in a complex biological world, and every one of us contains within us the *entire* biological world.

Now, if all my warnings have not scared you off, I welcome you to join me and other readers on this journey to a new awareness of Life, and its emerging realities.

PART ONE:
Walking in Two Worlds

Chapter 1

Science of Life and Knowledge of Life

Biological life is so diverse and complex that we can potentially learn endless information about any animal, plant, or bacteria we select. To date, however, our ability to access that information has been limited to studies endorsed by established academic standards. We can also make personal choices when selecting research methods depending on the researcher's priorities.

My intention, and the purpose of this book, is to suggest that there is a new potential source of knowledge that goes beyond the usual resources used by biological researchers. After many years in academic and public health agencies, I am aware that most of my colleagues have felt constrained in their explorations by two forces: doctrinaire limitations and inflexible ideas. The point of this book is to offer readers an intriguing new approach that can help them unlock the secrets of biological life.

Biology is the Science of Life

The word "biology" is derived from the words; *bios* ("life" in Greek) and *logos* ("study" or "science"). Yet, ironically, the definition of Life as a *biological* phenomenon is rarely discussed by professional biologists, whether working outdoors in natural conditions or indoors in laboratories. In fact, "What is Life?" is a question more frequently posed by philosophers, psychologists, and poets than biologists. One of the exceptions is a charming book written by Lynn Margulis and Dorion Sagan, titled *What is Life?*[1] It describes thoughts on such fascinating topics as the dynamics of the bacterial realm, the connection between sex and death, and theories of the origin of life.

A more widely read book with the same title was written in 1944 by physicist Erwin Schrödinger, titled *What Is Life? The Physical Aspect of the Living Cell.*[2] However, while Schrodinger's book had a significant impact among science lovers, it was written from a physicist's viewpoint and did not delve into theories about the essence of biological life.

Such disregard brings us to a question that may challenge readers of this book: "Should an attempt to unlock the mysteries of biological life be limited to the existing methodologies of modern science?" To answer this, we must first define the word "science." Commonly, "science" is defined as "the process of getting knowledge" from the Latin *scientia* (knowledge). In Hebrew, the word *mada,* meaning "science," shares the same root as *da'at* ("knowledge"). In fact, any meaning related to "science" refers to *knowledge that can be taught and learned.* At this point, however, we must admit that there are limits to the tools modern science can offer to help us explore the complexities of biological life.

Man created these limits; therefore, they can be removed by the same source. Biologists need to familiarize themselves with tectonic changes in research methodologies due to quantum physics, string theory, catastrophe theory, the science of complexity, and biosemiotics. So far, the impact of such advances has been limited in terms of biological research.

Scientists tend to be attracted to the latest scientific advances. Many who enter the profession of biology these days are less motivated to solve "the mysteries of life." Instead, they are more eager

---

[1] Lynn Margulis and Dorion Sagan, *What Is Life?* 2000, 288 pages.
[2] Erwin Schrodinger, What Is Life? The Physical Aspect of The Living Cell. 2001.

to learn new methodologies offered by molecular biology. These include detecting nucleic acid fragments, constructing chimeric genetic forms, creating genome sequencing, and other sophisticated techniques. While these new methodologies are intriguing, the science of biology is *far* more extensive than the study of molecules!

Chemistry is the branch of science that deals with molecules, other elements, and compounds. The branch of chemistry that focuses on studying compounds of living organisms is called "biochemistry." Therefore, when people discuss the molecular composition of a specific viral antigen or a specific fragment of the gene, they are referring to the "*chemistry* of life," not "the *biology* of life."

Organic molecules alone cannot explain the main properties of biological life -- order, reproduction, adaptation, growth and development, regulation, homeostasis, and sensitivity to environmental signals.

The Birth of Modern Science

Back in the 17th and 18th centuries, during the times of Galileo and Newton, a new period of science emerged. Known today as "modern science," it brought with it a new requirement: the need for experimental investigation and proof *before* one could claim truth and legitimacy to support scientific data and theories.

The culmination of the need to "prove" scientific knowledge arrived in the 20th century in the form of the "Falsification Principle" proposed by Karl Popper, an Austrian-British scientific philosopher. According to Popper's principle, a theory is not "scientific" until it is tested and proven either true or false.[3] The classic example related to biology is as follows: The theory that "all swans are white" will be considered "false" as soon as someone observes a *black* swan.

Another recent trend in modern science gives priority to "Hypothesis-Driven Research." Thus, researchers must answer

---

[3] Karl Popper, The Logic of Scientific Discovery. 1968, 48 pages.

specific, measurable, and answerable questions because a well-constructed hypothesis must be clear and "testable." As a result, many scientific biological journals only accept papers that present either proven or rejected results that are based on included experiments.

Let me be clear: I fully understand the importance of such qualifications for scientific acceptability, and most of my research projects have honored this requirement. However, if these principles are considered the *exclusive* source of assessing and obtaining scientific knowledge, we will eventually find ourselves seriously blocked when seeking important answers! Although direct observations still play a vital role in scientific progress, these days, the support for "descriptive research" is far less popular than that for hypothesis-driven research." One of the arguments for giving less importance to "descriptive research" is the difficulty in proving—or even repeating--such observations.

The Limits of Objective Knowledge

Modern science frequently declares that its goal is "to gain objective knowledge." This process requires a well-trained scientist to repeat an experiment on a specific natural object and hope to receive an observation similar or comparable to the original. The results of biological experiments can vary significantly within an established range due to the inherent "fuzziness" of biological life. Therefore, biological experiments must be repeated to establish the sample size and range of the expected variations.

The main reason experts insist on strict designs and accurate protocols goes back to the point that researchers must obtain and report results that can be replicated. What factors might cause different results? Perhaps the objects (animals, plants, microbes) tested were different. Perhaps the conditions under which the experiment was performed were different. Because of the stringent requirements involved, investigators are under pressure to standardize the conditions of their experiments as much as possible. For example, they might use animals of a genetically pure lineage for an experiment or create a cell culture that will grow bacteria under very specific conditions. When studying infections, which is one of the areas I have been involved in, I am aware that I must use "specific-pathogen-free" experimental animals. This is intended to minimize the influence of other agents of infection.

I agree that such requirements are fundamental--they define the quality of any biological laboratory. When I supervised a CDC laboratory, our adherence to such observations enabled us to deliver reliable data and valuable information. Even so, these observations alone do not reflect the diversity of biological life. Experimental conditions strictly controlled by a uniform design are rarely, if ever, observed in Nature! Therefore, such an experimental design can only partially reflect the environmental context of a natural biological system. Such a limitation is potentially crippling during this epoch of accelerating changes in environmental conditions.

It is clearly an advantage to have straightforward answers to well-formulated questions. But realistically, many aspects of inquiry are difficult to define. For example, contemporary experts are being challenged to define the word "health." Various organizations, from local public health agencies up to the federal level of the CDC and the International level of WHO, are now trying to rephrase their missions by using words like "health" or "well-being" instead of "disease." How, then, can we prove or disprove a statement when we have no specific hypotheses we can test?

I have also noticed that today's scientists tend to resist the question, "Why?" They feel much more comfortable working with questions like "How? Where? When?" The question, "Why" has now been relegated to the realm of philosophers, essayists, and metaphysicians. Philosophy has developed a sound training system to define an issue, test and find related words, and then use the arts of reasoning and argument to discuss the problems. Unfortunately, most academic programs in life sciences include little or no background in philosophical training. Indeed, science is not designed to have an answer for everything! Thus, it is unrealistic to assume that science is a "one-stop-shop" for solutions to every life issue. A far more genuine goal for scientists is to select and prioritize questions that *can* be answered.

Today, science is facing a new crisis: an increasing lack of public's faith in the ability of scientists to predict timing and nature of catastrophes, whether natural or social in origin. This attitude can jeopardize future advances aimed at delaying or even preventing the probable destruction of this planet's environment. British evolutionary biologist and educator Julian Huxley wrote that "Without the impersonal guidance and the efficient control provided by science, civilization will either stagnate or collapse, and human nature cannot make progress towards realizing its possible evolutionary destiny."[4]

Experience-based Knowledge

We must be aware of contradictions between science and other sources of knowledge, including our own personal and collective experiences. Consider the recent COVID-19 pandemic. In its early days, the public was urged to follow the CDC's recommendation for social isolation. Eventually, however, the decision to do so was supported by solid scientific evidence. It also involved individual choices people made for their unique situations, including personal experiences and plain common sense. During our lifetimes, we learn many things, but most of this information can hardly be classified as "scientific."

I found support for such a view in a book by Gary Zukav, in which he quoted his mentor, David Finkelstein, Director of the School of Physics at the Georgia Institute of Technology. Finkelstein stated: "Logos imitates but can never replace experience. It is a substitute for experience… The concepts initially formed by abstraction from a particular situation or experiential complexes acquire a life of their own… We have to live with the experience." When David Finkelstein

---

[4] Julian Huxley, *What Dare I Think?* 2011, 278 pages (originally published by Chatto & Windus, 1931).

was asked how he had communicated the experience, he answered: "You don't. But by telling how you make quanta [the plural of quantum] and how you measure them, you enable others to have it." [5]

Barry Lopez, the American nature writer/environmentalist, claimed that nobody could understand relationships within ecosystems unless they have experienced "direct and intimate" contact with Nature.[6] Practically all naturalists can support such a claim. The noted primatologist Jane Goodall has used the term "experience-near" as an empathic approach to investigating animal behavior of animals. She broke the norm by treating individual wild animals as "individuals."

Let me share another personal story to illustrate this point. Before I obtained my first permanent job in the world of academia, I worked as a zoologist in a large natural wildlife refuge. I often interacted with field zoologists and rangers who spent significant periods in the forest. By studying the footprints of wild mammals on snow, they could identify specific animals and follow their behavior. With the help of these field workers, I learned how to differentiate between tracks of wolves and big dogs, and between wild minks, polecats, sables, and other mustelids; all based on observations of size, paw shape, behavior patterns, and other signs. Such observations were recorded and later used in scientific reports. Some of the data received from reading animal tracks were also published and used to make practical recommendations.

How reliable were they? They were reliable enough to be used for future decisions! Were those methods "scientific" or not? Many experimental biologists would say, "not." Personally, I would resist debating them as "scientific" or "non-scientific." Though reports of animal tracks can hardly be defined as "scientific information," extensive observations of the tracks can be valuable. It can lead to new

---

[5] Gary Zukav, The Dancing Wu Li Masters: An Overview of the New Physics. 1979, pages 261-262.
[6] Barry Holstun Lopez, *Of Wolves and Men*, Charles Scribner's Sons, 1979, 320 pages.

knowledge of animal habitats, their distribution, food preferences, movement patterns, and more. Most important was our ability to correctly assess the information so it could be used when making future conservation decisions.

Here's another illustration of the "scientific" vs. "non-scientific" dilemma: our present inability to actually "see" a "biological species." However, we *can* see an animal or a bacterium that *belongs* to that species. But we can't see "disease," but similarly, we can see signs and symptoms, such as the redness of skin or an increase in body temperature. In other words, we cannot doubt an *observed* change in the color of skin or changes in temperature, but we *can* doubt that those characteristics represent a specific disease or *any* disease. Such individual observations are not abstract—*they are always concrete* because they refer to specific entities and repeated patterns of past research. These include the characteristics of biological objects and events or the changes that occurred under specific conditions. *Information about particulars is always acquired through personal perception.*

Experiential knowledge might not need to be obtained exclusively from personal experience. If related to human beings, it can also be sourced from the historical memory of a group to which the investigator belongs. A similar situation relates to so-called "traditional" or "indigenous" knowledge, which is specific to communities of native Americans, some tribes in Africa, and groups in other parts of the world. At times, the traditions of certain cultures may experience uneasy interactions between themselves and locals or "outsiders." However, there is no reason to create or maintain an unavoidable conflict between scientific and traditional systems. Far better for them to work together in a spirit of tolerance and mutual respect!

Folk medicine might be one example of specific information. When our team worked in Uganda to control the spread of bubonic plague, we soon realized that the locals had more trust in the use of herbal medicine or spiritual healing than in the Western concept of "germs." As a result, the afflicted community risked the lives of those who were infected with *Yersinia pestis* bacteria from flea bites. Instead of taking highly effective antibiotics that help prevent the spread of the disease, the natives chose to follow the treatment endorsed by their traditional healers. Fortunately, some of the healers began telling their patients that they needed their blood samples for diagnostic testing. Since then, it has become more common to include indigenous and scientific knowledge when supporting similar communities facing an urgent environmental crisis.[7]

Of course, considering indigenous knowledge only at the practical level would be limiting. Fulvio Mazzochi, a senior researcher at the Italian CNR, believes that an indigenous practice refers to a particular understanding of phenomena and order. For example, indigenous worldviews are not dualistic.[8] There is a growing need to understand how indigenous knowledge perceives the modern world's complexity in territories that have undergone rapid environmental changes. The difference in meanings and ways of life between Western science and indigenous knowledge makes their integration difficult. However, despite embracing different perspectives, it is still possible to relate to each other, mutually learn, and discover or develop shared meanings. In spite of different perspectives, there is still the possibility to relate to one another, to learn from each other, and to discover or develop shared meanings. There is still little doubt that modern science can gain a lot from such a dialogue, which has been highly efficient for those studying specific aspects of the natural world.

Pluralism About Knowledge

A keystone of modern science is the concept of "objectivity, which means there are "facts out there" in the outside world. The scientist's

---

[7] Jayalaxshmi Mistry and Andrea Berardi, *Bridging Indigenous and Scientific Knowledge*. Science, 2016, vol. 352, No.6291, pages 1274-1275.

[8] Fulvio Mazzocchi, *Why "Integrating" Western Science and Indigenous Knowledge Is Not an Easy Task: What Lessons Could Be Learned for the Future of Knowledge?* Journal of Futures Studies, 2018, vol. 22, no.3, page 21.

task is to discover and report these facts with "objectivity." Modern science has efficiently developed multiple methods to reduce the influence of personal biases, claiming that any efforts to reject objectivity will destroy science!

The problem is that everyone has personal biases. Thus, when we design a new research project, our experiences and preferences will definitely affect the choice of objectives and the methods to achieve these objectives. Scientific researchers must decide how to record and analyze data. Then, they must decide how to interpret and report the results. And ultimately, those involved in scientific research must be prepared to make *choices*! We cannot ignore or avoid the need to seriously reduce biases we hold, even if subconsciously. We must be aware of them, and, in certain situations, consciously decide to accept them. Philosopher Heather Douglas of Michigan State University described three areas where objectivity was needed concerning pluralism: 1) the interaction of humans with the world, 2) individual reasoning processes, and 3) social tendencies.[9]

It is always preferable to seek coexistence between scientific and experiential knowledge. Remember this when you read later chapters, where I will propose some intriguing new tools, you can use to access answers to questions involving biology. These tools were borrowed from traditions that might not commonly be perceived as "scientific" but were solidly based on experiential knowledge.

I mention the existence of experiential knowledge for two reasons. First, as a scientist, I have successfully used the knowledge I gained from personal experience. Second, we live in a world in which making discoveries is not limited to scientists! In fact, both scientific knowledge and experiential knowledge coexist wherever there is life. At times, conclusions derived from both sources can contradict each other. This creates a "paradoxical situation"—one that neither scientific thinking nor common sense can solve! In such a case, we

---

[9] Heather Douglas, *The Irreducible Complexity of Objectivity*. 2004, Synthese, vol. 138, no.3, pages 453-473.

must find answers that go *beyond* these limitations.

Under these circumstances, there is a strong appeal to turn to experiential knowledge, *including ancient sources,* to reveal secrets of biological life beyond the usual limits of scientific inquiry. At the same time, we must sidestep being limited by another possible duality: science versus experiential or traditional knowledge. We have seen the dangers of such duality when they emerged during the COVID-19 pandemic. We must also avoid giving credence to accusations between supporters of "proven" scientific facts and deniers of science's ability to support our health.

When we discuss the relationship between the methodology of modern science and the knowledge shaped by personal experience, one central question arises. How much, if at all, is the scientific aspect influenced by the investigator's personal experience? The investigator's influence was famously tied to the so-called "observer effect" in quantum physics. However, evidence of the investigator's effect is plentiful beyond the world of subatomic particles.

Modern science was predominantly based on the assumption that the knowledge we obtain is always "objective." Objectivity was considered to be faithfulness to facts and, ideally, "free of personal values." Simply speaking, it was believed that a scientist's personal predispositions should not interfere with attaining knowledge. This statement is undoubtedly true when conducting experiments and reporting obtained data. However, even in experimental work, the obtained information can only be partially unaffected by the investigator who designed the research project, including experimental work. Knowledge does not exist in an abstract space; it belongs to a person or a group of people, sometimes to a large group of people. Knowledge cannot be separated from the perception of a person and the ethical values accepted by society.

Without going deep into the history of science, we might acknowledge that the "fathers" of modern science – Isaac Newton, Galileo Galilei, Giordano Bruno, Gottfried Leibniz– couldn't imagine science free of human values and divine providence. Isaac

Newton probably wrote more about the Temple of Solomon and celestial prophecies than about the motions of solid physical bodies. The early Universities didn't have Faculties of Physics. Instead, they had Faculties of Philosophy, Theology, Law, and Medicine. None of these areas were free of values, including medicine at that time. But gradually, modern science began to attain "value-free" standards.

Biological life consists of both visible manifestations and invisible connections between them. The only way to learn about the invisible realm of biological life is to explore additional dimensions based on one's values. Accepting values in research leads to attaining flexibility when selecting a desired level of perception.

The bottom line is this: the challenge before us is not to seek a compromise between opposite sides. Instead, we must accept differences that can be developed into multi-dimensional, hierarchical practices to obtain a more excellent knowledge of life. Many researchers still have difficulty seeing the possibility of building a bridge between modern science and integrated knowledge. For them, the concept is contradictory and paradoxical. When knowledge from modern science does coexist with experiential knowledge, it confirms that such a paradox *does* exist. In fact, the existence of this paradox can work in our favor. When a challenge arises, we often find solid ground in traditions based on extensive collective knowledge.

Relationship Models between Two Knowledge Domains

To illustrate this point, let us conditionally name the two realms of knowledge. First is "value-free science" designed to accurately measure visible manifestations of natural phenomena. The second realm is the "integrative knowledge," which accepts the investigators' values and uncertainties when investigating the whole reality, including unfamiliar or hidden relationships. Below are some illustrations of models of the interaction between the domains.

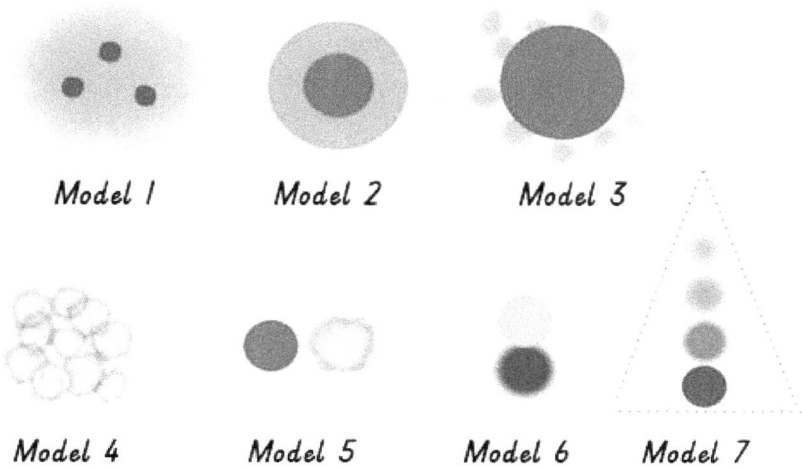

Figure 1. Interaction models between "value-free science" to determine the visible manifestations of natural phenomena (dark shapes) and "integrative knowledge" that accepts personal values and unknowns when investigating hidden realities (light shapes).

- **Model 1:** Overall, the surrounding reality is not well-formed and is primarily influenced by numerous invisible forces. Only a few objects (animals, people, stars) are more or less defined. This model dominated human societies before modern times and is still common among many indigenous peoples.

- **Model 2:** All manifestations of the visible world around us are governed by specific laws that correspond to the laws of the world above us. There is no contradiction between these levels; on the contrary, physical laws are governed by orders of a higher invisible world. Newton and many other great scientists of his time shared such a vision. Some scientists still share such a belief.

- **Model 3:** The world around us is material and exists in separate parts, ranging from elementary particles to galaxies. Scientists pretend that they can figure out the structure and movement of *all* material things. "Hard science" requires strict rules, allowing reliable information to be accumulated and verified. Some human activities, such as history, philosophy, metaphysics, religion, and art, could be exciting and vital but unrelated to "real" science. Modern science embraced this view by the end of the 19th century and it still dominates academic institutions worldwide.

- **Model 4:** In philosophy, "subjective idealism" (or empirical idealism) is the premise that *nothing exists except within a perceiving mind.* According to such a view, all our knowledge is formed by our predisposition to accept it. Marcel Kuntz warned about such a view, writing, *"... the danger of a postmodern approach to science, that seeks to include all points of view as equally valid, is that it slows down or prevents much needed scientific research."*[10]

- **Model 5:** There is a complete separation between worldly and spiritual domains. Scientists and other intellectuals who favor a rather materialistic approach to research problems at work yet follow religious or spiritual practices at home often adopt this attitude.

- **Model 6:** The "higher" (spiritual) realm with strict values is the only "true" reality, whereas the "lower" physical world is an illusory reflection of the higher one. Such a perception has roots in the Eastern esoteric traditions and is still common among spiritually inclined communities in Western countries.

- **Model 7:** There are different levels of reality concerning the action of specific general laws. Only one of them, the physical world, is visible and exists independently of the

---

[10] Marcel Kuntz, *The postmodern assault on science.* European Molecular Biology Organization, 2012,
vol.13, no.10, pages 885-889.

investigator's values. The other worlds, or realities, retain their connections, and all are rooted in the hierarchical system. The followers of such views belong to some schools of thinking, including, but not limited to, integrative science and transdisciplinary learning. The Kabbalistic tradition presents the most developed portal for absorbing, describing, and applying knowledge discovered in such a hierarchy.

Perceptions of Reality at Different Levels

Manfred Max-Neef from the Universidad Austral de Chile defined a "Level of Reality" as "a set of invariant systems (those that remain unchanged under a specified transformation) to the action of certain general laws."[11] The existence of different levels of perception of reality was formulated by Edmund Husserl, a philosopher/mathematician who established the School of Phenomenology.[12] The main features of the levels, according to Husserl, are contextualism, openness, and circularity. Each level mutually finds and determines the others. Physicist Werner Heisenberg proposed a similar concept of "Regions of Reality." Quantum entities are subordinated to quantum laws that radically differ from the laws of classical physics.

Max-Neef defined "transdisciplinarity" as the result of the coordination between all hierarchical levels. Basarab Nicolescu, who contributed significantly to the formulation of the transdisciplinary principles, pointed out: "...two different levels of reality are different if, while passing from one to the other, there is a break in the laws and a break in fundamental concepts, like, for example, causality."[13]

---

[11] Manfred Max-Neef, Foundations of transdisciplinarity. Ecological Economics. 2005, vol.53, pages 5-16.

[12] Bence Peter Marosan, *Levels of the absolute in Husserl.* Continental Philosophy Review, 2021, Springer.

[13] Basarab Nicolescu, *Transdisciplinarity and Complexity: Levels of Reality as Source of Indeterminacy.*
Bulletin Interactif du Centre International de Recherches et Études Transdisciplinaires, 2000, no.15.

The existence of different levels of reality with general laws specific to each level leads to the adaptation of additional "dimensions" from which we can obtain knowledge about Nature. The coexistence of different levels of biological life revealed by biological science coincides with fundamental principles developed by the Kabbalistic tradition.

Later in this book, I will present opportunities to use authentic Kabbalistic principles as a unique path combining ancient experiences with modern biological science. As we continue our journey together, it will become clear that this approach can advance, and possibly even ensure, the survival of humanity and our endangered planet.

Chapter 2

The Search for New Dimensions of Knowledge

Many argue that biology (with the exception of molecular genetics) has no laws. In fact, biological knowledge has very definite scientific laws, but modern science's criteria do not allow them to investigate *living* systems as an independent phenomenon. Therefore, if we genuinely wish to achieve knowledge of life, we must develop a new, multidimensional approach to this search. At present, our eyes can only perceive the outside world by identifying familiar geometric dimensions such as height, length, and width. However, we need additional tools to better perceive our *invisible* environment.

Dimensions shape our perception of the outside world and right now, human beings can only recognize the spatial dimensions available to us as our visual apparatus evolves. Prehistoric man had the ability to see, and thus found food to survive, moving from place to place while eluding predatory animal species. Because the imagery of the outside world is so obviously three-dimensional, we never even questioned the fact that there might be more dimensions!

However, dimensionality can go far beyond simple geometry. One of the definitions of the word "dimension" in the *Oxford Dictionary and Thesaurus* is "a measurable extent of any kind." A dimension is not only a measurement—it is a *direction* that can actually be measured. Scientists start with the need for measurements and then select a system to measure objects and the directions of their movement. Otherwise, how can one compare the properties of separate objects, whether alive or not? In fact, we possess the most perfect instrument to do so and Nature supplies it. I am referring to our thoughts!

Visualizing New Dimensions through Thought Experiments

Albert Einstein is usually associated with highly sophisticated theories that few humans can comprehend. However, the real source of Einstein's genius was his ability to use visualized "thought experiments" as a fundamental tool that eventually led to his significant scientific discoveries. We are fortunate to have Einstein as a reference; nobody can deny that he was one of the outstanding scientists in modern science history. At age 16, he chased beams of light with his mind. Later, he used moving trains and flashes of lightning to explain his most penetrating insights about special relativity. For general relativity, he used images of a person falling off a roof, elevators that sped up, or blind beetles crawling on curved surfaces.

"Thought experiments" are really devices of the imagination. The term was coined by Ernst Mach, a Moravian-born Austrian physicist and philosopher, in 1896. In his essay, *On Thought Experiments*, he claimed that it was not ideas that were innate in man but rather our tendency to experiment. When considering a problem that could not be resolved by actual physical experimentation, Mach constructed an imaginary apparatus in the mind's eye that both reflects the thesis of the problem and allows for the manipulation of circumstances that were impossible to create in the tangible world.[1]

While all experiments are guided by theory, not all experiments require lab testing. We can distinguish between the imaginary

---

[1] James Robert Brown, The Laboratory of the Mind: Thought Experiments in the Natural Sciences. London, Routledge, 2010.

scenarios featured in thought experiments vs. the narratives that establish the scenarios in others' minds. As the Stanford Encyclopedia of Philosophy stated: "Once a scenario is imagined, it may assume a life of its own, and this partly explains the creative power of a positive thought experiment."[2] To serve as a functionally equivalent device, any thought experiment should be designed just as thoroughly as a physical experiment.

There are numerous examples of productive results based on thought experiments in science. One of them became famous under the name "Maxwell's Demon."

1. James Clerk Maxwell intended to contradict the possibility of violating the second law of thermodynamics by "employing" an imagined creature. According to the proposed law, work is required to produce heat when energy flows from a cool body into a warmer one. Therefore, in 1871, Maxwell envisioned two vessels containing gas at equal temperatures, joined by a small hole. The door between the vessels was opened or closed by "Maxwell's Demon." This regulated the flow of energy by passing only fast-moving molecules from Vessel A to Vessel B and only slow-moving ones from Vessel B to Vessel A.

2. Another famous example is "Schrödinger's Cat," a product of the thought experiment imagined by the renowned physicist Erwin Schrödinger in 1935. Schrödinger posed the question, "When does a quantum system stop existing as a superposition of states and become one or the other?" To answer the question, Schrödinger mentally designed a surprisingly sophisticated experiment. It included a steel chamber, a Geiger counter, a tiny bit of radioactive substance, a hammer, a small flask of hydrocyanic acid, and even an experimental animal--the cat!

---

[2] James Robert Brown and Yiftach Fehige, *Thought Experiments*, in *Stanford Encyclopedia of Philosophy*, 2019, pages 1-2.

Michael Chandler, a medical doctor affiliated with the Weill Cornell Medical School of Medicine, claims the paradigm of the thought experiment can offer an instructive look at the value of contemporary medical education and practice.[3] Particularly, he demonstrated how medical diagnostic reasoning can be understood as the Basian statistical modeling blended with a "Bayesian-based thought experiment." Famed 19th-century German physician and pathologist Rudolph Virchow used visualization to stimulate the view that cells formed the basis of living organisms.[4]

A View from the 2D Flatland

The ability to conduct thorough experiments is imperative when considering the idea of additional dimensions, as you'll see in the following presentation. This one was introduced by a thought experiment in the form of a small book, *Flatland,* by Edwin Abbott in 1884.[3] Many readers are familiar with Abbott's book. Still, its role in presenting the "extra dimension" concept is important enough to deserve more explanation. The story, presented pseudonymously by "Mr. A. Square," described a two-dimensional world resembling a list of paper geometric figures: triangles, squares, and other polygons.

The culmination of the story arrived when the Square was visited by the Sphere, a *three-dimensional creature.* The Square was unable to see the Sphere as anything other than a circle that seemed to change size--and could even disappear! However, when the Sphere moved up and down in Flatland's environment, the Square could only see the circle either expand or contract.

The point is this: a Flatlander (an inhabitant of a two-dimensional world) couldn't visualize a three-dimensional object in its entirety. It was only visible to a Flatlander when the cross-sectional projection of this object penetrated the two-dimensional world (Figure 2). The Sphere intended to introduce the Square to "Solids"

[3] Michael Chandler, Visual Thinking and the Art of Medical Diagnosis, in *Design and Science,* Ed. Leslie Atzmon, 2023, page 288.
[4] Ibid., page 293.

and reveal exactly how they were constructed, "Behold this multitude of moveable square cards. See, I put one on another … Now a second, and now a third. See, I am building up a Solid with a multitude of Squares parallel to one another. Now the Solid is complete, being as high as it is long and broad, and we call it a Cube."[5]

This, however, was not a vision the Square could see! He replied, " … but to my eye, the appearance is as of an Irregular Figure whose inside is laid open to the view; in other words, methinks I see no Solid but a Plain."[3]

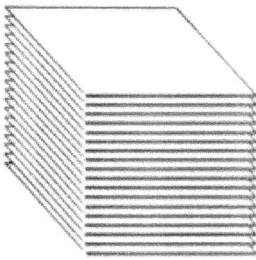

The view from the
point of Sphere

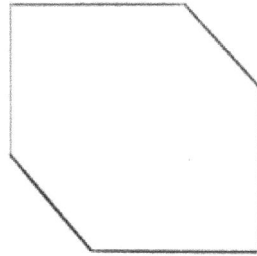

The view from the point
of Square

Figure 2. While a 3D subject ("Sphere") sees many parallel square cards that construct a "solid cube," a 2D subject ("Square") sees a plain, irregular figure (modified from Abbott's Flatland).

Further, Mr. A. Square was taken by the Sphere to visit Spaceland, the unreachable world of third dimensions. This journey evoked a harrowing emotional response from the Square, who could visualize only two-dimensional cross-sections of three-dimensional Spaceland. "Either this is madness, or it is Hell," cried the Square. "It is neither," replied the Sphere. "It is knowledge; it is three

---

[5] Edwin Abbott Abbott, *Flatland: A Romance in Many Dimensions.* 1884.

dimensions. Open your eyes once again and try to look steadily."[3]

I wish to emphasize two details from the Square's journey. First, he *could* see the "insides" of completely fenced enclosures of the Flatlands, including his own study room and even the highly secured Parliament building. Even more striking, the people of the three-dimensional worlds could see *inside* the bodies of lower-dimensional beings and view their internal organs.

The second note: after his awakening to the existence of another dimension, the Square became open to the idea of the existence of worlds with more than three dimensions! Ironically, the Sphere was not ready for such a perception. Inhabitants of each world (one, two, or three dimensions) were uncomfortable about going a step further into the domain of a potential new existence.

While Abbott's book became quite popular, a less popular publication was written the same year (1884) by Charles Howard Hinton, a British mathematician. Hinton's *What is the Fourth Dimension?* was not as entertaining as Abbott's book. Still, it demonstrated that the idea of an existing world of additional dimensions was "in the air" by the late 19th century.

Since Hinton was a mathematician (Abbott was not), he provided more geometrical illustrations that argued for the possibility of higher dimensions, particularly the fourth dimension. Hinton's world existed along the perimeter of a circle rather than on an infinite flat plane.[6] Hinton used it as a means of experimentation and creation, in contrast with Abbott, who treated the dimensionality analogy as a tool for description.

Later, Hinton extended the connection to Abbott's work with his book, *An Episode of Flatland: Or How a Plane Folk Discovered the Third Dimension.*[7] Hinton suggested that points moving around in three dimensions might be imagined as successive cross-sections of a

---

[6] Charles Howard Hinton. *The Fourth Dimension.* 1884 (From "Speculations on the Fourth Dimension, Selected Writings of Charles H. Hinton," Dover Publications, 1980. 270 pages).

[7] Charles Howard Hinton, *An Episode of Flatland: or, How a Plane Folk Discovered the Third Dimension.* 1907.

static, four-dimensional arrangement of lines passing through a three-dimensional plane. Hinton argued that, for us to gain an intuitive perception of higher space, we had to reject the concepts of right and left, up and down.

According to Linda Dalrymple Henderson, Professor in Art History Emeritus at the University of Texas at Austin, "The fourth dimension had become almost a household word by 1910."[8] Indeed, several animated films based on *Flatland* have since been released, including the film *Flatland,* released in 2007. Later, numerous imitations or sequels to *Flatland* were created. One was *Sphereland: A Fantasy About Curved Spaces and an Expanding Universe* by Dionys Burger in 1957[9].

In the novel, Mr. Square is ostracized by his community for introducing the idea of a third dimension. Later, society becomes more open to the ideas of "Spaceland" and begins accepting change and advancement. However, when a prominent surveyor discovers a Triangle that has more than 180 degrees, he is fired from his job, since such an object could not exist in Euclidean geometry. With help from the Sphere, the heroes can prove this theory. However, the established scientific community remains unable to comprehend the proposed ideas.

A more recent sequel is the book, *Flatterland,* written by Ian Stewart, professor of mathematics at the University of Warwick. According to the author, around 100 years after A. Square's experience, his great-great-granddaughter is visited not by Sphere but by the "Space Hopper," a creature with a grin, horns, and a spherical body. In contrast to the Sphere, the Space Hopper can travel beyond Spaceland to any imaginable world ("Mathiverse"). Here is Stewart's introduction to the travel phenomenon:

"There are lots of different Fourth Dimensions – not to

---

[8] Linda Dalrymple Henderson, *The Fourth Dimension and Non-Euclidean Geometry in Modern Art.* The MIT Press, 2018, revised edition, 760 pages.

[9] Dionys Burger. *Sphereland: A Fantasy About Curved Spaces and an Expanding Universe.* 1983. 208 pages.

mention Fifth, Sixth, or even a Hundred-and-First – many of which they experience in their daily lives, but fail to recognize … It's a question of being sensitive to what you are actually experiencing…They should look for a Fourth Dimension outside their own space." [10]

A Carrot's Journey Through a 2-D World

The *Flatland* story introduces a key tool for us to understand the following thought experiments. Michio Kaku, professor of theoretical physics at the City University of New York, and his co-writer, Jennifer Thompson, proposed using a thought model of a carrot passing through a two-dimensional world ("Flatland").[11] A carrot is an example of a biological object (a domesticated form of the wild carrot, *Daucus carota*).

Now, imagine that an inhabitant of a 2D world (a Flatlander) can see only successive slices of the carrot, which correspond to circles. The Flatlander sees an orange circle change sizes and later turn into a green circle corresponding to a green carrot top. Then, suddenly, the circles disappear just as mysteriously as they appeared (Figure 3).

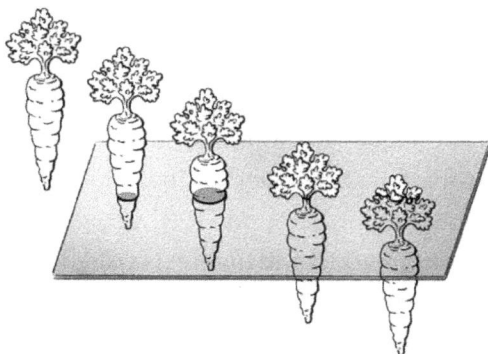

Figure 3. A carrot is passing through the Flatland.

Kaku and Thompson commented:

---

[10] Ian Stewart. *Flatterland: Like Flatland, Only More So.* 2002, 301 pages.
[11] Michio Kaku and Jennifer Thompson. *Beyond Einstein: The Cosmic Quest for the Theory of the Universe.* 1995, 224 pages.

"Likewise, if we were to encounter a higher-dimensional being, we might first see three spheres of flesh circulating ominously around us, getting closer and closer. As the spheres of flesh grabbed us and flung us into higher-dimensional space, we would see only three-dimensional cross-sections of the higher universe… Although we might understand that these various objects actually were part of the one higher-dimensional object, we would not be able to visualize this object completely or what life would be in higher-dimensional space". [12]

For my presentation at the Science & Nonduality Conference in San Jose, California, in 2014, I introduced another step in developing the "Carrot Model" by slicing leaves above the ground.[13] The new twist in this situation was that after several sequential stages, a Flatlander could see not only one but several separate green circles of different shapes and sizes.

As outsiders living in a 3D world, we know that all those 2D cross-sections are actually the manifestations of one carrot. However, while the Flatlander could not see the entire carrot, he could potentially learn how to become aware of it based on his observation of its cross-sections. This required being open to accepting the existence of additional dimensions, which required careful and consistent practice.

Now, let's take one more step. Imagine that the familiar Carrot is traveling side-by-side through Flatland with a friend: a Beet (Figure 4). Now, we can see pairs of circles of different colors changing their size. Suddenly, both figures begin to change their colors and shapes! But the most interesting thing happens afterward, you can see separate flat figures of different shapes and colors. Some of them are closer to each other, regardless of their shape. How many objects can

---

[12] Ibid., page 171.

[13] Michael Kosoy, *Entangled Life and Kabbalistic Biology*. The Science & Nonduality Conference, San Jose, California, October 2014, https://www.youtube.com/watch?v=8ZlFTJMMzOo

a Flatlander count? This is not an easy task, as at least three different characteristics must be considered: color (let's assume that the Flatlander is not color blind), shape, and the distance between figures that appeared in 2D.

Figure 4. A carrot and a beet in close proximity pass through the Flatland.

Here is the next thought experiment. This time, two carrots of approximately the same size travel through the 2D world (Figure 5). We see pairs of circles of orange color that represent cross-sections of the bodies of both carrots. These circles are replaced by many green, roundish cross-sections representing the carrots' tops. Eventually, it becomes impossible to determine to which carrot the green bodies belong. Does it matter? This is a thought experiment, but in real life, it can be biologically important to determine to which individual the observations belong. Let's imagine that one of these carrots is infected by bacteria of *Listeria monocytogenes*, which cause a serious food-borne infection.

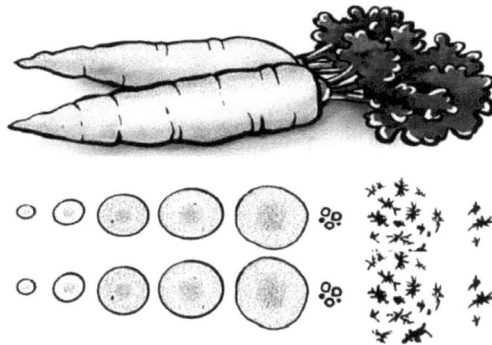

Figure 5: Two carrots in close proximity simultaneously pass-through Flatland.

With Flatlander Square, it was possible to imagine worlds with more than three dimensions. In the following years, some scientists seriously embraced this possibility.

Entering the Fourth Dimension

Hermann von Helmholtz, a prominent German physicist and physician, revolutionized the field of ophthalmology by inventing the "ophthalmoscope." He created this instrument to examine the inside of the human eye. Von Helmholtz compared the inability to "see" the fourth dimension to the inability of a blind man to imagine describing a color, such as 'redness.'[14] As mentioned earlier, prehistoric man had a particular survival advantage because of his ability to visualize finding and moving food. However, these primitives had no selective pressure placed on them to visualize a fourth dimension!

When Abbott, Hinton, and Helmholtz described the "fourth dimension," they referred to it as "an additional spatial dimension." However, I wish to clarify this description to avoid confusion with the view that considers Time as the "fourth dimension." In 1908, Hermann Minkowski, Einstein's mathematics professor, introduced an image of "space-time." He represented Time as "an additional

---

[14] Hermann von Helmholtz. *The Facts of Perception.* From *Selected Writings of Hermann von Helmholtz.* 1971 (originally written in 1878).

coordinate axis."

After Einstein's general theory of relativity was published, the concept of time as a fourth dimension became extremely popular. In fact, *Flatland* was mentioned in a letter entitled "Euclid, Newton and Einstein," published in Nature magazine in 1920. In this letter, Abbott is depicted as a prophet due to his intuition about the importance of time in explaining certain phenomena. However, the tendency to call "time" a "fourth dimension" should not confuse us. If we accept the possibility of more than three spatial dimensions, time can be considered either the fourth, seventh, or even *another* dimension.

Einstein experienced a serious dilemma when faced with the possibility of additional spatial dimensions. In 1919, he received a letter from Theodor Kaluza, a mathematician at the University of Konigsberg. In his letter, Kaluza proposed a five-dimensional theory of gravity. If this theory were correct, it would solve the fundamental theoretical problem that Einstein had been working on. Einstein was impressed by the elegance of mathematical argumentation, but because there was no experimental proof, he had a problem envisioning how this theory could be proven. Eventually, Einstein overcame his reluctance and promoted the publication of Kaluza's arguments. Despite this honor, however, most physicists continued to view Kaluza's arguments with skepticism.

Later, in 1926, Swedish mathematician Oskar Klein gave Kaluza's classical five-dimensional theory a quantum interpretation. The unified field theory of gravitation and electromagnetism was then designated the "Kaluza–Klein Theory." This theory was built around the idea of a fifth dimension *beyond* the standard 4D of space and time. In time, the Kaluza–Klein theory played a significant role in developing the so-called "string theory." According to this theory, the point-like particles of particle physics are replaced by one-dimensional objects called "strings." Overall, string theory tries to describe all the fundamental forces and forms of matter that exist in fundamental physics. The string theory has even been branded as the "theory of everything."

Strikingly, in string theory, spacetime is 10- or 11-dimensional! The theoretical models proposed in physics can help prepare us to accept the existence of additional dimensions. Thus far, however, we are still reluctant to accept the idea of additional dimensions to the real biological world.

Do not, however, be confused by a consistent use of the term "fourth dimension" by Russian mystic and writer Piotr Ouspensky. His first work, published in 1909, was titled *The Fourth Dimension*. However, for Ouspensky, the fourth dimension was *the realm of unseen higher spiritual things.*[15] He indeed used the term "fourth dimension" after being influenced by Charles Hinton. Ouspensky used the analogy of the dot, the line, the plane, and the cube similarly to Hinton and made an additional interesting point that I wish to cite: "We are ourselves beings of four dimensions, and we are turned toward the third dimension with only one of our sides, i.e., with only a small part of our being… The greater part of our being lives in the fourth-dimensional world".[13] However, Ouspensky's objective was utterly different from that of our scientific efforts. He was focused on the development of a seeker's inner world, not on a scientific exploration of the external world.

The Use of Dimensionality in Biology

Speaking of dimensions in biological investigations, we must be aware of drastically different ways we apply the concept of dimensionality in biology compared with applications in mathematics, physics, and chemistry. Historically, all measurements considered "authentic" in the exact sciences (physics? and chemistry) are precise or "exact," in keeping with the term "exact sciences." Based on "exact measurements," researchers are capable of accurate measurements, quantifiable predictions, and other rigorous methods of testing hypotheses. Take heed, however: *Dimensions and measurements are rarely straightforward!*

---

[15] Piotr Ouspensky, *The Fourth Dimension*. 2010, 52 pages.

This brings us to our investigation of living systems, which are different from those in the "exact sciences." The task of a biologist is to measure life by observing biological objects in their full diversity and biological processes in their constantly changing dynamics. However, some might argue that biology is not a "true science" because each biological object is unique. I have heard from a mathematician that biology is more akin to history, to which we cannot apply scientific rules/laws regarding its existence.

I believe that we should regard the dynamic variability of biological systems as an "exact rule" rather than as an annoying problem for those in the exact sciences! It is relatively easy to measure a variety of individual biological bodies, such as animals, by size, shape, weight, etc. In fact, a common process by which naturalists learn about biological organisms is first to capture an animal, for example, a lizard, and then measure its body length. The most common measurement herpetologists take of a lizard is its "snout–vent length"-- from the tip of the animal's face to the most posterior opening of the cloaca.

But even with this simple procedure, a professional zoologist may face challenging questions. For example, zoologists from Argentina, U.S.A., and UK proposed to improve the measurement of a lizard's body size by introducing what they called "multidimensional measure." This involved adding one more dimension, the lizard's weight, which has, in fact, turned out to be a significantly more productive measurement[16].

Biologists have developed the Shannon Index and other criteria to measure the diversity of organisms, but those criteria are more complex than measuring weight or length. Nevertheless, biologists *can* measure variability and diversity, which are essential characteristics of biological systems. Those indicators actually measure "directions" beyond the visible spatial dimensions.

---

[16] Daniel Pincheira-Donoso et al. *Body size dimensions in lizard ecological and evolutionary research.* Herpetological Journal, 2011, vol. 21, No.1.

The Dimensional Transformation from One Life Stage to Another

The measurement of dimensions in biology is even more challenging. Because we cannot even measure the essential parameters of living objects, how can we directly measure such intangibles as "evolution" or "adaptation to the environment?" How can we measure animal behavior and physiology? While biologists can use indicators that reflect those processes, these indicators only analyze phenomena *indirectly*.

The word "dimensionality" is not commonly found in most biologists' vocabularies. During the last decade, however, the use of these and related words has become more common. Here are several examples: multidimensional patterns in microbial communities[17], multiple dimensions of biodiversity[18], multidimensional plasticity in insect development[19], and multidimensional dimensionality in cardiac modeling[20].

At the "4D Biology Workshop for Health and Diseases" in Brussels in 2020, the workshop focused on finding organizing principles for large-scale genomics and proteomics. It also sought to set the vision for handling large-scale biological and medical data. Recognizing the need to develop analytical approaches in modern neuroscience, Alexander Gorban of the University of Leicester proposed a guide to "the path from the *curse of dimensionality* to the *blessing of dimensionality.*"[21]

Jean-Pierre Eckmann of the University of Geneva and Tsvi Tlusty of the Center for Soft and Living Matter in Korea describe a

---

[17] George Golovko et al. *Identification of multidimensional Boolean patterns in microbial communities.* Microbiome, 2020, vol. 8, no.1, page 131.

[18] S. Naeem et al. *Biodiversity as a multidimensional construct: a review, framework, and case study of herbivory's impact on plant biodiversity.* Proceedings Biological Sciences, 2016, vol.283.

[19] Nadja Verspagen et al. *Multidimensional plasticity in the Glanville fritillary butterfly: larval performance is temperature, host, and family specific.* Proceedings Biological Sciences, 2020, vol.287.

[20] Alan Garny et al. *Dimensionality in cardiac modeling. Progress in Biophysics and Molecular Biology.* 2005, vol.87, no.1, pages 47-66.

[21] Alexander Gorban et al. *High-Dimensional Brain in a High-Dimensional World: Blessing of Dimensionality.* Entropy, 2020, vol.22, no.1, page 82; the italics are mine.

dimension as *"the number of independent parameters that describe a [biological] system."*[22] The researchers proposed that living systems achieve "dimensional reduction" by their capacity to learn through evolution. They also cite the ability of individual genotypes to produce different phenotypical plasticity.

There is a recent trend to accept the *concept* of multidimensionality and develop 3D/4D models in biology. However, using dimensionality to measure transformational processes in biology is still being investigated, and its effectiveness is yet to be proven. The main obstacle to accepting higher dimensional levels is our attachment to the conventional 3D perspective.

To study animal behavior, for example, investigators must continuously observe and record specific patterns of the animal's food searches. It is far more difficult to define one or more parameters that truly represent an animal species' adaptability when searching for food under *changed* natural conditions. Similarly, investigators can present data on point mutations in the bacterial genome, or they can record morphological variations observed in the natural populations of plants. Those investigators can then claim that their observations indicate evidence of evolution. However, the truth is that they *cannot* measure the evolution of bacteria or plants. Biological processes are far too complex to be measured by any specific "ruler."

Two prominent Kabbalists, Rabbi Yitzchak Ginsburgh and Rabbi Moshe Genuth, noticed an analogy between embryonic development and the switch of dimensionality stages. They called the proposed model the "Point-Line-Area." This general model shows a development occurring in three stages--from a singular ("point") to a multidimensional area. Ginsburgh and Genuth illustrated the development of a fetus using the following model:

> "Before being inseminated by a sperm cell, the ovum, the female egg is in a point-like state. In this state, it is safe from

---

[22] Jean-Pierre Eckmann and Tsvi Tlusty. *Dimensional reduction in complex living systems: Where, why, and how.* BioEssays, 2021, vol.43, e2100062.

any mishaps, but at the same time, it cannot develop further. To develop, its point-like state needs to be distributed and permeated ... At the moment of insemination, the creation of the fetus begins, and the ovum proceeds to a line-like state. The line-state is characterized by continuous development occurring at a steady pace ... Because of the relative lack of complexity (or inter-inclusion), every entity or system that is in a line-like state suffers from the risk of instability until it passes into the more complex and mature aria-like state ... Only towards the end of the pregnancy term does the fetus mature enough to enter a state that is area-like ... Thus, independence and freedom characterize the area-like state."[23]

The "Point-Line-Area" imagery is represented throughout the early Kabbalistic texts by three Hebrew letters: *Yud* (a suspended point), *Vav* (the line-like letter expressing connection), and *Hey* (the letter expressing space). The interaction between these three letters can be thought of as a hidden code that unlocks all possible dimensional transformations! However, we can talk more about this only after learning some Kabbalah principles in Part 3.

Another illustration of transformation in terms of dimensionality is that of an insect's complete metamorphosis from the young (larva) to adult individuals differentiated by their forms and lifestyles. Let's examine the Western North American Monarch (*Danaus plexippus*), a common butterfly found west of the Rocky Mountains. This description is presented intentionally as a schematic to emphasize a single aspect: the *difference in perception between the stages of one insect species.* Adult females of the Western Monarch lay eggs on milkweed plants. These eggs are 1.2 mm high and 0.9 mm wide and are attached to the bottom of a leaf. After about four days, the eggs transform into the larval stage, called "caterpillar." There are several stages between larval molts called "instars."

The main objective of a caterpillar is to find food to eat. A

---

[23] Yitzchak Ginsburgh and Moshe Genuth. *Wisdom: Integrating Torah & Science.* 2018, pages 114-115, bold is mine.

caterpillar lives and moves in a two-dimensional world, while a butterfly actively explores the three-dimensional world. After the transformation from a caterpillar to a butterfly through the chrysalis (pupa) stage, the adult butterfly needs an entirely different dimension for orientation. The main goal is to mate and lay eggs. Some species of adult butterflies feed on nectar from flowers. Their search for such a food requires completely different criteria from the caterpillar's need to eat leaves.

I wish to cite the words of Ginsburgh and Genuth again when using the proposed "point-like-area" model to describe the life cycle of a Monarch butterfly:

> "Butterflies start in the point-like state of an egg… Once the egg hatches, the insect emerges as a caterpillar, representing its line-like state. This is its most precarious stage of life, and few caterpillars survive long enough to spin a cocoon where they go through the complete metamorphosis necessary for their transformation into a butterfly, their final and most mature area-like state."[24].

When thinking about such a metamorphosis, our usual priority is to observe changes that occur in tissues, organs, and physical appearance. Consider the collapse of dimensionality during the metamorphosis from a caterpillar to a chrysalis and the sudden "appearance" of an additional dimension during the process of transformation from the chrysalis to the butterfly. Richard Bach, an American writer, has written, "*What the caterpillar calls the end of the world, the master calls a butterfly.*"[25]

Dimensionality in Biological Organization

There is a marked similarity in the changing dimensions of any animal during its development through its consecutive life stages. Think about the dimensions required for an embryo to orient inside

---

[24] Ibid., page 115, bold is mine.
[25] Richard Bach, *Illusions: The Adventures of a Reluctant Messiah.* 1977.

its mother, or for a young offspring to leave the nest, or for an adult animal of the same species to find food. It's easy to understand that animals at these life stages perceive the outside world very differently and react to different stimuli.

All living things are hierarchically arranged according to their "biological levels" of organization. These levels correspond to biological categories such as cells, organs, organisms, populations, communities, and ecosystems. At each level of organization, we can expect entirely different dimensions by which they can be measured.

In the case of cells, we must consider their size, their specialization for specific tissues, and their responses to certain chemicals. Organisms require very different parameters for individual development, their search for food, behavioral repertoire, and many other activities. Qualities and measurements required to characterize unique populations are also very different and are not restricted to characteristics specific to individuals. These parameters include diversity, birth rate, mortality, selective pressures of the environment, etc. Each higher level of biological organization *must* be invisibly present at the lower level. For example, when an organism experiences stress, this stress affects not only the organism but also, indirectly, every cell and organ.

Let's consider the changes in the role of each element as "a journey through dimensionality." Take the image of a rabbit, which "travels" from the "cell world" through the "organismal" and "populational" worlds to the "ecosystem world." It's the same rabbit but reflects different levels of dimensionality:

- At the cellular level, we have a conglomerate of the rabbit's cells.
- At the organismal level, we have a complex of interconnected tissues, organs, and circulatory systems belonging to the individual rabbit.
- At the population level, the same rabbit is a part of the wider rabbit population that occupies a specific location.

- At the ecosystem level, the rabbit plays the role of a grass-eater, a prey for foxes, a host of tularemia bacteria, and many others.

Different levels are required for different tasks. I was pleased to discover an elegant and useful illustration of the transformed biological role of animals as seen through the lens of different dimensional systems (Figure 6). It appeared in a book written not by a biologist, but by a mathematician, Ian Stewart, whom I mentioned earlier. In *The Annotated Flatland*, he wrote:

> "The "new" world here contains the old one … In theoretical ecology, for instance, the populations of various creatures are considered as coordinates in a plane. If there are r rabbits and f foxes, then the state of the wood is represented by the point with coordinates (r, f) … The introduction of a new species – say, pheasants, whose population is represented by new coordinate, p changes the rules of the ecosystem because foxes can now prey on pheasants as well as on rabbits. The new state is represented by a point (r, f, p) in three-dimensional space. **A new dimension has appeared**: the dimension "pheasant population." Just as the Sphere propels A. Square into a new dimension, so the introduction of pheasants propels the woodland ecology into a new dimension."[26]

---

[26] Ian Stewart, *Introduction and Notes to The Annotated Flatland.* 2002, page 156; the bold type is mine

The "fox-rabbit ecosystem" before the introduction of pheasants

The "fox-rabbit-pheasant ecosystem" after the introduction of pheasants

Figure 6. The introduction of pheasants represents the appearance of a "new dimension" for foxes (using the words of Ian Stewart).

A Journey Through the Viral Dimensions

Let us make one more thought experiment. This time, imagine a situation in our three-dimensional world where you can see changes in "slices," one after another. This would be similar to a series of connected images strung sequentially to create a visual effect of movement during a film montage. First, the images are of specific viruses that invade the cells of a specific species of bats somewhere in Asia. Later, you will observe some changes in the RNA sequences of those viruses. (Remember, you have no technical limitations in an imaginary experiment!) However, the form of the viruses remains the same.

During subsequent cross-sectional observations, you can recognize more changes in viral RNA, slight changes in the speed of viral replication, and the occupation of other bat species by this virus.

After many such "sliced" observations, you will suddenly realize that individuals of one primate species have died (let's call them "people"). You don't know if there is any connection between viruses in bats and the death of primates. However, as you continue to observe, you can see that the number of deaths exponentially increases as cases appear on different continents.

After many cross-sections, you witness that these primates have begun to kill animals belonging to other species of mammals (let's call them "minks"). Is there any connection between viruses in bats, the deaths of people, and the killing of the minks? You can continue to observe more and more changes, including the behavior of the primates. At this point, you know where I'm heading. Remember the carrot leaves? Are those separate entities? We know that they belonged to one carrot because we left it in 3D, but a Flatlander can see only separate figures of different shapes.

The Laws of Nature in Higher Dimensions

We must be very aware of what we measure. Otherwise, we will see more examples of the miscommunications between scientists that plague modern science. Of course, it's normal to have different perspectives and interpretations of scientific observations. However, claims that someone has actually "measured evolution" or any other complex biological process are baseless and result in time-wasting debates. Instead, we must admit that to measure such complex processes, we need to apply additional dimensions and, in particular, external (outside) perspectives.

In his book *Hyperspace,* Michio Kaku[27], the theoretical physicist, stated, "The [additional] dimension pushed boundaries of modern science to their limit. It was more scientific than the scientists." Claiming this, he cited an interesting remark made by Peter Freund, a professor of theoretical physics at the University of Chicago: "The laws of nature become simpler and elegant when expressed in higher dimensions, which is their natural home." Freund used the following

---

[27] Michio Kaku, *Hyperspace.* 1994, page 63.

illustration:

> "Think, for a moment, of a cheetah … In its natural habitat, it is a magnificent animal … unsurpassed in speed or grace by any other animals. Now, think of a cheetah that has been captured and thrown into a miserable cage in a zoo … Now, we see only the broken-spirited cheetah in the cage, not its original power and elegance. The cheetah can be compared to the laws of physics, which are beautiful in their natural setting. The natural habitat of the laws … is higher-dimensional space-time. However, we can only measure the laws … when they have been broken and placed on display in a cage, which is our three-dimensional laboratory."[28]

This scenario is much more accurate regarding the laws of biology!

The Kabbalah and the Dimensions of Life

Some might complain that the images we discussed here are not real, but simply imaginary. Indeed, Flatland is an image! Samuel Ben-Or Avital, the Founder of Le Centre du Silence in Boulder and my Kabbalah teacher, often points to the difference between "the view from the helicopter and from the boat." He wrote: "… with this observer awareness, you can see the whole from different angles. Any time you need to make a decision, it is very beneficial to explore both of these ways of seeing".[29]

This difference is not only about the range of vision. From a higher perspective, we can see relationships between separate objects, which you cannot observe from a lower point of view. The Kabbalistic tradition supplies us with a unique experience: using it, we can switch from one dimension to another during the "thought" connections between the worlds (*olamot*). We will start preparing ourselves for such a journey in Part 3. This experience is important because it shows that such a switch of dimensions is not limited to the imagination, but can actually be used as a practical research tool!

---

[28] Ibid., page 12.
[29] Samuel Avital. *From Ecstasy to Lunch*. 2020, 367 pages.

Sarah Yehudit (Susan) Schneider, a biologist at the University of Colorado in Boulder who became the foremost teacher of Kabbalah in Israel, refers to the "carrot" model of Flatland to introduce Kabbalah. In her small, but fascinating book, *Evolutionary Creationism: Kabbala Solves the Riddle of the Missing Links*, Schneider wrote:

> "A ten-dimensional picture, with its pieces already intact, translates itself, slice by slice, into four dimensions. The [biological] species are not actually arising from within the four-dimensional frame itself. They are being transferred from one state to another, like a globe flattening into a map. In this model, the species are related conceptually, though not ancestrally. They fit together like puzzle pieces, yet arise independently through some as yet unidentified process by which ten dimensions transfigured into four."[30]

My late teacher, Joel Bakst, once told his students that reading *Flatland* is a prerequisite for studying Kabbalah. He wrote in his last book:

> "The greatest difficulty in wrapping one's mind around the texture of the Kabbalah in general … is an unfamiliarity with the concept of dimensionality … our immediate, 3D space-time reality, is embedded in a four-directional fabric of a higher and more encompassing dimension … the Kabbalists throughout the generations, regularly refer to our limited, linear world as Olam haShafel, which easily translated as Flat Land".[31]

In another of his books, *Beyond Kabbalah: The Teaching That Cannot Be Taught*, Bakst strongly emphasized the interconnection between Kabbalah and a multi-dimensional approach to thinking. He

---

[30] Sarah Yehudit Schneider. *Evolutionary Creationism: Kabbala Solves the Riddle of the Missing Links.* 2005, 103 pages.

[31] Joel David Bakst, *Kabbalah of the Adamic Messiah.* 2020, 346 pages.

asserts that the most critical concept is the model of dimensionality, wherein the tangible world is "embedded within a hidden reality, a missing dimension."[32] From the perspective of *Beyond Kabbalah*, our major dilemma arises because we are unfamiliar with the concept of both higher and lower dimensions.

You'll recall that the word "dimensions" is present in the title of this book. In this case, however, I am not referring to geometric or physical dimensions. My most urgent message is this: We must change our perspective of research so we can approach additional dimensions by using thought experimentation. To me, the word "dimensions" refers to directions that *enable us to understand and measure knowledge about the outside world.* By doing so, we can evaluate our relationship with the universe. From a biologist's perspective, we can use dimensions to orient ourselves in the complex and challenging world of biological life.

In the next chapter, we will talk about the complexity of biological life and by what means we can recognize it. To most effectively approach how we will be able to explore complex and dynamic biological processes, we need to search for additional orientation marks. I propose to find these in the realm of the Kabbalistic tradition.

Brian Greene, a theoretical physicist and professor at Columbia University, coined the phrase "unification in Higher dimensions."[33] Green formulated such a statement based on Superstring theory, but it matches precisely the Kabbalistic principle! The Kabbalistic system of thinking leads us to the same conclusion. It describes a higher realm of existence based on more essential laws than one would apply to a lower realm.

The extreme complexity of biological life requires us to search for clear, basic principles that would orient us in a complex world, one that remains mostly invisible when explored through a three-dimensional lens. Fortunately, "Kabbalistic Biology" principles are here to serve us in that search.

---

[32] Joel David Bakst, *Beyond Kabbalah: The Teaching That Cannot Be Taught.* 2012, 418 pages.

[33] Brian Greene, *The Elegant Universe.* 1999, page 196.

Chapter 3

Life at the Edge of Disorder

In the second part of the 20th century, advances in modern science were heralded by the development of new technologies, such as IT and biomolecular sequencing. This era also introduced the interconnected, transdisciplinary platforms known as the "Science of Complexity," "Chaos Theory," "Dynamic Systems Behavior," "Catastrophe Theory," "Synergetics," and more. There was a common theme in these groundbreaking platforms: each addressed a complex system consisting of many simple parts that interacted in a non-linear way. The result in all cases was unpredictable structural and behavioral patterns.

In this chapter, I will focus on how important it is to begin to "think differently" about such complex systems. This refocusing is not just desirable; it is imperative if we are to successfully deal with unstable ecosystems, environmental crises, and rapidly changing innovations. When we shift our perception and are open to new avenues of information, we can be far more effective in both our investigations and our management of any existing complex, dynamic system! These include the economy, social movements, climate activity, and environmental transformation.

The Science of Complex Systems
An investigation of biological life would be strengthened using laws

typically applied to complex and rapidly changing biological processes. "Complexity science" does not feature a single theory because it encompasses more than one interdisciplinary theoretical framework. Its mission is to analyze complex and changeable systems. Recent advances in this science were accompanied by developing sophisticated mathematical models+ for scientists to analyze.

Founded in 1984, the Santa Fe Institute in New Mexico became the first research institute dedicated to the study of complex adaptive systems. There, researchers continue to target functional similarities and differences among complex dynamic systems in a variety of processes. These include the origin of organic life, biological evolution, the behavior of primates, the operation of the immune system, and the dynamics of epidemics. They also include non-biological systems such as financial markets, climate, social changes, and other complex and dynamic systems.

I became intrigued by the science of complexity when I recognized a profound similarity between some approaches taken by the science of complex systems and the principal tenets of Kabbalah. *The critical point here is to search for a dimension that will embrace the whole spectrum of complexity-simplicity.* After we take a short excursion into investigating chaotic and complex systems, we will be ready to face the challenge of moving from analyzing complex and unpredictable processes to making simple decisions. As we will see later, Kabbalah provides us with invaluable practical experience in switching back and forth between complexity and simplicity.

The Complexity of Biological Life

Let's admit the obvious: biological life is amazingly complex! No two identical organisms exist among the more than eight million species on Earth. In addition, every organism is constantly changing. What we might consider an individual organism one day can be very different the next day! According to biologists Ron Sender and Ron Milo of the Weizmann Institute of Science in Israel, about 330 billion human cells per person are replaced daily.[1] That's the equivalent of one percent of all body cells in each human organism. In each cell, complex networks of interactions occur between thousands of nucleic acid molecules, proteins, and metabolites.[2]

However, although biological life *appears* chaotic, there is a clear order in the organization of life. Every mammal starts with four limbs, every bird has two legs, every insect has six legs, and every arachnid has eight legs. An enormous number of biological attributes confirm the presence of order in the biological world. For example, researchers have discovered a strict order in the stages of embryonic development. Embryonic development is relatively stable under most environmental conditions. The evidence of embryonic stability and vulnerability in an always-changing world is an excellent confirmation of the order existing in the biological world. Watching the interplay between relative stability and unstoppable variability in biological life is fascinating.

Yet, all biological structures and functions constantly change, revealing their sensitivity to environmental conditions. Despite the actual stability of developmental stages, living beings in their early development are more sensitive to environmentally adverse influences than at any other time in their lives.[3] Margulis and Sagan described living things as "islands of order surrounded by an ocean of chaos."[4] Their ability to create order around a biological system is a significant criterion that discriminates between the environment's biotic (living) and abiotic (non-living) parts.

Entropy and Life

The Second Law of Thermodynamics states that "all closed physical systems are moving toward a state of equilibrium." The main consequence of such movement is an increase in disorder, otherwise

---

[1] Ron Sender and Ron Milo, *The distribution of cellular turnover in the human body*. Nature Medicine, 2021, vol. 27, no.1, pages 45-48.

[2] Uri Alon, *Simplicity in biology*. Nature, 2007, vol.446, page 497.

[3] James Wilson, *Environment and Birth Defects*. 1973, 305 pages.

[4] Lynn Margulis and Dorion Sagan, *What is Life?* 1995, 288 pages.

called "an increase in entropy." Entropy is a measurable physical property most commonly associated with states of disorder, randomness, or uncertainty. The concept and term are used in diverse fields, from classical thermodynamics, where it was first recognized, to statistical physics and information theory.

The natural tendency of the universe toward increased entropy or disorder seems to be defied by living organisms. There is some truth to this claim since biological systems do not actually violate the Second Law of Thermodynamics; instead, they operate within its constraints in a fascinating, complex way. Biological systems such as living organisms are not closed systems; they are open systems that constantly exchange energy and matter with their surroundings.

This distinction is crucial to understanding how living organisms can maintain and even decrease their internal entropy. By consuming energy from food or sunlight, organisms create ordered structures, grow, and reproduce. As a result of energy acquisition, complex molecular structures and processes, such as protein synthesis and DNA replication, are created and maintained. These processes also generate waste heat and byproducts, which are expelled into the environment. As a result, waste heat and byproducts increase entropy in the surrounding environment.

While alive, each biological organism maintains a stable, precise internal environment that differs from its surrounding environment. It achieves this by acquiring energy from nutrients coming from outside itself. It also survives through the process of "metabolism," by which the organism converts nutrients into energy. As Erwin Schrödinger said, "...an organism feeds on negative entropy."[5] Negative entropy ("negentropy") refers to a system becoming less disordered or more ordered. For example, in an organism, the entropy production due to the heat released from the difference between acquired energy and stored energy compensates for the negative entropy. This is so designed to reproduce itself by the special

---

[5] Erwin Schrodinger, *What is Life?* 1944.

arrangement of DNA nucleotide bases.[6]

In everyday language and according to most conventional dictionaries, "chaos" means disorder and randomness. However, in chaos theory and complex systems, the terms "chaos" and "disorder" are not the same. While both a chaotic system and a disordered system may initially appear to share common ground, the fundamental disparity lies in their capacity for self-organization. Despite its outward appearance of randomness, a chaotic system possesses the remarkable capability to spontaneously and unpredictably reconfigure itself into a state of order and organization. Fully accepting the difference between these terms, in the subsequent text, I use "chaos," "disorder," and "randomness" to describe the direction away from order and organization.

Chaos as a Precursor of Order

The apparent lack of pattern or predictability in events is common in many biological systems.[7] David Bohm has declared that "randomness and order are parts of the same spectrum; they exist side by side, related by context."[8] I would rephrase Bohm's statement as follows: "Order and disorder are parts of the same spectrum and exist side by side."

Let me explain how my perspective differs from that of Bohm. In today's society, the words "chaos" and "disorder" tend to elicit a strong negative response, and why wouldn't they? After all, death, wars, epidemics, the collapse of populations, and environmental destruction are examples of chaos and disorder. The challenge, however, is to see the reverse side of chaos: it is actually the stage preceding the emergence of a new level of order! Indeed, the order of living things can be destroyed: every living organism will eventually die; a biological population can be crushed; an entire species can become extinct. *But all order starts with disorder!* In fact, after

---

[6] Jinya Otsuka, *The Negative Entropy in Organisms; Its Maintenance and Extension.* Journal of Modern Physics, 2018, vol.9, pages 2156-2169.

[7] Yaron Ilan, *Order Through Disorder: The Characteristic Variability of Systems.* Frontiers in Cell Developmental Biology, 2020, vol.8, page 186.

[8] David Bohm, *Wholeness and the Implicate Order.* 2002, 284 pages.

dramatic environmental changes, it is quite common to see new biological species emerge.

"Chaos theory" is a new scientific paradigm based on sophisticated mathematics. In his book *Chaos: Making a New Science,* James Gleick, an American science historian, described the efforts of dozens of scientists who contributed to this developing field.[9] They explained that chaos serves as an unlimited source of new information. I now offer two questions related to chaos. The first is, "Why is it essential that chaos be regarded as the precursor of order in biology?" The second question is, "What is the relationship between chaos and the Kabbalistic tradition?"

In their article *"Chaos in Ecology: Is Mother Nature a Strange Attractor?"* ecologist Alan Hastings, a professor at the University of California in Davis, and his colleagues demonstrated the role of chaos in ecology. They concluded that "the study of chaos is important to ecologists because the lessons of nonlinear dynamics will provide very different answers than the linear models traditionally emphasized by ecologists."[10] James Weiss of the UCLA School of Medicine and his colleagues identified various phenomena exhibiting chaotic behavior in the biomedical field. These included the response of cardiac and neural tissues, fluctuations in leukocyte counts, variations in renal blood flow, and the epidemiology of measles. At least, they appear chaotic in the absence of sufficient information.

The Concept of Chaos in Kabbalah

Kabbalistic tradition considers the role of chaos not only positive, but essential. At the very beginning of Genesis, we meet the word *ha-Tohu,* commonly translated as "chaos," "disorder," "desolation," "formlessness," and "waste" (uninhabitable space). However, Kabbalistic scholars warn us not to interpret the word "chaos" as negative because all order comes from disorder!

---

[9] James Gleick, *Chaos: Making a New Science.* 1987, 362 pages.
[10] Alan Hastings et al. *Chaos in ecology: Is Mother Nature a Strange Attractor?* Annual Review of Ecology and Systematics, 1993, vol.2, pages 1-33.

In Kabbalistic teachings, *Tohu* (chaos) brings greater power and the potential for developing a new order. Since the power of Tohu was so intense and robust, *kelim* (plural of *kli;* literally, "vessels") were shattered. Kabbalah uses the word "*kli*" as a prototype for any manifested object that serves a particular purpose. Elementary pieces of shattered *kelim* create a new order (*Tikkun*). Embracing chaos (*Tohu*) and order (*Tikkun*) is a prerequisite for the existence of Life (*Haim*).

The *Zohar,* the Kabbalah's primary book, is based on the teachings of Shimon bar Yochai, a 2nd-century sage in ancient Judea. It vividly describes how Chaos (*Tohu*) unites with Order ("Dry Land" — *Yabbashah* in Hebrew) to transform Earth's "Dry Land" into a fertile Earth (*Malkhut*) that brings forth new life.[11] The Breaking of the Vessels (*Shevirat ha-Kelim*) is, according to Kabbalah, a clearing of the decks, a fresh start, and a challenge to the existing structures.[12]

Aron Moss, the Rabbi of the Nefesh Community in Sydney, Australia, illustrated the struggle between Chaos (*Tohu*) and Order (*Tikkun*) by using the imagery of twin brothers: Esau (the first-born, strong, a symbol of chaos) and Jacob (a symbol of order and balance). The relations between the brothers culminated in their joining in an embrace.[13]

Life at the Edge of Chaos

Strikingly, the main concepts of the Science of Complexity, which emerged in the second part of the 20th century, are similar to the views presented by the ancient sources of Kabbalah. To Stuart Kauffman, one of the most influential theorists of complex systems, "*Life exists at the edge of chaos.*" Kauffman states, "The fate of all complex adapting systems in the biosphere ... is to evolve to a natural state

---

[11] *The Zohar, Haqdamat,* commentary by Daniel Matt Pritzker Edition, vol. 1, Stanford University Press, Stanford, California, page 83.

[12] Sanford Drob. *Kabbalistic Metaphors.* 2001, page 122.

[13] Genesis, 33:4.

between order and chaos, a grand compromise between structure and surprise."[14]

Everyone understands the words "complex" and "complexity. " They are the opposite of "simple" and "simplicity." Norman Packard coined the phrase "edge of chaos." If "chaos" is on the right side and "order" is on the left side, the central pillar is the edge that not only separates, but also unites both. Roger Lewin cites Chris Langton in *Complexity: Life at the Edge of Chaos*:

> "If connectedness among [elements] within the system is low, then the effects of the initial perturbation will soon [run] out. This is when the system is near the frozen state. With high connectedness, any single change is likely to propagate hectically throughout the system… This is a chaotic state. At the intermediate state, the edge of chaos – with internal and between-species interactions carefully tuned – some perturbations provoke small cascades of change."[15]

There is a delicate balance between rigid order and adaptability in biological systems. When rigidity is too high, it can be hard to adapt to changing conditions and ineffective when responding to them. In contrast, too much disorder can lead to chaos and the breakdown of biological systems. In biology, rigid order is critical under some conditions, particularly in the organization and function of living organisms. Here are a few examples:

- The nucleotide sequence of DNA is highly specific and rigidly ordered; a change in a single nucleotide can lead to unpredictable changes.

- Cell membranes are composed of phospholipids arranged in a specific order, with hydrophobic tails facing inward and hydrophilic heads facing outward. This order creates a barrier that regulates the movement of molecules in and out of the cell.

---

[14] Kauffman, Stuart. *The Origins of Order: Self Organization and Selection in Evolution.* 1993, 734 pages.

[15] Roger Lewin. *Complexity: Life at the Edge of Chaos.* 1999. 235 pages.

- The order and organization of metabolic pathways are particular and rigidly ordered, with any disruption to the pathway resulting in metabolic dysfunction.

- An imbalance between insulin and glucose in the body results in high blood glucose levels. Type 2 diabetes could be caused by an imbalance in the "effect of insulin" and glucose in the case of insulin insensitivity.

- In cancer cells, the normal regulation of gene expression is disrupted, resulting in uncontrolled cell division and growth.

Biological systems can become highly complex in the transitional space between rigid order and disorder. This complexity arises from the interactions and relationships between various components and factors. Each of these is subject to predictable order and unpredictable disorder at different scales. Complexity in biological systems can be seen in how different components interact and influence each other.

Ecosystems are good examples of complex biological systems in which various species interact with the environment. For example, in a forest ecosystem, relatively simple and predictable interactions between tree species, herbivores, and predators can lead to complex patterns of competition, predation, and mutualism. Over time, these interactions can lead to complex patterns of adaptation and coevolution among different species. Inevitably, they will develop traits and behaviors to better exploit or defend against their environment and other species.

These patterns can be highly variable and change rapidly in response to disturbances such as climate change, habitat loss, or the introduction of invasive species. A previously uninhabitable space can become livable depending on the interaction between organizing and disorganizing forces. Hydroponics and aeroponics, for example, have allowed plants to grow in previously "uninhabitable" regions, such as deserts and Arctic regions.

The space of complex biological forms can be described as multidimensional, including various biological forms, each with unique characteristics. This space is one of high complexity, with emergent properties and behaviors arising from the interactions and relationships between various components and factors. Dimensions of this space may include size, shape, internal organization, genetic makeup, and environmental influences. Biological forms in Nature are created after a complex interaction of these factors (Figure 7).

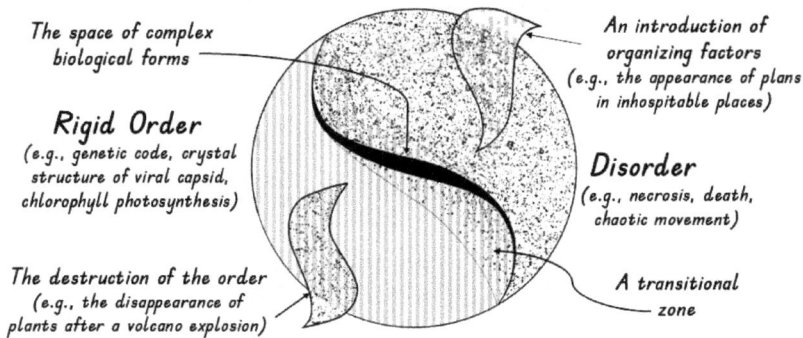

Figure 7. The complexity of biological life can emerge in the transitional space between rigid order and disorder.

In contrast to common belief, chaos is *also* about order, but in a way contrary to its name. Ultimately, the sublime order that lies beneath the apparent disorder *appears* to be out of control. Rabbi Joel David Bakst wrote about chaos in *Beyond Kabbalah*:

> "Tangled, splintered, jagged, twisted, fissured, twined, contorted, convoluted, serpentine, intricate, serrated, craggy, and fractured. Yet, simultaneously, there is an undeniable recurring pattern, especially a pattern that appears within the same structure at the same time, only at different scales. Every fracture of the shape replicates a fraction of itself on a smaller or larger scale, continuing infinitely. Infinite chaos reveals an organizing principle of beatific, transcendent order!"[16]

---

[16] Joel David Bakst, *Beyond Kabbalah: The Teaching That Cannot Be Taught.* 2012.

There is a zone between chaos and order within which life develops. The founder of the Visa credit card association, Dee Hock, coined the term "chaord" to describe an orderly and chaotic behavior that is simultaneously exhibited in a system.[17] A "chaordic" system is dynamic, self-organizing, adaptive, and resilient, balanced between order and chaos. Like any complex system, a biological system blends stochastic and deterministic features and displays properties of both. The space "at the edge of chaos" is where life innovates, and elements are not hard-wired but are flexible enough for new connections and innovations. Various levels of order become possible out of chaos.

One of the main points of the Science of Complexity is the recognition that it is inadequate to analyze separate properties of elements of *any* biological system, such as molecules and cells. Why not? *Because this new science is unable to accept the entire, complex system of biology as being indivisible.* According to a frequently cited metaphor, "a complex system is more than just the sum of its parts." Therefore, this new investigative model requires a complete shift in perspective.

Any complex system should include the following requisites: many simple elements, multiple levels of interaction between the elements, the openness of the system to exchange energy and information with its environment, and finally, a dynamic structure. Complex systems demonstrate several principal properties: emergence, irreducibility, nonlinearity, direct and indirect feedback loops, self-organization, and unpredictability. Let's break these down for a clearer understanding.

Emergence and "The Butterfly Effect"

We begin with a phenomenon called "emergence," which refers to the appearance of new order structures and functionalities that arise from the interaction of the entities. The word "emergence" is widespread in many areas of human activity: political emergence, economic emergence, etc. Epidemics are often associated with emergence.

---

[17] Dee Hock, *Birth of the Chaordic Age.* 2000, 345 pages.

"Emergent infectious diseases" is commonly used to identify some new diseases.

We must acknowledge the *reality* of emergence in our history. The arrival of the Emergence Theory was hardly accidental: it was preceded by two world wars, the Holocaust, the Great Depression, avant-garde movements in art and literature, and quantum revolutions in physics. These were massive shifts, and stable systems can become both unpredictable and chaotic even with minimal influences.

Edward Lorenz, an American mathematician and meteorologist at MIT, formulated the idea that chaotic systems are characterized by extreme sensitivity to their initial conditions. In the early 1960s, he developed a model of weather conditions that demonstrated that a slight change in starting conditions could lead to a much greater change later. In one of his presentations, Lorenz used a catchy metaphor called *The Butterfly Effect,* which became very popular. He said: "A flap of a butterfly's wings in Brazil could set off a tornado in Texas."

Such a statement initially sounds far-fetched, so let's explore it further. I will rephrase Lorenz's saying as more of a biological metaphor: "A change in insect behavior could set off a die-off of rabbits in Texas." Does it still sound as if it's unrealistic? Yes, it does. However, it's a fact that a chain of interconnected processes can be *initiated* by an improbable event, like a change in the mating behavior of butterflies in northern Brazil (Figure 8).

Suppose that intermediate processes include the following activities:

1. The black-headed grosbeaks (Latin name *Pheucticus melanocephalus*) developed an insensitivity to secondary plant poisons, allowing these birds to ingest monarch butterflies without vomiting.

2. The new food resource increased the bird population density.

3. The birds changed their migratory paths from Brazil to

central Mexico.

4. The dropping of bird feces near a livestock farm resulted in a change in the gut microbiome of cows.

5. The modification of the cows' microbiome transformed a previously harmless symbiotic bacterium into a virulent bacterial strain.

6. The delivery of modified bacteria with increased virulence to cottontail rabbits eventually caused a massive die-off of these animals in Texas.

The above scenario is entirely hypothetical and extremely unlikely. However, we still cannot say it's impossible, even if such a possibility were very low. Several years ago, my colleagues from Texas showed me a picture of dead rabbits presumably dying from tularemia, the disease caused by the bacteria *Francisella tularensis*. It is still not known what caused this outbreak of tularemia to occur.

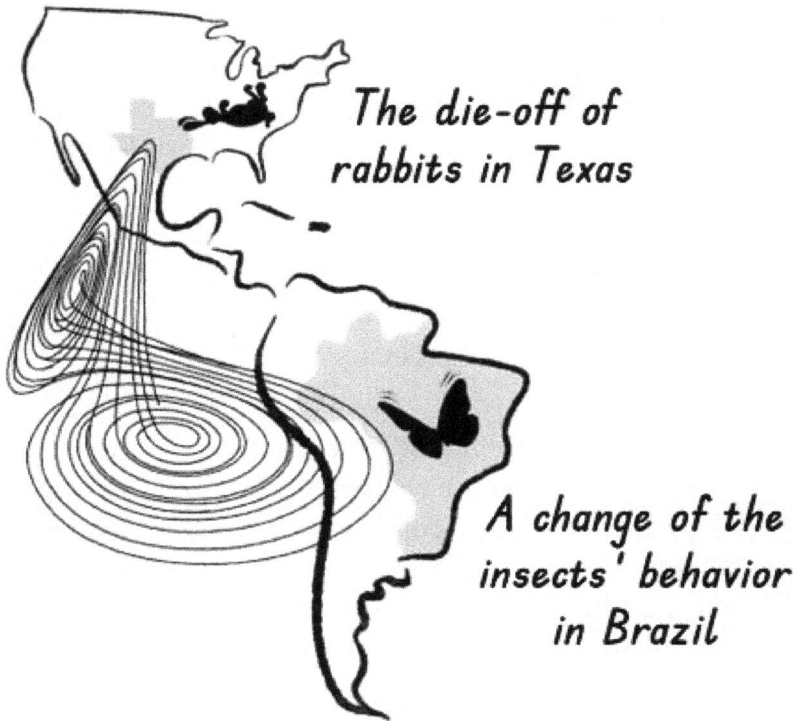

Figure 8. A change in a butterfly's behavior in Brazil could set off a die-off of rabbits in Texas.

While such a hypothesis still has no ground, would it be outrageous if it were rephrased as "a flap of a bat's wings in China set off a crash of the stock market in New York" amid the COVID-19 pandemic? Let's consider a scenario wherein a bat of a particular species in a cave in Yunnan, China, carried a mutation of a particular coronavirus that led to the deaths of millions of people, triggering a deep economic crisis and social unrest on all continents. Could anybody predict such a scenario before 2019? It's highly unlikely!

Entropy and Epidemic Forecasts

A "complex adaptive system" is a collection of individual agents free to act unpredictably. This is one of the basics of such systems. The word "unpredictably" refers to science's inability to predict the stage of emergence when analyzing only "simple" or separate variables. Such a focus calls into question the accuracy of any long-range prediction.

In 2013, I wrote an article for the journal *Entropy*, describing increasing contrast between the surge of information about elements of infectious systems and the scientists' very limited understanding of the causality of epidemic. "The situation is paradoxical because the expectation is that new information should lead to a better understanding and more accurate predictions of the distribution and dynamics of infectious diseases, and, overall, to a decrease in entropy."[18]

The above may sound discouraging in terms of making forecasts, but I prefer to call it "challenging." As James Gleick commented, "…had Lorenz stopped with the Butterfly Effect, an image of predictability giving way to pure randomness, then he would have produced no more than a piece of very bad news."[19] However, Lorenz saw more than the unavoidable unpredictability of climate behavior. His model offered new directions for taking into account the sensitivity of weather to local initial conditions. Similarly, scientists investigating complex epidemic systems should not restrict themselves to acknowledging the unpredictability of the systems' behavior. Instead, they invent ways to handle their investigation of such systems.

When considering the chaotic nature of the COVID-19 pandemic, Chris Piotrowski, a researcher at the University of West Florida, offered an optimistic conclusion. He stated that "Chaos theory predicts that individual and collective efforts towards functionality override short-term upheavals, so as to strive towards stability. Thus, a state of normality and societal equilibrium is

---

[18] Michael Kosoy. *Deepening the Conception of Functional Information in the Description of Zoonotic Infectious Diseases.* Entropy, 2013, vol.15, page 1930.

[19] James Gleick, *Chaos: Making a New Science.* 2008, page 22.

ultimately achieved and maintained."[20]

Every epidemic system shares common properties, such as (1) self-regulation to support the existence of a whole system of a particular infection and (2) emergence phenomena expressed in a high increase of intensity (epidemics). Both infectious agents and hosts can express emergent properties arising from interactions, whether among their components at each level or from environmental factors.

New, more realistic investigative models require a dynamic shift from emphasizing separate components (the infectious agents, the susceptible hosts, and the environmental factors) to focusing on the whole system and back. Because their actions are interconnected, one agent's behavior can change the context for various other agents. Systems are "complex" when they represent many independent agents interacting with each other. The richness of these interactions allows the system to undergo self-organization as a whole.[21]

Self-Organization and Levels of Analysis

"Complexity science" emphasizes the importance of "levels of analysis." New phenomena at the level of the organism cannot be fully explained by observations made at cellular or macromolecular levels. Complex systems exist at different levels of organization, ranging from molecular to ecosystem levels. The levels include molecules, cells, tissues, organisms, communities, and ecosystems.

Self-organizing systems will spontaneously arrange their components, which enables them to adapt to quickly changing environmental conditions.[22] Complex systems can reorganize their internal structures without the intervention of an external agent (such as a virus). Instead, they operate under conditions far from equilibrium, and as Dean Rickles and his colleagues from the

---

[20] Chris Piotrowski. *Covid-19 Pandemic and Chaos Theory: Applications based on a Bibliometric Analysis.* Journal of Projective Psychology & Mental Health. 2020, vol. 27, pages 1-5.

[21] M. Mitchell Waldrop, *Complexity: The Emerging Science at the Edge of Order and Chaos.* 1992.

[22] Donald Coffey, *Self-organization, complexity, and chaos: The new biology for medicine.* Nature Medicine, 1998, vol.4, pages 882–885.

University of Calgary noted, are "embedded in the context of their own histories."[23]

Self-organizing systems are also identified by the involvement of "non-linearity" and "feedback loops." The latter regulates and amplifies fluctuations in complex systems that lead to new forms of organization. "Negative feedback loops" are mechanisms that keep things stable. "Positive feedback loops" increase the change or output: the result of a reaction is amplified to make it occur more quickly. Examples of negative feedback loops are regulations of body temperature, blood pressure, regulation of blood sugar, production of red blood cells, and many other processes. Some examples of positive feedback include contractions in childbirth and the ripening of fruits.

Jeffrey Goldstein, a professor of Philosophy at Adelphi University, compared inner self-organizational transformation to the process of ancient alchemy.[24] In alchemy, chemical compounds are challenged by various procedures to bring out their hidden essences. Because each essence is locked inside each of the compounds, the purpose of alchemy is to accelerate the progress of the inner transformation of each of the compounds. Like alchemy, self-organization is a process of transformation whereby hidden potentials are unleashed and actualized for an authentic change.

The Potential Power of Nonlinearity

A critical step toward describing self-organizing systems was the theory of "dissipative structures" developed by physical chemist and Nobel laureate, Ilya Prigogine. Prigogine was intrigued by the ability of live organisms to survive under conditions of "nonequilibrium" (a state of imbalance between opposing forces or processes). According to Prigogine's theory, dissipative structures can not only maintain themselves in a stable state far from balance, but they can also create

---

[23] Dean Rickles et al. *A simple guide to chaos and complexity.* Journal of Epidemiology Community Health, 2007, vol.61, pages 933–937.

[24] Jeffrey Goldstein. *The Unshackled Organization:*
Facing the Challenge of Unpredictability Through Spontaneous Reorganization. 1994, page 3.

a new order and evolve.[25] Another essential property of dissipative systems is the number of critical points of instability, called *bifurcation points*--from Latin: *furca* (fork). Each bifurcation point represents a point in the system's behavior where instability suddenly forces a change in the system's behavior, a "point of no return." This change can lead to a loss of stability. It might also lead to the emergence of a new scenario with different properties.

Prigogine recognized the link between "nonequilibrium" and "nonlinearity." In mathematics, "nonlinearity" describes a situation where the relationship between an independent variable and a dependent variable cannot be predicted from a straight line. When we describe complex biological systems, we must distinguish the unique features of nonlinear systems by contrasting them with linear systems. The features of nonlinear systems may look strange and illogical, but it is precisely those strange features that indicate the enormous power of transformation locked within nonlinear systems.

Here are a few more features of nonlinearity. In linear systems, the whole is merely the sum of its parts. However, in nonlinear systems, the whole is greater than the sum of its parts. In linear systems, interactions are restricted to one-way influence. In nonlinear systems, interactions are multidirectional and multidimensional. Linear systems, when in a state of equilibrium, remain roughly the same. However, nonlinear systems can go through the transformation during conditions called "far-from-equilibrium." Prigogine's nonequilibrium thermodynamics states that such conditions are required to create and maintain life. Not all scientists agree with this definition, and some may argue that life is made up of many reactions in the "near-equilibrium" range.[26]

To demonstrate the need for nonlinearity in an analysis of ecological systems, let's consider an example of the dynamics of

---

[25] Ilya Prigogine and Isabelle Stengers. *Orders Out of Chaos: Man's New Dialogue with Nature.* 1984, 349 pages.

[26] Georgi Gladyshev, *The Second Law of Thermodynamics and the Evolution of Living Systems.* Journal of Human Thermodynamics, 2005, vol. 1, Issue 7, pages 68-81.

biological systems in which two species interact: one as a predator and the other as prey. Consider the case of a group of rabbits and a group of foxes coexisting in one forest. The foxes are the predators, and the rabbits are their prey. When foxes have lots of food, they multiply over time. As the population of foxes goes up, the population of rabbits will, naturally, go down. As more and more rabbits are eaten, there is less food for the foxes. Gradually, the population of foxes goes down, too. With fewer foxes, the rabbits have fewer predators and start to multiply. Once again, there is more food for foxes, and they stop starving and start multiplying again. This scenario was originally described by Alfred Lotka by a pair of *nonlinear* differential equations, known widely as the Lotka–Volterra equations.[27]

One of the main advantages of this new nonlinear scientific approach is that researchers can better approximate the irregular and organic behavior of complex biological systems. As Goldstein stated, "...we no longer need to understand these systems by putting them into the procrustean bed of linear and equilibrium abstractions."[28] If linear systems have predictable outcomes, nonlinear systems may have unpredictable outcomes.

Nature *Does* Make Jumps!

Change is gradual and cumulative in linear systems, whereas in non-linear systems, change can be unanticipated and revolutionary. I will illustrate the differences between linear and nonlinear approaches by taking on the most critical question in biology: How did life evolve? Many questions about evolutionary pathways remain unanswered because of our incomplete fossil record. Scientists are often unable to determine how quickly a species has evolved.

Gottfried Leibniz's principle, *Natura non facit saltus* (Latin for "Nature does not make jumps"), has turned out to be one of the central claims of Darwinism. It sees the process of evolution as being

[27] Brauer and Castillo-Chavez. *Mathematical Models in Population Biology and Epidemiology.* 2000.
[28] Jeffrey Goldstein. *The Unshackled Organization: Facing the Challenge of Unpredictability Through Spontaneous Reorganization.* 1994, page 19.

steady, slow, and continuous. All significant evolutionary changes happen during a long geological time frame. Darwin believed that small changes in a species slowly add to significant changes over very long periods of time. Therefore, such a view is called "gradualism."

However, one of the problems with this dominant view is the absence of fossils "in-between" many taxonomic groups of known organisms. In 1972, paleontologists Stephen Gould of Harvard University and Niles Eldredge of the American Museum of Natural History challenged the assumption that evolutionary change was continuous and gradual. They called their concept, "punctuated equilibrium."[29] This hypothesis states that evolutionary changes transforming old biological forms into new ones occur in short bursts separated by long periods of stability, or *stasis*. The most significant point here is the observation that *species can change very rapidly over a few generations, and then settle down again to a period of little change.* The theory holds that species evolve rapidly when isolated and are under pressure to adapt to new environments. According to Gould and Eldredge, biological evolution is *not* linear.

Paleontological arguments for "punctual equilibrium" are numerous and varied. Michael Benton and Paul Pearson of the Department of Earth Sciences, University of Bristol, analyzed the results of 58 studies of speciation patterns in fossil records published between 1972 and 1995. Their analysis demonstrated the widespread occurrence of stasis in fossil records. Benton and Pearson stated, "It therefore seems clear that stasis is common and had not been predicted from modern genetic studies."[30]

Such "punctuated equilibrium" represents an excellent example of how disorder caused by a new situation can potentially lead to sudden, new biological forms. I will demonstrate a dilemma between punctuated versus gradual evolution using the origin of modern horses as an example. The origin of horses served as evidence for

---

[29] Niles Eldredge and Steven Jay Gould. *The Pattern of Evolution.* 1999, vol.219, pages 140-145.

[30] M.J. Benton and P.N. Pearson, *Speciation in the fossil record.* Trends in Ecology & Evolution, 2001, vol.16, pages 405–411.

evolutionary "gradualism" for 150 years after Othniel Marsh's "Horse Series."[31]

Figure 9 illustrates five alleged evolutionary steps (species). These are:

1. *"Hyracotherium,"* renamed by Othniel Marsh to *Eohippus*, was an animal approximately the size of a fox with multiple small hooves. It lived approximately 52 million years ago.

2. *"Mesohippus"* lived around 34 to 32 million years ago. It had longer legs than Eohippus and stood about 60 cm tall. It had three toes, the central one longer and larger than the others, but all toes touched the ground and carried weight.

3. *"Merychippus"* lived from 16 to 5 million years ago. Though the animal retained the primitive character of three toes, it looked more like a horse. Moreover, it was the first horse known to have grazed.

4. *"Pliohippus"* lived around 15-12 million years ago. It had evolved into the first one-toed, hoofed horse, while the two side toes had been reduced to splint bones.

5. Finally, the modern horse (*Equus ferus caballus*) has a single toe/hoof and was domesticated about 4,000 years ago.

---

[31] Othniel Charles Marsh (1831–1899) was an American professor of paleontology at Yale College and president of the National Academy of Sciences. He discovered 30 kinds of supposed fossil horses in Wyoming and Nebraska in the 1870s. In 1879, he arranged these in an evolutionary sequence and displayed them at Yale University's Peabody Museum. The exhibit has been duplicated in countless museums and books.

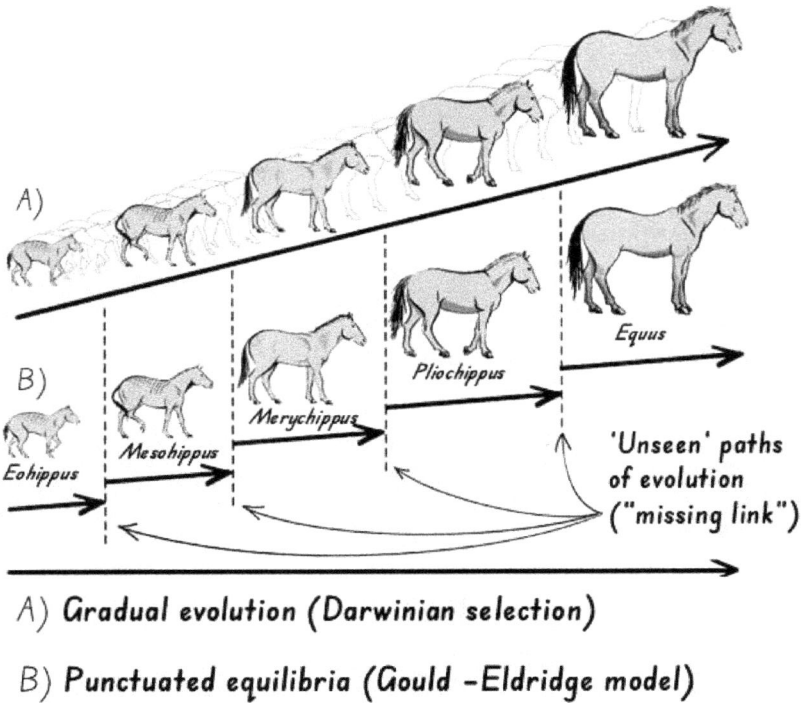

A) **Gradual evolution (Darwinian selection)**

B) **Punctuated equilibria (Gould -Eldridge model)**

Figure 9. The concept of punctuated evolution of horses challenges the concept of gradual evolution.

The evolution of horses was a frequent example of linear evolution. The problem was that all the major changes in horses that paleontologists fixed upon (size, tooth height/type, and number of toes) could be understood as adaptations to changing environments. In the actual fossil record, some earlier horses were larger than later ones. Additionally, the fossils portrayed in the supposed lineage spanned different continents, separated by vast ocean expanses.

An alternative view holds that horse fossils do not represent the direct descent of modern horses, but instead represent branches of a tree. Jonathan Wells tackled this problem in the *Icons of Evolution*: "By the 1920s, it was already becoming clear that the evolution of the horse was much more complicated than Marsh's linear picture

74

implied."[32] Wells noted that horse evolution is often portrayed as a straightforward linear process, even in contemporary discussions. This is with minimal consideration of the group's prior diversity. Why does this tendency persist? In science, there exists a tendency to be drawn toward recent developments, where our curiosity leans toward unraveling the linear narrative that led us to our current state.

The "Cambrian Explosion" is a classic example of a massive punctuation in the history of life on our planet. It refers to a period of rapid diversification and evolution of animal life that occurred around 541 million years ago, during the Cambrian period.[33] Before the Cambrian explosion, most life on Earth consisted of simple, single-celled organisms such as bacteria and algae. However, during the Cambrian period, there was a sudden proliferation of complex, multicellular animals with diverse body plans and anatomical features, including the first appearance of arthropods, mollusks, and chordates. The causes of the Cambrian explosion are still debated, but it is believed to have been driven by a combination of factors, including increases in atmospheric oxygen levels, changes in ocean chemistry and temperature, and the evolution of new developmental and genetic mechanisms that allowed for the rapid diversification of body plans.

Another example of evolution's "jump" is a phenomenon known as "pre-adaptation." When environments change, traits initially developed for one purpose can be co-opted for another. An example of pre-adaptation is the feathers of birds. It is believed that feathers first evolved in dinosaurs as insulation or display. However, as birds evolved from dinosaurs, feathers took on a new function: flight. A feather's structure makes it well-suited to powered flight. Therefore, feathers that evolved for insulation or display were pre-adapted for

---

[32] Jonathan Wells, "Icons of Evolution: Science or Myth?
Why Much of What We Teach About Evolution Is Wrong." Regnery Publishing, 2002, page 1999.
[33] James Valentine et al., *Morphological complexity increase in metazoans.* Paleobiology, 1994, vol.20, pages 131-142.

birds' new functions.[34]

Thinking in Complexity

One of the critical lessons from developing the "Science of Complexity" is the emergence of nonlinear thinking, or "thinking in complexity." In his book *Thinking in Complexity*, Klaus Mainzer, a German philosopher and scientist, wrote: "Linear thinking may be dangerous in a nonlinear complex reality... Linear thinking may fail to yield a successful diagnosis".[35]

Kurt Richardson and his colleagues at the Institute for the Study of Coherence and Emergence in Boston believe that complexity thinking provides "a clear warning as to the danger of uncritically adopting any 'black and white' theoretical position."[36] Tackling modern-day crises like the coronavirus pandemic, Fulvio Mazzochi of the National Research Council in Rome, Italy, argued that there is a need for "a substantial mindset shift" if we want properly trace causality.[37] Conceptually, he argued, we should pay much more attention to connecting patterns and structures that foster thinking that "goes beyond a linear and simplistic idea of causality." What would really be helpful to better cope with complex situations like the pandemic? Mazzochi answers this question, "refining our way of reading the world."

Helena Knyazeva, a professor at the National Research University in Moscow, summarized the following set of ideas illustrating the concept of nonlinearity:

1. Under certain conditions, nonlinearity can strengthen fluctuation.

---

[34] Richard Prum and Alan Brush, *The evolutionary origin and diversification of feathers.* The Quarterly Review of Biology, 2002, vol.77, pages 261-295.

[35] Klaus Mainzer, *Thinking in Complexity: The Computational Dynamics of Matter, Mind, and Mankind.* 2004, page 15.

[36] Kurt Richardson, Paul Gilliers, and Michael Lissack. *Complexity Thinking: a Middle Way for Analysts.* 2004.

[37] Fulvio Mazzocchi, *Tackling modern-day crises: Why understanding multilevel interconnectivity is vital.* Bioessays, 2021, vol.43, no.3, e2000294.

2. Open, nonlinear systems demonstrate the existence of thresholds of sensitivity. Thus, below a threshold, everything diminishes and disappears, while above a threshold, everything excessively increases.

3. Nonlinearity provides the discreteness of the evolutionary path of its systems. It means that not every evolutionary path is possible; only a potential spectrum of evolutionary paths is available for use.

4. Nonlinearity stresses the possibility of emergent (unexpected) changes.

5. Nonlinearity implies the possibility of sudden and rapid development at certain evolutionary stages.[38]

Gabriel Burstein and Constantin Negoita of CUNY Hunter College have developed what they call a "Kabbalistic System Theory."[39] Their concept acts as a unifying science to create a common language for psychological, social, economic, and cultural systems and their current multi-faced crises.[40] This system is based on modeling the Tree of Life as a hierarchical three-level feedback control system corresponding to the cognitive, behavioral/emotional, and activity levels.[41] I mention the research of these scientists because it was a successful effort to nourish the system theory with the richness of Kabbalistic thinking.

The practical value of accepting biological objects and processes as complex systems is underscored by the development of "systems biology." Significantly, the systems view was introduced into biology when an avalanche of genomic data threatened the ability of

---

[38] Helena Knyzeva, *Nonlinearity of time in the complex world.* Philosophy, 2012, vol.21, page 214.

[39] Gabriel Burstein and Constantin Virgil Negoita.
A Kabbalah System Theory of Ontological and Knowledge Engineering for Knowledge Based Systems. International Journal of Advanced, 2013, vol.2, No.2.

[40] Gabriel Burstein and Constantin Virgil Negoita. *A Kabbalah System Theory Modeling Framework for Knowledge-Based Behavioral Economics and Finance.*
Computational Models of Complex Systems, 2014, pages 5-23.

[41] Gabriel Burstein and Constantin Virgil Negoita. *Foundations of a postmodern cybernetics words*

researchers to interpret the complexity revealed by the sequencing of whole genomes of many organisms. The systems approach is a new way to combine available data about a particular biological process while emphasizing the uniqueness of the living organism.

A similar crisis was created by the rapid growth of ecological information. Hannaha and Weinberg raised the question as to whether progress would consist of "adding further layers of complexity to a scientific literature that is already complex almost beyond measure".[42]

Here, in fact, is a major paradox: How can information about complex systems be focused on making straightforward decisions? The ignorance of complexity leads to limited views and eventually to dogmatism and incorrect actions. An unlimited accumulation of complex information can result in getting lost in a forest of complexity. How, then, does one simplify the principles of a complex world?

The Complexity of Epidemics

While discussing complexity in the evolutionary ecology of infectious diseases, my son, Roman, and I proposed investigating a "simplicity-complexity" continuum.[43] Analyzing numerous variables related to the status of either microbes or infected persons can be perfectly appropriate for building mathematical models. However, it could be less useful when a clear and convincing definition is required to find an immediate solution during epidemics.

Which pathogen was responsible for the emerging outbreak when multiple infectious agents were reported? What range of variability of the virus can we accept to define it as a pathogen? What specific animal species can we define as a reservoir host? From the layman's point of view, it is not helpful to be presented with all the complexities of viral ecology when in an Ebola fever virus outbreak.

---

[42] Mihjalo Mesarovic et al. *Search for organizing principles: understanding in system biology.* System Biology, 2004, vol.1, no.1, pages 19-27.

[43] Michael Kosoy and Roman Kosoy. *Complexity and biosemiotics in evolutionary ecology of zoonotic infectious agents.* Evolutionary Applications. 2017, page 5.

The public expects and requires a clear definition of the source of infections and risk factors. Between such extreme situations, there is an unlimited number of intermediate positions representing potential probabilities for the expression of epidemics.

When studying epidemics, the perspectives within the "simplicity-complexity" dimension might change in a variety of ways:

- From different virulence factors to multi-sided descriptions of the pathogen.

- From well-defined pathogens to broad microbial communities.

- From clinically manifested disease cases to various infectious patterns ranging from latent infections to deadly epidemics.

- From single environmental factors, such as temperature, as it affects a host-parasite relation to the broad ecological context.[44]

Simplexity: The Combination of Simplicity and Complexity

While choosing how to keep a balance between "simplicity" and "complexity," we can refer to the recently proposed term "simplexity." Jack Cohen and Ian Stewart used this term in their book *The Collapse of Chaos: Discovering Simplicity in a Complex World*. The authors argue that scientists have been focusing on a system's complexities when they should be focusing on its simplicity.[45]

An essential point about "simplexities" is that their presence is guaranteed once we know the rules. In reality, simplicity is incomplete without complexity, and complexity is not complete without simplicity. Philippe Compain of the University de Strasbourg defined "simplexity" as "the combination of simplicity and complexity within the context of the dynamic relationship

---

[44] Michael Kosoy. *Deepening the Conception of Functional Information in the Description of Zoonotic Infectious Diseases.* Entropy, 2013, vol.15, page 1949.

[45] Jack Cohen and Ian Stewart, *The Collapse of Chaos: Discovering Simplicity in a Complex World.* 1994, page 411.

between means and ends."[46]

Alain Berthoz, professor of physiology at the College de France, chose the word "simplexity" for the title of his book.[47] He argues that in a complex world, solutions are not simple because there can be no simple solutions – they are necessarily "simplex." Berthoz introduced the concept of simplexity as "the set of solutions that living organisms have found despite the complexity of natural processes."

Alas, simplification in a complex world is challenging! Nature requires that we choose, refuse, connect, and imagine to enable us to act in the best possible manner. Jack Cohen and Ian Stewart proposed not only the term "simplexity" but also "complicity" as the state or condition of being an accomplice. "Simplexity" indicates the easy way for different rules to generate similar or identical features. It works when the rules themselves are very similar. The term "complicity," in the words of Cohen and Stewart, represents the situation when "totally different rules converge to produce similar features, and so exhibit the same large-scale structural patterns."[48]

Here is an example of complicity: malaria is caused by a plasmodium parasite that invades the red cells of human blood. Female *Anopheles* mosquitoes transmit such parasites. When mosquitos suck a sick person's blood, they ingest the parasite. After that, the mosquitos can feed on another person and pass the parasite on through the bite. There is no plasmodium parasite in human blood if there are no mosquitoes. Blood was not created to provide food for mosquitoes. Separately, mosquitoes have no plasmodium if there are no sick people around. The ability of mosquitoes to fly did not originate so that they could transmit malaria.

However, when all these species – plasmodium parasites of a specific species, susceptible people, and mosquitoes of a specific genus interact, they provide the potential for entirely new

---

[46] Philippe Compain, *The challenge of simplexity. The simple and complex in organic synthesis*. Actualité Chimique, 2003, vol.265, pages 129-134.

[47] Alain Berthoz, *Simplexity*. 2012. 265 pages.

[48] Jack Cohen and Ian Stewart, *The Collapse of Chaos: Discovering Simplicity in a Complex World*. 1994, page 414.

possibilities. Once the malaria transmission cycle starts, it leaves, figuratively, its history behind. In the words of Cohen and Stewart: "Simplexity merely explores a fixed space of the possible. Complicity enlarges it. And both processes collapse the underlying chaos, producing stable features from a sea of complexity and randomness."[49]

Think about the dimensions that we discussed earlier. A separated plasmodium parasite doesn't live in a "world of malaria." A separate mosquito doesn't live in this world either, nor does a healthy person live in this "world." It is a new world manifested in a new dimension! Other parasites, mosquitoes, and other people don't enter this world, and the dimensions of this world do not apply to them!

In *Hyperspace*, Michio Kaku declares that "the laws of Nature simplify and unify in higher dimensions." The question is, how can we use high-dimensional space as a unifying principle? The ability to traverse higher dimensions is the best opportunity that Kabbalistic practice can provide! We will explore some tools provided by Kabbalah in Part 3. At this point, however, we will need to become oriented in the "complexity-simplicity" ocean.

The "Cosmic Accordion" Practice

I cannot think of a better illustration of the "complexity-simplicity" practice than the image used by my teacher, Samuel Avital; he calls this practice "the journey between simplicity and complexity," or "the Cosmic Accordion."[50] He wrote that the expression "Cosmic Accordion" was born out of necessity since this natural law occurs repeatedly in a variety of situations. Avital explains that the Cosmic Accordion is a principle found throughout Kabbalistic literature. It represents a journey from the infinitely simple to the infinitely complex (Figure 10).

---

[49] Ibid., page 415.
[50] Samuel Avital, *The Invisible Stairway*. 1982, pages 235-237.

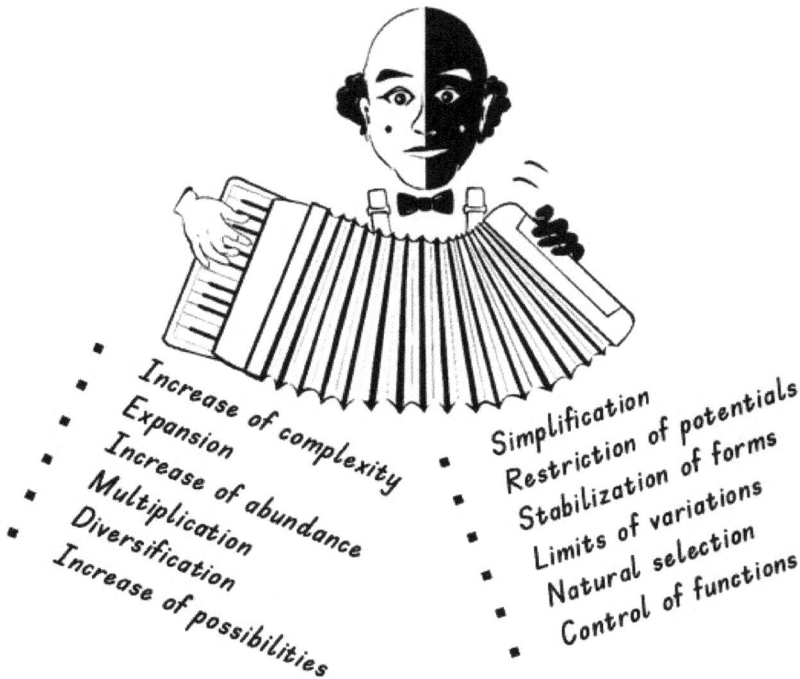

Figure 10. The "Cosmic Accordion" represents the journey between complexity and simplicity (according to the teaching of Samuel Avital).

Avital uses the breath as the simplest example of the productive opposition between the "complexity-simplicity" interplay:

"After inhaling comes exhaling, after contraction, expansion. The journey of this existence through simplicity and complexity, a sort of ball bouncing "up" and "down," between the "ceiling" and the "floor" … between two worlds, the "here" and "there," the "visible" and the "invisible" … We really cannot perceive things unless we separate them… Division precedes uniting… Different manifestations, one essence… An important part of our work is learning to be harmonious with opposites or becoming, as we say, "lovers of paradox." … We have to express weakness with strength. Our point of reference, the body, is a structure made of both the hard and the soft." [51]

---

[51] Samuel Avital, *The Body Speak Manual*. 2001, page 191.

In biology, each process consists of an interplay between these two tendencies: *from simple to complex* and *from complex to simple*. Here are several examples of movements from simple to complex in biology:

- An expansion of a biological species to new habitats.

- An increase in the diversity of the plant community.

- An increase in the variability of animals within a particular population.

- An appearance of new variants of microbes.

- An emergence of new social patterns in bird behavior.

- A diversification of butterfly wing patterns.

- An intensification of genes' recombination in response to foreign antigens.

Given this opportunity for the expansion and multiplication of biological systems (populations, species, and ecosystems), an increase in complexity could be accompanied by reduced stability ("order") and an inevitable risk that these systems might not survive. As biological systems become more complex, they often become more diverse and interconnected, with many interacting components vulnerable to the impact of small changes.

This can lead to greater adaptability and stability, but it can also make the system more vulnerable to disturbances and disruptions. For example, an increase in the complexity of an ecosystem, with more species and interactions among them, can increase its overall productivity and resilience. At the same time, it can make the ecosystem more susceptible to disruptions, such as habitat loss, pollution, or the introduction of invasive species. Overspecialization leads to reduced complexity and vulnerability, so higher complexity allows for more generalization and is more stable.

Similarly, an increase in a population's complexity, with more genetic variations and diversity, can enhance its ability to adapt to changing conditions. However, it can also make it more vulnerable to disease or environmental stressors.

Alternatively, while the trend toward simplifying biological systems can sometimes result in increased stability, this is only one possible outcome. There are several other reasons why simplification can also have disadvantages. For example, simplification can lead to reduced   genetic variation and diversity, limiting a population's ability to adapt to changing environmental conditions. It can also reduce the potential for development of new traits or functions.

Moreover, simplification is often associated with a loss of complexity and interconnectivity among different system components. Simplification in parasites increases their vulnerability. Another example is the loss of keystone species, those on which other species in an ecosystem largely depend. This is such that if the key species were removed, the ecosystem would change dramatically.

We must get through these trends of modern thought to illustrate how necessary it is to balance biological life's complexity and simplicity. Alas, the science of complexity does not provide such an orientation. It may be very effective in describing levels of complexity, but orientation marks require a system of coordination outside the described biological system. *Kabbalistic Biology* offers readers an explanation via the system of orientation.

Chapter 4

Beyond the Traps of Formal Logic

Logic is often defined as "the study of the laws of thought, correct reasoning, valid inference, or logical truth." Over time, logic has attained the stature of a science and, as such, has developed a formal language that is not easily understood by non-specialists. Because biologists investigate organisms and biological processes, the question arises: Should they learn the science of logic to enable them to proceed with greater understanding?

Here's my response: When my colleagues and I are working with a variety of living creatures, we need not study the formal rules of the science of logic. However, we must be confident that our interpretations of our observations and experiments are accurate. We must also admit that the *way* we think drastically changes how we interpret scientific results. Because we must consider two important aspects, the multi-dimensionality of biological life and the simplicity-complexity spectrum, *how* we interpret our observations will be of critical importance.

Aristotelian, or "classical," logic has had an unparalleled influence on the history of Western thought and continues to lead the reasoning applied to modern science. In biology, classical logic is unquestionably dominant.

Going Beyond Classical Logic
In this chapter, I wish to emphasize that the *style* of thinking dictated

by classical (or formal) logic is not the only available instrument to measure reasoning. We must also consider alternative perspectives. But first, let's learn more about "classical logic." In classical logic, the main assumption is that each component of a problem can be assessed as either "true" or "false." This "true/false dichotomy," when applied to each component of the statement, justifies creating a "branch" to add to one's "decision tree." A "dichotomy" divides a whole into two separate parts. Every component of the whole must belong to one part or the other. *Nothing can belong simultaneously to both parts.* The dilemma, "To be or not to be?" from William Shakespeare's play, Hamlet, is probably the best illustration of such logic, followed perhaps by the conclusion that a statement is "either true or false." According to classical logic, it cannot be both.

According to Aristotelian logic, there are four kinds of logical opposition between "universal" and "particular" propositions. If we compare two classes, *A* and *B*, they can be "contrariety," "contradiction," "subcontrariety," or "subalternation." For the first proposition, *A* and *B* are "contrary" to each other if they cannot be *true* together. For the second proposition, *A* and *B* are "contradictory" to each other if they cannot be *true* together and cannot be *false* together. For the third proposition, *A* and *B* are "subcontrary" to each other if they cannot be *false* together. For the fourth proposition, *B* is a "subaltern" to *A* if B cannot be *false* whenever *A* is *true.*

The second proposition, called in classic logic "the law of excluded middle," states that for every proposition, either this proposition or its negation is true. Aristotle formulated by Aristotle in *Metaphysics*, saying that it is necessary in every case to affirm or deny and that nothing can exist between the two parts of a contradiction.[1] Gottfried Leibniz formulated the most common expression of this principle as: "*Every judgment is either true or*

---

[1] Aristotle, *Metaphysics*. W.D. In *Great Books of the Western World*, Encyclopædia Britannica, Inc., Robert Maynard Hutchins (ed.), vol. 8, 1952.

false."[2] Thus, we learn that we must select a correct answer from two or more plausible but mutually exclusive options. We assume that we must choose Option A or Option Not-A. In biology, however, this conclusion can be misleading.

The law of the excluded middle has been deeply ingrained in our minds since we learned from pictures in children's books that one picture shows a rabbit and another shows a fox. Aristotelian logic is the logic of classes or categories. A simple example is the following statement: "All dogs are mammals." This statement indicates that there is a relationship between the class of dogs and the class of mammals. Thus, the class of dogs is a subset of the class of mammals, or equivalently, all members of the class of dogs are also members of the class of mammals. To demonstrate how this principle of classical logic is ineffective in biology, I will provide a few simple examples.

"Don't Cry Wolf ... When You Meet a Dog"

When we identify a biological species, we select some parameters to label it as belonging to one or another species. In most situations, this process works well... except when it doesn't! In the latter case, biologists must seek additional criteria and assign their observations to a specific class or species. Ultimately, each selection will lead to a choice between which of the two classes best fits the observation. If biologists still cannot assign a conclusion to one of the categories, they can admit to an "intermediate status" between two or more categories. This status can either be labeled as a "new category," or it can be assigned to one of the known categories after more sophisticated methods reveal additional information.

Let's take a very simple example. Everybody can distinguish a dog from a cat, right? Beyond the obvious signs, there are biological characteristics, including morphological, genetic, and behavioral. These present a spectrum of criteria that allows a professional zoologist to assign a dog to the species *Canis familiaris* and a cat to

---

[2] Andrei Kolmogorov, *On the principle of excluded middle*. From Jean van Heijenoort (From *Frege to Gödel, A Source Book in Mathematical Logic*, 1967, page 421.

the species *Felis catus*. The most reliable criterion that distinguishes biological species is the animal's reproductive uniqueness. Various mechanisms of reproductive uniqueness prevent different species of animals from interbreeding.

It could be more complicated to distinguish a domestic dog from a wolf (*Canis lupus*), rather than from a cat. Biological similarities between domestic dogs and wolves have prompted many zoologists to consider the domestic dog a subspecies of *Canis lupus,* namely, *Canis lupus* subspecies *familiaris*. However, other zoologists consider the domestic dog to be a separate species (*Canis familiaris*).

Cats present a similar situation. *Felis catus* is the scientific name for the domestic cat species. However, the cat can also be referred to as a subspecies of the species *Felis silvestris,* which goes beyond domestic cats (*Felis silvestris* subspecies *catus*) and includes the African wildcat (*Felis silvestris* subspecies, *lybica*) and the European wildcat (*Felis silvestris* subspecies, *silvestris*).

The situation becomes even more complicated by a wide variety of hybrids after crossing either a wildcat species with a domestic cat or breeding a wolf with a domestic dog. Is your pet a dog or a dog-wolf hybrid? The answer could have significant implications. Should a species be protected as "endangered," or should it be controlled as "dangerous?"

What is the distinction between cats and dogs? That depends on *how* we think. Both cats and dogs belong to the order Carnivora. This order consists of lower taxonomic groups, or "families." One of the families is Canidae, which includes wolves, dogs, and dog-like species, while another family belonging to this order is Felidae, which includes cat-like mammals. Therefore, both cats and dogs share various features specific to carnivores (animals belonging to the order *Carnivora*).

There are over 260 species of Carnivora. What does it mean to be a carnivore? Carnivores are predators and get most of their food by killing and eating other animals. All of them tend to establish territories. Carnivores have well-developed senses, especially sight, hearing, and smell, in order to acquire food to survive. Professional zoologists use certain anatomical features to determine how mammals rank as carnivores. All carnivores generally have four "canine" teeth (fangs or tusks) with a conical shape used to kill prey. On the front legs are toes with five digits. There is a long list of other features that define carnivores, which we need not go into here.

The point is that we can imagine a carnivore "prototype" -- an imaginary animal that exhibits the essential features of this order. Likely, all carnivores have common ancestors, called "miacids," that once inhabited Eurasia and North America. When we assign a cat, a dog, or another mammalian predator to the order *Carnivora*, we don't compare these animals with each other. Instead, we compare them with the "carnivoran prototype."

And here is the paradoxical conclusion: *to some degree, a cat is still partially a dog, and a dog is partially a cat.* This is indeed a paradox, and we must acknowledge it. The question, "Is it a cat or not a cat?" is like the question, "Is it A or not-A?" or "To be or not to be?" To answer it, we must consciously switch to a different system of reasoning.

Don't Cry Plague ... When It's Not

Similar debates are widespread in animal taxonomy, and even more so when we enter the worlds of plants and microbes. Here is an example from the world of microbes, one in which I was directly involved. The pathogen of plague, a disease that kills more people than any other, is *Yersinia pestis*. This bacterium is very similar to another bacterium called *"Yersinia pseudotuberculosis,"* which causes enteric illness after someone has consumed contaminated food or water. Both *Yersinia pestis* and *Yersinia pseudotuberculosis* are so like each other that some scientists argue they should be considered one species.

However, on a practical level, this conclusion would not work. *Yersinia pestis* is the etiological agent of plague and is considered a bioterrorism agent of the highest risk (Category A). To treat this

89

bacterium as an especially dangerous pathogen, it must be well-defined. Is it *Yersinia pestis* or *not-Yersinia pestis*? Is it plague or not plague? Definitions of particular etiological agents can go above or below the bacterial species level. Another example is *Borrelia burgdorferi*, a well-recognized pathogen of tick-borne Lyme disease. The so-called "*Borrelia burgdorferi sensu lato complex*" (from Latin *sēnsū lātō* - "in the broad sense") is a diverse group of bacteria found worldwide that includes both named species and yet-to-be-named variants. Which species/variant within the species *Borrelia burgdorferi* can be identified as an agent of Lyme disease? The answer to this question is widely disputed.

Another example of how we may be limited in defining a *single* source of a disease is the phenomenon of cancer. After a normal cell accumulates a certain number of crucial mutations, it "becomes cancerous." However, there is no clear indication as to *when* the cell might have become cancerous. Classical reasoning makes a binary distinction ("yes" or "no") between the cancerous and non-cancerous.

One proposed solution to the above situation is "probabilistic logic." Nils Nilsson, a professor of computer science at Stanford University, first formulated the term. In a paper published in 1986, he argued that statements of "truths" are, in fact, only probabilities.[3] In the case of cancer, probability reasoning can provide information about the *likelihood* of tissue becoming cancerous at a specific age of the patient. Such factors would include the presence of specific genes, behavioral patterns such as smoking, and so on. However, the ultimate conclusion comes down to classical logic: at any specific point, the cell is either cancerous or non-cancerous.

Walter Elsasser, a physicist turned theoretical biologist, wrote an article titled "A Form of Logic Suited for Biology." Here he compared the difference between physics and biology: "Since physics deals primarily with extension, its chief tool of description is the

---

[3] Nils Nilsson, *Probabilistic Logic*. Artificial Intelligence, 1986, vol.28, pages 71-87.

continuum. On the other hand, biology starts with taxonomy; correspondingly, the dominant concept in biology is that of a *class*".[4]

Traps of Traditional Logic

Robert Horn, an American political scientist who taught at Harvard, Columbia, and Sheffield universities, suggests that some flaws of traditional logic may be more common and insidious than expected. Why? To begin with, these concepts can be limited by traditional ideas about change and stability.

In his paper, "Traps of traditional logic & dialectics: what they are and how to avoid them," Horn presented descriptions of seven traditional logic traps:

1. *The Forever Changeless Trap.* In this trap, we believe that the current condition will remain the same forever.

2. *The Process-Event Trap.* This trap appears when we begin thinking of objects as "events," whereas we would do better to think of objects as "processes."

3. *The Solve-It-by-Redefining-It Trap.* This could be called "The Definition- Can- Do- It Trap," which attempts to solve problems by simply redefining them.

4. *The Independent Self Trap.* In this trap, we separate organisms from their environment, and ourselves from our interdependence with others.

5. *The Isolated Problem Trap.* In the grip of this trap, we regard problems as being unconnected to their wider contexts.

6. *The Single Effect Trap.* In this trap, we think that we can cause a single effect with no "side effects."

7. The Exclusive Alternatives Trap. Traditional logic tends to make us think in terms either-or analyses. However, many

---

[4] Walter Elsasser, *A Form of Logic Suited for Biology.* Progress in Theoretical Biology, 1981, vol. 6, pages 23-62, italics in this quote are the author's to indicate the distinction between the methods of physics and those of biology.

situations demand that we juggle more than two alternatives.

Considering the traps of classical reasoning, some logicians have proposed many non-classical" formal systems that differ significantly from standard logical systems such as "propositional" and "predicated" logic. Among those, one branch of logic is called "many-valued logic." Such logic rejects bivalence, allowing for truth values other than "true" and "false." One of these directions is "three-valued logic" developed by logician and philosopher, Jan Łukasiewicz. He replaced the old bivalent logic with the new trivalent one, using statements such as "true," "false," or "indeterminate."

Robert Anton Wilson, in *The New Inquisition*, proposed a non-Aristotelian system of classification. Here, propositions can be assigned one of seven values: true, false, indeterminate, meaningless, self-referential, game rule, or strange loop. Wilson did not devise a formal system to manipulate propositions once they were classified but suggested that we can clarify our thinking by not restricting ourselves to simplistic true/false binaries.

Alternative ways of thinking

To better understand biological life's complex phenomena, we require alternatives to formal logic that adhere to the principle of non-contradiction, where either "A" or "not A" is true. The paradoxical situations illuminated in the "dog-not-dog" and "plague-not-plague" relationships allow for the simultaneous existence of "A and "not A." According to this concept, two contradictory statements can coexist, offering Hamlet an alternative "to be *and* not to be."

Alternative forms of logic can provide new perspectives on complex, dynamic, and paradoxical phenomena. Quantum mechanics, for instance, involves particles existing in a superposition where they are simultaneously in multiple states, like Schrödinger's cat, which is both alive and dead until it is observed.

The concept of dialectic logic or paraconsistent logic can be related to this alternative way of thinking. Dialectical logic views reality as a dynamic process in which contradictions and tensions can drive change and development. According to Hegel's dialectical method, the interaction between opposing ideas (thesis and antithesis) creates a synthesis, which then becomes a new thesis, creating an ongoing intellectual process. Paraconsistent logic, on the other hand, provides a formal method to accommodate contradictions while avoiding inconsistency. Paraconsistent logic lets contradictions exist *without* making every statement true.

Now, let's turn our attention to another direction: the infinitely valued logic known as "fuzzy logic." This is the logic used to solve complex problems, including those of modern biology. Fuzzy logic was formulated in the mid-1960s by Lotfi Zadeh, a professor at the University of California, Berkeley.

"Fuzziness" is not a respected term in science and is commonly associated with the quality of being" vague" or "without sharp outlines." A better word to equate with fuzziness is "multivalence." The Merriam-Webster Dictionary defines multivalence as "the quality or state of having many values, meanings, or appeals." Fuzzy, or multivalent, logic has a deep connection to the "incompleteness theorem" developed by the famous logician and mathematician Kurt Friedrich Gödel. Gödel's theorem claims that science contains true statements that cannot be proven. One of the popular formulations of this theorem is: "There is more to the *truth* than can be caught by *proof.*"

Compared to classical binary sets where variables may take on "true" or "false" values, fuzzy logic variables permit any real number between 0 and 1 to be considered a "truth value." *Fuzzy logic offers the concept of partial truth,* where the truth levels may range between completely true and completely false. Fuzzy logic *accepts* inaccuracy, uncertainty, and unpredictability. In a broad sense, fuzzy logic refers to "fuzzy sets" with blurred boundaries. In a narrow sense, *fuzzy logic is a logical system that tries to formalize approximate reasoning.* It is defined as "a form of knowledge suitable for notions that cannot be defined precisely, but which depend upon their context."[5]

---

[5] Joseph Bih, *Paradigm Shit – an Introduction to Fuzzy Logic.* EEE Potentials, 2006, pages 6-21.

Sets that are fuzzy or multivalent break the "laws of the excluded middle" to some degree.[6] Objects belong only partially to a fuzzy set. They may also belong to several sets. Relationships between sets show the paradox at the heart of fuzzy logic. In classical logic, an object either does or doesn't belong to a set. A glass of water is either empty or full. Fuzzy sets cover a continuum of partial sets, such as glasses only half full or one-third full. The example with the empty and full glasses is symbolic but very real. There are unlimited possible variants between an empty glass and a full glass. (Figure 11).

Figure 11. There are unlimited possible variants between an empty glass and a full glass.

Although the example with glasses (Fig.11) is simple, it still illustrates a "fuzziness." Claims that the glass is empty or full can also be questioned. The glass could be empty and dry or empty with drops of water still in it. It may be empty of water, but not of air. The claim about a full glass leads to even more questions. When is a glass full enough? What are the criteria for "fullness?"

The examples of an empty glass and a full glass can represent a wide range of situations, from biological objects like organisms to more abstract concepts like knowledge about evolution. With the "glass" metaphor, I illustrate that many things in life exist on a continuum rather than as binary opposites. The metaphor suggests that any given situation can fall somewhere between an utterly empty state and a complete state. For example, health and illness exist on a spectrum rather than being completely healthy or ill. It is challenging to define a health status between these extreme situations.

---

[6] Robert Anton Wilson, *The New Inquisition: Irrational Rationalism and the Citadel of Science.* 2020, 350 pages.

Consider my warnings when using the "empty-full" glasses example. First of all, the situation with glasses as physical objects implies a linear progression from one physical state (empty) to another (full). It is possible to arrive at different states through multiple paths or trajectories. In 1921, Albert Einstein made a very relevant statement to the discussed problem, "So far as laws of mathematics refer to reality, they are not certain; and as far as they are certain, they do not refer to reality."[7]

More important is the fact that complex biological systems cannot be reduced to a simple continuum, so do not take this metaphor literally! A biological system may follow multiple paths or trajectories, depending on interacting factors, rather than following a linear path. For example, the dynamics of a population of organisms can be influenced by a wide range of factors, such as changes in the environment, the availability of resources, and interactions with other species. Complex and unpredictable interactions between these factors can result in endless possible outcomes that do not neatly fit into a continuum between two states. These factors may include genetic variation, developmental processes, or commensal interactions.

Life is *Not* Black-or-White

"Fuzzy logic" means different things to different people. For some, according to Petr Hájek, a professor at Charles University in Prague, it is a philosophy of life: "a way to break the stranglehold that black-and-white thinking of the tradition has upon us."[8] For others, it is a more accurate way of describing our ordinary language "in which we do not think that everything is either true or false, but where we recognize the shade of gray that populates our thoughts and linguistic communication."

Bart Kosko, a professor at the University of Southern California's Viterbi School of Engineering and a former student of Zadeh, in his

---

[7] Albert Einstein gave an address on 27 January 1921 at the Prussian Academy of Sciences in Berlin. He chose as his topic Geometry and Experience. The lecture was published by Methuen & Co. Ltd, London, in 1922.

[8] Petr Hájek. *Metamathematics of Fuzzy Logic*. Trends in Logic, vol. 4. 1998, 297 pages.

book, *Fuzzy Thinking*, formulated a so-called "mismatch problem." The world is gray, but science is black and white.[9] One of the examples that Kosko provided was the statement that "grass is green." Such a statement isn't 100% true, however, because "the blade of green grass turns brown." Merlin Sheldrake, an investigator of fungi, claimed in *Entangled Life*, "Biological realities are never black-and-white."[10]

The leading expert on Talmud and Kabbalah, Adin Steinsaltz, wrote:

> "Life is not a black-or-white matter; on the contrary, every event and occurrence in our lives is somewhere on the gray scale. In the scientific realm, too, there great advantage in discerning the shades. In fact, a major difference between the scientist and the layman is in their ability to see minute differences between objects. Understanding small differences and how they are created is the most basic tool for developing in-depth knowledge... While the health intellectual sees the shades of gray as interim stages between the black and the white, the rotten intellectual sees neither the black not the white: he sees only the shades of gray."[11]

Is a Zebra Vertically Striped?

The answer to the simple question, "Are zebras vertically or horizontally striped?" might vary. In general, people say that zebras have primarily vertical stripes. This pattern of stripes is thought to provide the animal with several benefits, including confusing its predators and serving as a form of visual communication with other zebras. Some parts of the zebra's body have vertical stripes, while others have diagonal or horizontal stripes. In addition, there are three species of zebra, each with a different pattern of stripes. For example, the Grévy's (or imperial) zebra has predominantly vertical stripes,

---

[9] Bart Kosko and Satoru Isaka, *Fuzzy Logic*. Scientific American, 1993, pages 76-81.

[10] Merlin Sheldrake, *Entangled Life*. 2020, page 42.

[11] Ron Goldschlager and Adin Steinsaltz, The Mystery of You: A Journey through the Paradoxes of Life. 2011, page 345.

whereas the mountain zebra has more diagonal ones (Figure 12). The example with zebra strips is purposely oversimplified. However, this example represents a frequent dilemma when describing biological systems, including animals.

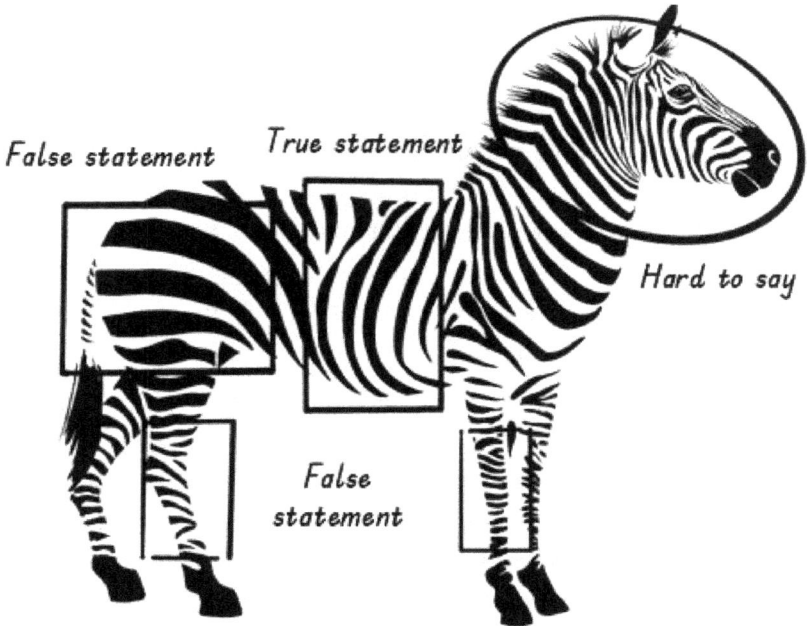

Figure 12. The question of whether zebra is vertically striped has different answers.

Some might argue that the example of zebra stripes is unimportant in biological practice. Let me, instead, take a situation commonly experienced by zoologists working with wild rodents in North America. Two of the most abundant species of mice are the deer mouse (*Peromyscus maniculatus*) and the white-footed mouse (*Peromyscus leucopus*). Mice of these species look very much alike to a non-specialist. However, experienced zoologists can distinguish those species in most cases. One of the common characteristics used for identification is tail color. In deer mice, the tail is bicolored, the darker top half and the lighter bottom are sharply differentiated. That differs from white-footed mice, in which the line separating the upper and lower portions of the tail is less distinct and more gradual. These characteristics vary geographically, however, and in some areas, the two species are difficult to distinguish based on external morphology. In addition, young mice do not express such an evident distinction between species. In some cases, identifying these species is difficult without special genetic testing conducted in laboratory settings.

A decision about species identification depends on many ecological and geographic factors. For example, in some regions, either deer mice or white-footed mice are absent, and in this case, zoologists could easily make their choice. Identification of mouse species could have practical importance. For example, deer and white-footed mice carry different types of hantaviruses, and the ability to identify their species could be critical for identifying a source of human infection. Circumstances and the significance of observations do affect the way we think!

An argument could be made that the example with the mouse tail color will only interest a small group of zoologists. Would forecasting epidemics be vitally important? I ask this question at a time when the COVID-19 pandemic continues to spread. I can explain how epidemiological predictions are made, and why they almost all fail! If you wish to learn, be prepared for more specific terms. If not, feel free to skip the following two paragraphs.

Nearly all epidemiological predictions are based on the so-called "SIR model" developed by Roy Anderson and Robert May, two distinguished scientists from the University of Oxford around 40 years ago.[12] In this model, "S" stands for the number of "susceptible hosts," "I" for the number of "infectious hosts," and finally, "R" for the number of "recovered hosts" (those are individuals who became immune and did not participate in the future spread of the infection, at least for some time). Since then, many modifications have been proposed to fine-tune the model. For example, they introduced

---

[12] Roy Anderson and Robert May, *Infectious Diseases of Human: Dynamics and Controls.* 1991, 757 pages.

"exposed hosts" (SEIR model), "carrier status" (SICR model), "maternally-derived immunity" (MCIR model), and more.

These models assume that each host is either infected or not, either susceptible or not, and either recovered or did not. Right? Wrong, I argued in my article published in *Entropy* in 2013.[13] In reality, host-pathogen relations are much more complex, reaching the point that you cannot define any parameter by using "yes" or "no" statements. The individual host is infected. But infected by what? Maybe it is a closely related but non-pathogenic strain. Maybe the detected microbe was inhaled briefly and is gone from the host organism. Maybe the microbes can stay in the host organism for a long time but are inert in their ability to produce the infection. A status of "susceptibility" and "recovery" is not easier to define by "yes" or "no" statements. The problem is not in the lack of information. We have tons of information. The problem is our reliance on our "dualistic" logic ("yes" or "no").

We need a different form of statistics to analyze a complex biological system where context and significance are inescapable components. "Fuzzy statistics" usually refers to a combination of "fuzzy set theory" the treatment of ambiguous, imprecise, or subjective data and traditional statistical methods. James Buckley, professor of mathematics at the University of Alabama at Birmingham, prepared a series of books on fuzzy statistics, including *An Introduction to Fuzzy Logic and Fuzzy Sets, Fuzzy Probability and Statistics,* and others.

Fuzzy logic and fuzzy sets are increasingly being used in biology. Here is a list of the titles of some selected scientific articles: "Fuzzy logic for biological and agricultural systems," "Fuzzy epidemics," "Medical diagnosis, diagnostic spaces, and fuzzy systems," "Employing fuzzy logic in the diagnosis of a clinical case," "Fuzzy logic-based approaches for gene regulatory network inference," and

---

[13] Michael Kosoy, *Deepening the Conception of Functional Information in the Description of Zoonotic Infectious Diseases.* Entropy, 2013, vol.15, pages 1929-1962.

"Application of fuzzy logic in oral cancer risk assessment."

Still, most biologists are entirely unaware of such logical thinking. I have heard remarks from my colleagues about "fuzzy logic" given in an offhand manner as if implying vagueness and unreliability. Kosko recognizes that the difficulty with accepting non-classical logic is rooted in cultural tradition. "The only barrier remaining to wider use of fuzzy logic is the philosophical resistance of the West."[14]

Kosko added, "Fuzziness begins where Western logic ends." The Japanese scientific community embraced fuzzy logic earlier than did scientists in the West. Some attribute this to Japan's cultural environment. Kosko commented that "the yin-yang symbol is the emblem of fuzziness." Fuzzy logic introduces human subjectivity into the realm of objective science and uses human senses as they are, without adding the complication of abstraction. As a result, Japanese companies have become leaders in fuzzy logic-based research, and in transforming that research into industrial applications. It is now in use around the globe in business, finance, management, and even in defense areas (such as the report, *Short Range Missile Autopilot Design using Fuzzy Control Technique* given at the Conference on Aerospace Sciences and Aviation Technology in 2018).

Constantin Virgil Negoita, a professor of Computer Science at Hunter College, City University of New York, has written several books on fuzzy sets. In one of his books, *Expert Systems and Fuzzy Systems*, Negoita explored the effects of semantic systems on decision support systems. In it, he asserted that using fuzzy set theory can help an expert computer system draw from its knowledge base more efficiently, and therefore make more accurate and reliable decisions.[15] One of the important claims made by Negoita is that "The *description is fuzzy, not the things… Inexactness is not a liability. On the contrary, it is a blessing given the sufficient*

---

[14] Bart Kosko. *Fuzzy Thinking.* 1993, 318 pages.
[15] Constantin Negoita, Expert Systems and Fuzzy Systems. 1985, 190 pages.

information it can convey with less effort."[16]

From Fuzzy Logic to Kabbalistic Logic

To explain a fuzzy set, Negoita invites us to re-enter the two-dimensional world of Flatland; here is his explanation:

> "Each Flatlander is incarcerated in a flat set. If we fling a Flatlander into our three-dimensional world, he can see only two-dimensional cross-sections of our world, a family of crisp sets. Simply put; by adding another dimension, we can capture more features. *This is what a fuzzy set does. It adds a new dimension: our evaluation of the membership.* Using classical flat mathematics, a fuzzy set can be represented by a family of crisp sets, projected on the Flatland."[17]

Negoita sees a deep connection between "Fuzzy Set Theory" and "Postmodernism," the late-20th-century style in the arts, literature, and science that departs from modernistic traditions. According to Negoita, "The fuzzy set theory …encouraged the acceptance of uncertainty as a condition of everyday life … helped to accelerate the shift from the Enlightenment ideal of the two-valued logic to a postmodern preoccupation with many degrees of truth."[18] To continue the line of connection between fuzzy logic and postmodern thinking, Negoita's close associate, Gabriel Burstein, developed a very interesting theoretical framework. He wrote a series of publications in which he and his co-authors credited Kabbalah with providing the foundation of the postmodern fuzzy logic theory.[19]

Fuzzy logic addresses complex and uncertain information through the lens of human knowledge and subjectivity. However, more progress is needed when we explore the multi-modal aspect of

---

[16] Constantin Negoita, Remembering the Beginnings. International Journal of Computers, 2011, vol. 6, no. 3, page 460.

[17] Ibid., page 458, italics are mine.

[18] Constantin Negoita, *Postmodernism, cybernetics and fuzzy set theory.* Kybernetes, 2002, vol.31, pages 1043-1049.

[19] Gabriel Burstein and Constantin Negoita. *Foundations of a postmodern cybernetics based on Kabbalah.* Kybernetes, 2011, vol. 40, pages 1331-1353.

fuzzy logic. [20] There are multiple angles of assessing the truth due to different cognitive, emotional, and behavioral aspects. Burstein and his colleagues formulated a fractal multi-modal logic of Kabbalah that integrates the cognitive, emotional, and behavioral levels of humanistic systems.[21]

Fuzzy logic deals with possibilities and linguistic variables. Bruce Friedman, a physicist whose research included fuzzy logic and who is also knowledgeable about Kabbalah, makes a distinction between the notions of "probability" and "possibility," a helpful tool for applying possibility theory.[22] The rules for the application of possibilities can be significantly different from those for the application of probabilities. Friedman wrote, "You have a choice of a number of possible things that you can do at a given time. When you make that choice all of the possibilities vanish except for that one possibility which is transmuted into actuality… The state of all possibilities is not considered to be nothing, but is deemed to be everything. The contraction is then taken to result in nothing which eliminates "all possibilities" except the one possibility that is made real."[23]

Now let's be clear: Fuzzy logic doesn't mean Kabbalistic logic! Nevertheless, an essential similarity between the two is that they both allow for multiple levels of interpretation and meaning. As a result, our system of thinking is increasingly enhanced with each passing level of reality and perception. Both fuzzy and Kabbalistic types of reasoning recognize that strict binary distinctions may not capture the full complexity of reality and that multiple levels of interpretation may be necessary to gain a deeper understanding. They both allow for more nuanced and flexible approaches to understanding complex systems and ideas.

---

[20] Gabriel Burstein et al. *Kabbalah Logic and Semantic Foundations for a Postmodern Fuzzy Set and Fuzzy Logic Theory.* Applied Mathematics, 2014, vol.5, pages 1375-1385.

[21] Gabriel Burstein and Constantine Negoita, *A Kabbalah System Theory of Ontological and Knowledge Engineering for Knowledge Based Systems.* International Journal of Advanced Research in Artificial Intelligence, 2013, vol.2, page 10.

[22] Bruce Friedman, *Mystery of Black Fire.* White Fire. 2016, page 252.

[23] Ibid., page 253.

Placing the words "fuzzy thinking" and "Kabbalah" on one line, I must make a very important claim. As it addresses issues of higher dimensions, the language in Kabbalah becomes direct, confident, and free of any "fuzziness." Unlike fuzzy logic, Kabbalistic reasoning recognizes the existence of true values. *In Kabbalah, truth is viewed as the unity of all things in one Source.* Kabbalah recognizes that truth is often beyond human comprehension, and can only be approached through images, metaphors, and concepts, such as the Tree of Life, which we will discuss in Chapter 10.

One way in which Kabbalah explores these mysteries is through the use of "spiral logic." Unlike traditional linear logic based on binary distinctions and linear progressions, spiral logic allows for a more nuanced and dynamic approach to understanding complex systems and ideas. *Spiral logic is based on the idea that truth values can be represented as points on a spiral curve,* rather than as discrete true or false values on a linear scale.

Complex relationships between obtained information and concepts can be represented instead of simple binary distinctions. Kabbalah uses complex correspondence systems to explain reasoning, such as the associations between the Sefirot and various aspects of the natural world or the human psyche. In symbolic diagrams, such as the Tree of Life, these systems are spiraled outward from a central point to encompass a wide range of concepts. Nevertheless, Kabbalah does not exclusively use spiral logic, but instead uses a traditional system of thought to uncover more profound truths about reality.

When the concepts of multi-dimensionality, complexity, uncertainty, and nonlinearity first appeared, they forced scientists to rethink the definition of "science." The process of exploring the manifestation of biological forms and actions requires a multi-valued system of thinking and flexibility in expressing language. At the same time, the famous physicist, Michio Kaku, wrote, "The laws of nature become simpler and more elegant when expressed in higher dimensions, which is their natural home."[24] Through our knowledge of higher-dimensional space, we can enhance our abilities to learn more about biological life. The challenge is *knowing how to travel through different dimensions to obtain this knowledge.* Such a practice would require the ability to accept unfamiliar levels of complexity and uncertainty and to be open to alternative points of view.

We have a choice: we can either ignore the challenges illuminated by the multidimensionality of biological life, or we can accept paradoxes as natural parts of our thinking practice. Paradoxes are often perceived as challenging and frustrating in thinking and problem-solving. While paradoxes may not be solvable in the traditional sense, we *can learn to live with them.* In this way, paradoxes can be seen as powerful tools for expanding our thinking and pushing us to consider new perspectives. To do so, however, we must be ready to begin practicing alternative ways of thinking.

---

[24] Michio Kaku, *Hyperspace.* 1994, page 12.

PART TWO:
Entering the Invisible World of Biological Life

Chapter 5

Information is *NOT* Knowledge

In previous chapters, we discussed the importance of dimensionality and complexity—both powerful tools that can add to our knowledge of biological life. In this chapter, I propose that modern science's preoccupation with collecting massive data and processing all available information will not automatically bring us insight into biological processes.

The biological world is neither linear nor straightforward; hence, we need multi-sided and multi-dimensional viewpoints. No dimension is "right" or "bad," but it can be functioning at a "higher" or "lower" level, as we will discuss in this chapter. Similarly, levels of "simplicity" do not have a "good" or "bad" side; instead, they can move in either direction, from simple to complex. Researchers can choose a question and define a perspective to look at a biological system.

The Power of Perception

I am convinced that the main reason for most disagreements between biologists is not based on a lack of information or the validity of the concept. The fact is that scientists may, at times, regard a biological object from *different points of view*. The solution? Setting up a system that will allow *all* perspectives to be factored into the search for answers. Biologists must also be aware of the levels of dimensionality and complexity. Which method will be most appropriate for a specific investigation? Can researchers depend on personal preferences to get optimum results? It takes the ability to consider a multidimensional approach when selecting a particular perspective.

Let's take a look at the situation created by the spread of the coronavirus during the recent COVID-19 pandemic. When the pandemic began, many TV channels and blogging websites offered interviews with biologists and epidemiologists familiar with the behavior of viruses. Many were trusted experts, including molecular biologists, immunologists, virologists, epidemiologists, and infectious diseases doctors. However, the challenge for the guest experts was daunting: they were all being asked questions they could not definitively answer!

TV interviewers tossed out questions to their guests on how quickly the epidemics would spread. What level of group immunity might stop the pandemic? How could viewers avoid infection? These were reasonable questions from a public obsessed with the emerging threat of an out-of-control pandemic. In its early days, only a handful of scientists had studied the transmission of coronaviruses in human populations. Even epidemiologists who had worked during an epidemic caused by a similar virus could not answer every question. To discover why they were unable to do so, let's take a look at the challenges scientists face in such situations.

First of all, most experts are specialists in a particular discipline. They might know details about a viral protein's structure but have minimal knowledge about its potential to cause a full-blown epidemic. In another possible scenario, doctors specializing in infectious diseases might be able to see actual and potential developments in clinical studies. However, their previous knowledge of viruses might have been limited to a class in virology taken years ago in medical school. In such a case, the expert's ability to reconstruct a general picture of the epidemic could be impaired by the lack of information about a more recent outbreak.

Second, epidemics are being recorded in very diverse environments and social conditions. Information obtained from observations in one place will not necessarily apply to different environments and conditions.

Third, the epidemic process is *very* complex. Scientists have information about a few elements within this system, especially regarding new emergent diseases. However, the main problem for scientists is the imbalance between the limited information available and the breadth of knowledge required to handle such complex systems as epidemics.

"Unknown Unknowns"

When experts with varying experiences are asked questions, their training and experience will certainly affect their answers. There are two other challenges, as well. The first is a choice between giving a simple or a complex answer. If the answer is too technical, most questioners will be unable to understand it; conversely, if the answer is too general, it could come across as vague or simplified. Another challenge is assessing the *level* of unknowns regarding the COVID-19 infection under discussion. This can be a dilemma for scientists who dislike admitting, "I don't know," if they lack confidence in their ability to give an accurate answer. A question also arises when experts are aware of "what we don't know, but may know *someday*" versus "what we *do not know,* period." My teacher, Samuel Avital, prefers to say "Not Yet Known" instead of "Unknown." He explains that "unknown" sounds as if all possibilities are closed. In contrast, the words "Not Yet Known" open the door to infinite potential for discoveries.

This can arise when recognizing the reality of the small proportion of "known" data versus the possibly "unknown–hidden–invisible" data. There is a saying: "There are known knowns; those are things we know we know. We also know there are known *unknowns;* that is to say, we know there are some things we do not know. But there are also *unknown unknowns*—the ones we don't know we don't know." The idea of "unknown unknowns" was conceived in 1955 by two American psychologists, Joseph Luft and Harrington

Ingham. They were in the process of developing a technique to help people better understand their relationship with themselves, as well as with others.

We'll return to this idea later to learn how it can be valuable when we are searching for fundamental knowledge. Ralph Gomory, the President of the Alfred P. Sloan Foundation, wrote in *Scientific American* in 1995: "We are all taught what is known, but we rarely learn about what is not known, and we almost never learn about the unknowable. That bias can lead to misconceptions about the world around us."[25]

Data is the "Food" of Modern Science

"In the beginning, there was the data" would be an appropriate statement to start discussing the hierarchy of knowledge. Without data, scientists would have no empirical basis for their theories or hypotheses and would be unable to test or refine their ideas. One could compare data to "food" that can be digested by being chewed. The end product of "chewing" is information. Just as food is essential for human survival and growth, data is essential for the growth and advancement of scientific knowledge.

Data are a discrete and unprocessed raw set of values received from observations. Values can be quantitative, such as the number of rodents captured in 100 traps during a night, or qualitative, such as the length of a mouse. Every observation and measurement of a biological object provides data about a specific object, such as an animal or an animal cell. Data merely exists. By itself, it lacks meaning or significance. For example, the measured deer mouse's length turned out to be 180 mm. In the future, a zoologist may conclude that this mouse is large or small based on the data, but the data is always the data. The most critical measurements are correct when within a defined system of reference.

Each branch of biology has its specific methods of research. Each

---

[25] Ralf Gomory, *The Known, the Unknown and the Unknowable.* Scientific American, 1995, page 120.

biologist uses methods appropriate to explore different levels of biological organization, whether cells, tissues, organisms, species, ecosystems, etc. Clearly, different research methods result in different data types. For example, different types of microscopes can provide different measurements of the structure of individual cells and their components. The selection of methods will be affected by various factors, including personal preferences, available resources, institutional history, and many others. As a result, the data will also be different. That is normal. Keeping it in mind is all we need to do. While the data obtained from different methods may not always be aligned, this does not mean that one method is inherently better or worse. Instead, each method may provide a unique perspective of the system being studied, and it is up to the scientist to evaluate and interpret the data meaningfully.

We should always anticipate errors in measurement. If measurements are repeated, the data should remain-- if not the same--close enough. If the same object could be measured by a more sophisticated tool, the data could then be more accurate. It's essential to assess the *quality* of the data. *But data remain simply data until they are not proven to be false.*

Data is objective in the sense that each element of data represents a real "object." I would be more cautious with the definition of objectivity of data presently taken for granted. Some factors can consciously or unconsciously influence the choice of a specific biological object or the selection of a particular way to measure it. For example, when an ornithologist chooses to investigate one bird from a flock of the same species, the investigated bird may have been selected because it looked healthy. Such a choice can be justified under certain conditions, but it might also be subjective. This is an example of so-called "observer bias." Observer bias is the tendency of investigators to look beyond what they are seeing. Instead, they see what they expect or want to see. It's a very common-- and serious-- problem when collecting data in any research. Some practical tools can reduce, if not omit, observer biases. These include making a random selection of objects or using blind and double-blind experiments.

The term "data" comes from the Latin *datum,* that means *given.* An interesting remark about the meaning of data was made by Irun Cohen, professor of immunology at the Weizmann Institute of Science, who was awarded the prestigious Robert Koch Prize for his work on autoimmunology. Cohen noted, "Data are not to be taken for granted. Data, despite the name, are not so much *givens* as they are *takens.*"[26] The continuation of Cohen's thought is remarkable, "*Data, like the Torah, are given to all but received only by those prepared to see what the data mean.*"[27] Scientists may not recognize data as "facts" if they are not prepared to accept or interpret them!

More Is Not Automatically Better

There is a mistaken belief that so-called "Big Data" arriving in increasing volumes will automatically lead to complete and valuable information. This belief has been strengthened by the development of powerful computers that can process and store enormous amounts of data. "Big Data" refers to data sets that are too large or complex for traditional methods of processing. These sets have the potential to transform many areas of biology into new ways of doing research.

Human genomics is one area where Big Data storage is exploding. Exceptionally, few readers will be able to comprehend such an amount (I cannot), but below are some numbers from the Human Assembly and Gene Annotation Database.[28] This reference site provides a data set that includes >32 billion base pairs partitioned into >20 thousand protein-coding and >22 thousand non-protein-coding gene annotations transcribed into >200 thousand transcripts. I provide these numbers specifically to illustrate the massive amount of data available in this field for a specific resource.

A common expectation from Big Data is that their analysis will

---

[26] Irun Cohen, *Rain and Resurrection.* 2010, page 125; italics by Cohen.

[27] Ibid., page 125.

[28] www.ensembl.org/Homo_sapiens/Info/Annotation

allow 1) higher potential for inclusiveness, 2) reduce our desire for exactitude (precision), 3) and put a strong emphasis on correlations, that is, relations between phenomena or variables.[29] The role of correlations has been well accepted as the starting point in investigations. However, the central premise of Big Data analysis is the priority of relationships over causal explanation (the explanation of the internal mechanisms of a phenomenon). The expectation is that the analysis of Big Data will defy or supersede conventional knowledge.

However, this perception of Big Data creates a significant challenge for biologists, thanks to the sudden arrival of massive genetic, phenotypic, and environmental data and information. With the development of modern technologies, including molecular detection and genome sequencing, the ability to generate new biological data has increased tremendously. As a result, analyses of "big data" might bring helpful information but they might not! *More is not automatically better!*[30] Data by themselves are meaningless. *"The idea that with enough data, the numbers speak for themselves hardly makes sense."*[31]

Christopher Ponting, a British computational biologist and the Chair of Medical Bioinformatics at the University of Edinburgh, warns of overexcitement about *"big knowledge from big data."* Ponting wrote: "This is because sequence annotations do not always implicate function, and molecular interactions are often irrelevant to a cell's or organism's survival or propagation. Merely correlative relationships found in big data fail to provide answers to the why questions of human biology."[32] In conclusion, Ponting said, "Life

---

[29] Viktor Mayer-Schönberger and Kenneth Cukier, *Big Data: A Revolution That Will Transform How We Live, Work, and Think.* 2014, 272 pages.

[30] Michael Kosoy and Roman Kosoy, *Complexity and biosemiotics in evolutionary ecology of zoonotic infectious agents.* Evolutionary Applications, 2017, page 1.

[31] Fulvio Mazzocchi, *On Big Data: How should we make sense of them?* EMBO Reports, 2015, vol. 16, page 1253.

[32] Chris Ponting, *Big knowledge from big data in functional genomics.* Emerging Topics in Life Sciences, 2017; vol. 1, no. 3, page 245.

sciences are awash with data, but relatively bereft of knowledge."[33]

Climbing the DIKW Pyramid

To arrive at the process of data transformation, I will use an imaginary model that has become quite popular in recent years. The idea of a hierarchy from data to wisdom was formulated back in 1989 by Russell Ackoff, the founder and head of the Institute for Interactive Management in Philadelphia. This model is commonly called a "DIKW Pyramid." Here, "D" stands for *data*, "I" for *information*, "K" for *knowledge*, and "W" for *wisdom* (Figure 13).

Notably, Ackoff proposed one more level, placing "understanding" above "knowledge" but below "wisdom." Thus, there can be no understanding without knowledge and no wisdom without understanding. For the whole model, the more appropriate name should be "The DIKUW Pyramid." However, Ackoff's perceptive statement has been ignored by many of his followers. Why? I suggest it is due to the confusion between conventional ideas of "understanding" and "wisdom" in modern societies.

Let's examine each level of the DIKW Pyramid. I'll explain in the following chapter why it's crucial for developing new dimensions, particularly from the perspective of the Kabbalistic tradition. The DIKW pyramid, incidentally, has been seriously discussed and applied by representatives of sectors such as science, business management, public health, and education.

Before I explain more about the DIKW Pyramid, I would like to share the following quote from Ackoff: "An ounce of information is worth a pound of data. An ounce of knowledge is worth a pound of information. An ounce of understanding is worth a pound of knowledge." These inspiring words were offered by an expert whose research, consulting, and education have impressed over 250 corporations and 50 governmental agencies in the U.S. and abroad. Ackoff explains that each of the higher levels in this hierarchy includes the categories that fall below it. For example, there can be no

---

[33] Ibid., page 247.

information without data and no knowledge without information. Curiously, Ackoff's impression is as follows: "... on the average, about forty percent of the content of human minds consists of data, thirty percent information, twenty percent knowledge, ten percent understanding, and virtually no wisdom."[34]

There are different questions for each DIKW world. Since data does not have an independent meaning, the main questions are "How was the data collected and presented and, most importantly, how reliable was it?" There are many questions about information, like how much data to use, where and when data was collected, who reported the data, etc. The following questions are about generating or selecting datasets, structuring presented data, and, most importantly, discovering what meanings can be drawn from this data.

The main question in the world of knowledge is about the *value* of available information. How can this information help us solve our most important problems, like stopping the epidemic? As we discovered, some information can be disregarded if it's not helpful or relevant. The main question about understanding, which is inseparable from wisdom, is "why?" (Figure 13).

---

[34] Russell Ackoff, *From Data to Wisdom:*
Presidential Address to ISGSR. Journal of Applied Systems Analysis, 1989, vol. 16, page 3.

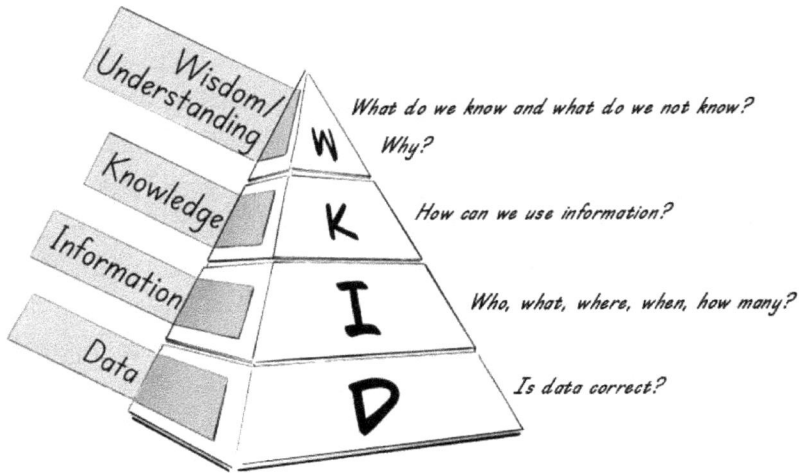

Figure 13. The D-I-K-W pyramid reflects the idea of a hierarchy from data to wisdom. Adequate questions are asked at different levels of the Knowledge Hierarchy.

Data is *Not* Information

If defining "data" is quite simple and straightforward, defining "information" is a different story. The word "information" has multiple meanings and definitions and could be used in different contexts. For example, in the field of communication, information may refer to the content of a message. In biology, it may refer to genetic messages or the signals exchanged between cells or animals. In this chapter, I use the word "information" as a fundamental step of gaining knowledge.

Data are symbols that represent properties of particular objects (e.g., animals or bacteria), processes (e.g., movements of animals from one point to another), and environments (e.g., soil temperature). Information gives data meaning by processing and also provides a context in which data makes sense. The word "*in-formation*" indicates that data is *in the process of* being formed or shaped, or that it exists within a formation process.

The ability to actually interpret the data obtained to gain information about biological systems could create drastic differences. The ability to distinguish relevant from irrelevant data, as well as to recognize patterns and trends, may only *sometimes* be evident.

116

Perception also plays a role in formulating hypotheses and designing experiments. After all, scientists must use their perception to identify potential variables and factors that could influence the biological system being studied. In addition, the perception of scientists can also influence the interpretation of data and the conclusions drawn from studies. Scientists may bring their biases and assumptions to the research process and interpret data in ways that confirm their preconceptions. Therefore, scientists must remain open-minded and approach their research critically, constantly evaluating and reevaluating their conclusions based on new data.

To turn data into information, it must be organized and interpreted until it is meaningful to the audience. This process may involve techniques such as data analysis, visualization, and communication. The critical point is that data is *not* information! Missing this point can be a source of potential misinformation. *Data has no meaning or value without context and interpretation.* The transformation of data into useful information is a creative process that requires both effort and expertise.

Here is one of the differences between data and information: to become information, data must be prepared, organized, and structured--in other words, "formed" (*in*-formed). Such a difference is functional, not structural, but when transmitted into "information," the data scale itself is usually reduced. The resulting "information" is used to describe and answer such questions as: *who, what, where, when,* and *how many?* Information is always inferred from data, and information systems generate, store, retrieve, and process data.[35]

Since our focus is on receiving knowledge about complex biological systems, we may receive two separate kinds of information. The first kind represents patterns showing how the data is organized. In this case, information is narrowly defined as "a degree of non-randomness" (an increase in entropy), such as a sequence of characters in the text or in a piece of DNA. For the second kind of

---

[35] Jennifer Rowley, *The wisdom hierarchy: representation of the DIKW hierarchy.* Journal of Information Science, 2007, vol. 33, pages 163-180.

information, I follow Gregory Bateson's definition: "Information— the elementary unit of information—is *a difference that makes a difference.*"[36] This kind of information is called "functional information"—a concept formulated by Alexei Sharov of the National Institute on Aging.[37] Therefore, information is not only about amounts of coded data. *Information carries value.* There are a range of external and internal needs biologists seek to meet. These building blocks fall into four different categories: functional, emotional, life-changing, and social impact. The main question is: value to whom, and for what purpose?

To explain the situation, let me start with a simple example not directly related to biology. Each of us receives information with accelerated intensity. Information comes from both friends and acquaintances, financial institutions, utility services, political reporters, and philanthropic organizations, to list a few sources. Do we need all this incoming information? Everybody agrees that we do not. Some information may be important; some may be very important, but most information is simply annoying. The same is true of information directed at animals, plants, and microbes. Some information is vitally important for the survival of organisms; other information can be perceived without it being considered critically important.

Information is not only about the object of our investigations. *The ability to receive and transmit information is the most critical parameter of its value.* However, such a value is not always evident in processed data. The receiver must be open and able to absorb the information! The same signal can carry valuable information to one recipient and result in "information noise" to another. The same is true of information that scientists receive from scientific literature.

Here, I wish to point out a significant distinction between the

---

[36] Gregory Bateson, *Steps to an Ecology of Mind; Collected Essays in Anthropology, Psychiatry, Evolution, and Epistemology.* 1972.

[37] Alexei Sharov. *Functional Information: Towards Synthesis of Biosemiotics and Cybernetics.* Entropy, 2010, vol.27, no.12, pages 1050-1070.

quality of information to "make a difference" and one's ability to consciously choose to select valuable information. If the former belongs to the realm of information, the latter belongs to the realm of Knowledge.

Information is *Not* Knowledge

If definitions of information create controversies, they are even more relevant to the next level of the DIKW Pyramid: "Knowledge." Knowledge is the core element in learning about the biological world—and applies to any world inside or outside of us. The DIKW pyramid is sometimes called a "Knowledge Hierarchy."

The central theme of this chapter is this: *Information is NOT knowledge*! The confusion between "information" and "knowledge" is actually the biggest roadblock to developing real knowledge in modern science. The notion, "Information is NOT knowledge," is commonly attributed to Albert Einstein. The quote ends with, "...the only source of knowledge is experience." Another quote related to the subject is by John Naisbitt, "We are drowning in information but starved for knowledge."[38]

The problem arises when seekers ask inadequate questions. If information asks the questions, "Who," "Where," "When," "Why," and "How Many?" then knowledge asks only one question: "Why Does it Matter?" The power of knowledge lies in its ability to generate original ideas, open doors to new opportunities, and resolve challenging problems.

Because of the pivotal role of knowledge in our search for information about biological life, I can offer a simple example. Remember the world of Flatland in Chapter 2, where Mr. A. Square lived? If you recall, Flatland is flat. However, many polygons (triangles, squares, pentagons, hexagons, and other figures) live in Flatland. In this imaginary world, the polygons are alive and can move freely across the two-dimensional plane. Polygons (*poly*) "many" and (*gon*) "sides" are closed figures of line segments (not curves) and

---

[38] John Naisbitt, *Megatrends: Ten New Directions Transforming Our Lives.* 1984.

possess as many angles as they have sides. For example, a triangle has three sides and three angles; a pentagon has five sides and five angles, and so on. Each corner has a certain measure of angles.

The question is: How can the polygons recognize each other and explore their world? Mr. Square offers extensive explanations of this process. Here is a sample from his explanation: "... you will suppose that we could at least distinguish by sight the Triangles, Squares, and other figures... On the contrary, we could see nothing of the kind, not at least to distinguish one figure from another. Nothing was visible, nor could be visible, to us, except Straight Line."[39]

The Square could see a straight line and nothing more. From one point, it could only discern the line of a particular size. The closer this polygon comes to the viewer, the larger the line becomes. Similarly, polygons can recognize elements of their own environment, such as houses. Thus, "data" in Flatland is limited to the size of a line observed from a particular point of view. Inhabitants of Flatland can perceive images of other objects by connecting data points. Those reconstructed (formed) images become the source of "information." Based on the available information, the inhabitants can recognize polygons as triangles, squares, pentagons, and other figures.

So, what does this mean for them? A lot! Here is Mr. Square again, "Our Soldiers and Lowest Classes of Workmen are Triangles with two equal sides... Our Middle Class consists of Equilateral or Equal-Sided Triangles. Our Professional Men and Gentlemen are Squares ... and Five-Sides Figures... Next above these come the Nobility, of whom there are several degrees, beginning at Six-Sided Figures."[40] "Information" about the shapes of the flat figures is being transformed into "knowledge" of the social structure of Flatland. What, then, happened to Mr. Square when he traveled to the Three-Dimensional world? This experience provided him with knowledge about the organization of Flatland as it appeared from Above. Now, he could manipulate objects inside his house without entering

---

[39] Edwin Abbott. *The Annotated Flatland: A Romance of Many Dimensions.* 2002, page 33.
[40] Ibid., page 33.

through the doors.

Mr. Square had reached the level of "Wisdom" by recognizing the existence of Higher Dimensions. Interestingly, the Sphere, with a more developed understanding of the Third Dimension, could *not* reach the level of "Wisdom" of Higher Dimensions!

Yes, the Flatland model is simple. But we, humans, study the world outside us, including the biological world, precisely in the same manner! We observe separate pieces of "Data" that don't mean anything. We can also combine and sort out these pieces of data to obtain "Information" about biological objects from the collected data. We then can proceed to develop a system of "Knowledge" that assesses the value of available information about biological systems. Knowledge signifies that one has formed perspectives.

Richard Boland of Case Western Reserve University and Ram Tenkasi of the Center for Effective Organizations suggest that success will depend on how effectively diverse individuals can organize and develop their unique knowledge competencies. These authors also emphasize that success depends on "how effectively they can integrate and synergistically utilize their distinctive knowledge through a process of *perspective taking*."[41] In another one of his papers, Professor Boland noted that changing a practice "includes not only the learning of new practices but also unlearning old and outmoded knowledge." Thus, knowledge can regulate itself, either by "unlearning" or by choosing to place "mental blocks" on the acceptance of new knowledge, based on one's specific preference. An individual's formation of knowledge is a highly creative process that requires one's constant awareness.

Ignorance as an Intentional Blockage of Knowledge

Since we are discussing the ability to put "mental blocks" on the acceptance of certain information, let's address a couple of issues that strongly affect how information is transformed into knowledge. I will label the first issue "ignorance." We usually give the word "ignorance"

---

[41] Richard Boland and Ram Tenkasi. *Perspective Making and Perspective Taking in Communities of Knowing.* In *Organization Science*, 1995, vol.6, no.4, pages 350-372; italics are mine.

a negative connotation. It means "a lack of knowledge, understanding, or information." We can find numerous examples of how ignorance of obvious knowledge can negatively impact scientific progress. Francois Jacob, the famous French biologist who was awarded the 1965 Nobel Prize in Medicine, discussed this in his article, "*Understanding Science and Knowing Ignorance.*" As an example of the dangerous consequences of ignorance, he reminded us how badly the science of genetics was damaged by ignorance in the Soviet Union.

So, what is the meaning of the saying: "Ignorance is bliss?" In some cases, people might intentionally choose to ignore new information if facing reality brings change or necessitates taking responsibility for one's actions. It could be more comfortable to maintain the status quo by staying uninformed and avoiding the impact that knowledge might have on one's beliefs, values, or actions. However, it's critical to recognize that ignorance can also be a consequence of various external factors, such as inadequate access to information, lack of education, institutional pressure, or societal norms.

"It is always advisable to clearly perceive our ignorance," wrote Charles Darwin in *The Expression of the Emotions in Man and Animals.* Daniel Boorstin, nominated as Librarian of U.S. Congress in 1975, wrote, "The greatest obstacle to discovery is not ignorance – it is the illusion of knowledge."[42] Commenting on Boorstin's saying, Estelle Frankel wrote in the Introduction to *The Wisdom of Not Knowing* about the willingness of "(to) *not know, to trade the certainty of the known for the unknown.*"[43]

There is also another side to ignorance in science. Professor Stuart Firestein, Chair of the Department of Biological Sciences at Columbia University, argued that the ignorance of a subject could be a motivating force. In his book, Firestein shows how scientists use ignorance to program their work and identify what should be done,

---

[42] Danie Boorstin, *The Discoverers: The History of Man's Search to Know His World and Himself.* 1985, 768 pages.
[43] Estelle Frankel, *The Wisdom of Not Knowing.* 2017, page 1.

what the next steps are, and where scientists should focus their energies. There is a false impression of science as a deliberate, step-by-step method to finding things out and getting things done. In fact, says Firestein, more often than not, science can be compared to looking for a black cat in a dark room and there may not even *be* a cat in the room! He wrote, "In science, there are dark rooms everywhere that have been found to be completely empty."[44]

The first thing to recognize is that ignorance fails to describe the breadth of its subject. It is crucially important to decide "against the enormous background of the unknown, what particular part of the darkness he or she will inhabit," said Firestein.[45] Adam Frank, a professor at the University of Rochester and a theoretical/computational astrophysicist, once commented to the NPR blog 13.7: Cosmos & Culture that "to fight ignorance, we must start by admitting our own."[46]

In recent decades, we have seen a dramatic increase in creative approaches to overcoming scientific ignorance and uncertainty. These new approaches can be traced in part to a realization that ignorance and uncertainty cannot always be reduced or banished from science. Michael Smithson, professor of Behavioral Sciences at James Cook University, Queensland, Australia, wrote that "the fact that ignorance is negotiable and yet fundamental to scientific work poses several important dilemmas and prospects. We may be participating in a shift from the traditional research strategies of reducing or banishing ignorance toward a deeper understanding of and greater capacity to cope with ignorance."[47]

Brazilian Rabbi Nilton Bonder proclaimed: "Even a naïve or irrational person may be less ignorant than those who perceive reality through the distortion of their conditionings."[48] Illustrating his point,

---

[44] Stuart Firestein, *Ignorance: How It Drives Science.* 2012, page 65.

[45] Ibid., page 58.

[46] Adam Frank, https://www.npr.org/sections/13.7/2018/01/12/577356257/science-says-that-to-fight-ignorance-we-must-start-by-admitting-our-own

[47] Michael Smithson, *Ignorance and Science: Dilemmas, Perspectives, and Prospects.* 1993, vol.15, no.2.

[48] Nilton Bonder, *Yiddishe Kop.* 1999, page 17.

Bonder asked the question, "Which is better, a fast horse or a slow horse?" The answer that he proposed is, "It all depends on whether you're headed in the right direction or not."[49] Bonder also wrote that "Awareness of ignorance is fundamental in any type of research."[50] He reasoned that we are easily diverted from our goals without realizing it. That is why mapping out our ignorance can be an effort-saving strategy. Bonder retells the story of the great mystic Bal Shem Tov. When one of his visitors complained about his ignorance, Bal Shem Tov consoled him by saying that if he had realized himself as an ignorant man, that would be a great accomplishment!

The Streetlight Effect

The main question is: "How do scientists use ignorance, consciously or unconsciously?" Here, I will introduce another related issue as a famous metaphor for knowledge and ignorance. It's known as "The Streetlight Effect." This story is traced back to Nasreddin Hodja, a legendary folk character in the Middle East and Central Asia in the 13th century. The story has many modern versions. Here is a short one: A man is searching for something under a streetlight. A passerby stops and asks him what he lost. The man replies that he lost his keys, and both men begin to look under the streetlight together. After a while, the second man asks the first one if he's sure that he lost his keys here. The response is: "No, but the light is better here."

While this scenario is a joke, similar situations are actually common in science! Milton Friedman, a cancer specialist and professor of radiology at the New York University School of Medicine, called his presidential address to the American Radium Society: "The Light is Better Here!"[51] It is very common for scientists to choose their objectives and methods for new projects based on the availability of resources, comfortable conditions, support of university leadership (and funding), and many other factors that will

---

[49] Ibid., page 22.
[50] Ibid., page 21.
[51] Milton Friedman, *The Light is Better Here!* The American Journal of Roentgenology, Radium Therapy and Nuclear Medicine, 1968, vol. 102, pages 3-7.

provide a "better light."

In science, "The Streetlight Effect" story refers to the question: "Do we measure what needs to be measured, or do we measure what is *easily* measured, even if it is invalid or irrelevant."[52] Abraham Kaplan, the first American philosopher to systematically examine the behavioral sciences, in his book, *The Conduct of Inquiry*, defined "The Streetlight Effect" as: "…the tendency for researchers to focus on particular questions, cases, and variables for reasons of convenience or data availability rather than broader relevance, policy import, or construct validity."[53] I found Kaplan's definition in the article written by Cullen Hendrix, a professor at the University of Denver. He demonstrated "The Streetlight Effect" in climate change research on Africa.[54]

Manuela Battaglia of the Diabetes Research Institute in Milan, Italy, together with Mark Atkinson of the University of Florida, demonstrated how "The Streetlight Effect" dimmed knowledge when studying Type 1 Diabetes.[55] In a bid to play it safe, researchers preferred to replicate previous research instead of covering new territory. Such an attitude inhibits deep research and solutions, especially since it is easier to publish studies that can be verified quickly or are less contradictory.

Sources classified under "Streetlight Effects" have often shown limited viability when it comes to big data. For instance, using big data for the Ebola epidemic in West Africa led to an overestimation of the spread of the disease--and an underestimation of local initiatives' potential. Considering how modern technologies (proteomics, transcriptomics, and genomics) contribute to understanding snake venom, Choo Hock Tan warns about falling

[52] William Shadish et al., *Experimental and Quasi-Experimental Designs for Generalized Causal Inference*. 2001, cited from Hittner et al., *The Third Cognitive Revolution*, 2019.
[53] Abraham Kaplan, *The Conduct of Inquiry: Methodology for Behavioral Science*. 1998, 452 pages.
[54] Cullen Hendrix, *The streetlight effect in climate change research on Africa*. Global Environmental Change, 2017, vol. 43, pages 137-147.
[55] Manuela Battaglia and Mark Atkinson, *The Streetlight Effect in Type 1 Diabetes*. Diabetes, vol. 64, pages 1081-1090.

into the "Streetlight Effect," which symbolizes, in his words, "cognitive availability bias."[56]

I am not accusing any of my colleagues of following "The Streetlight Effect." I have been in their shoes far too often myself! However, I am urging researchers to be aware of how these factors influence their research. We could learn more from "The Streetlight Effect" metaphor, which highlights our tendency to focus on what is visible or easily observable while neglecting what is not readily apparent or difficult to observe.

Therefore, it is crucial to investigate any possible opportunities to learn about the "invisible" (dark) parts of biological systems, as they can offer insights into how the visible aspects of life are formed. Just as the bulk of an iceberg is hidden below the surface of the water, there may be unknown or hidden aspects of biological systems that can significantly impact our health and well-being. By exploring these hidden aspects of biology, we can gain a more comprehensive understanding of biological systems and identify new avenues for treatment and intervention.

While it's important to be aware of the potential pitfalls of "The Streetlight Effect," it is also essential to recognize that beginning research in well-lit areas can be a valid strategy. It can work as long as researchers are willing to explore the shadows in search of the optimal solution. If there is a willingness to go further and push into the shadows in the right direction, searching under the "streetlight" could be an appropriate place to start investigations. In the paper mentioned above by Coo Hock Tan, the author offers a productive way of looking at the streetlight metaphor, " ... one should therefore acknowledge what potentially lies outside the edge of light (did I drop my keys in the dark?), and one should be vigilant enough to tell apart the real from fake found under the light (Which keys are mine, or rather which are not mine?)."[57]

---

[56] Choo Hock Tan, *Snake Venomics: Fundamentals, Recent Updates, and a Look to the Next Decade.* Toxins, 2022, vol. 14, no. 247.

[57] Ibid., page 2.

I believe we can use another metaphor, namely "The Flashlight Effect," to illustrate the need to explore hidden aspects of biological systems to avoid "The Streetlight Effect." Just as exploring the dark corners of a street with a flashlight can reveal unexpected and potentially dangerous obstacles, examining the hidden aspects of biology can also uncover new and potentially challenging aspects of biological life.

For example, exploring network analyses of complex diseases may reveal previously unknown genetic mutations or epigenetic factors contributing to disease risk. However, illuminating these hidden aspects of biology may not be easy. Just as a flashlight requires charged batteries and basic experience to operate effectively, exploring biology's hidden aspects requires specialized tools and expertise (Figure 14). The decision to apply new techniques, analytical skills, concepts, or ideas can act as a "flashlight" to illuminate new areas for research in exciting, new ways.

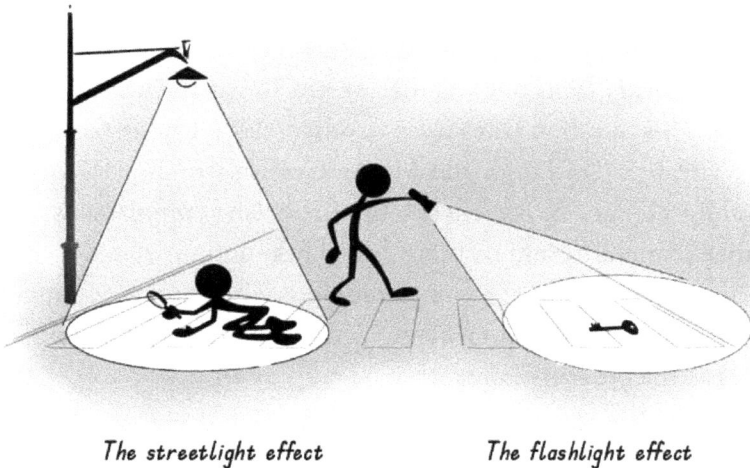

The streetlight effect   The flashlight effect

Figure 14. Scientists often choose objectives and methods that provide "better light." It could be a "streetlight" with the tendency to focus on particular questions for convenience, rather than broader relevance. Conversely, it could be a "flashlight," one you carry with you to find answers to well-defined questions.

Nilton Bonder cited the words of a Hasidic master who warned his disciple, "You are like someone walking with a guide through the forest on a dark night. As soon as the guide moves away from you, darkness takes over. But if you carry your own light, you need not fear the darkness."[58]

Knowledge can dramatically alter the way we perceive information. Many people quote a saying attributed to the American writer Anais Nin that illustrates this idea. She stated in her autobiographical novel, *Seduction of the Minotaur*: "We don't see things as they are, we see them as we are."[59] Nin did not claim authorship of this aphorism, instead, she attributed these words to the Babylonian Talmud. More precisely, in the Talmudic tractate, Berakhot, 55b, Rabbi Samuel ben Nahman said: "A man is shown in a dream only what is suggested by his own thoughts."

Knowledge is *Not* Wisdom

One essential attribute of knowledge is completeness. Certain information can estimate the "incompleteness" of data, but the completeness of knowledge is verified only by practice. The Tree of Knowledge bears fruit, but whether or not to eat them is another question. This question takes us to another level--"Understanding." Rabbis Yitzchak Ginsburgh and Moshe Genuth wrote in *Wisdom: Integrating Torah & Science:* "In Kabbalah *completeness* is indicative of maturity and rectification." These authors state: "…the most important quality of completeness is also known as inter-inclusion, whereby every element in a structure holographically includes all the other elements".[60]

Among the definitions of "Wisdom" I have seen are: "accumulated knowledge," "ability to increase effectiveness," "adds values, which require … judgment," and "the highest level of abstraction." To me, these words can refer to both "understanding"

---

[58] Nilton Bonder, *Yiddishe Kop.* 1999, page 89.

[59] Anais Nin, *Seduction of the Minotaur.* 2014.

[60] Yitzchak Ginsburgh and Moshe Genuth. *Wisdom: Integrating Torah & Science.* 2018, page 114.

and "wisdom." Ackoff himself defined wisdom as "evaluated understanding." He emphasized that "there can be no wisdom without understanding and no understanding without knowledge."[61]

"Understanding" requires having a perspective from outside the world in which an object stands ("under-standing"). There is no explanation for wisdom: it either is or it isn't. While Knowledge asks, "How?" Understanding asks, "Why?" Wisdom asks, "What can we know, and what can't we know yet?" Valerie Ahl, a psychologist at the University of Wisconsin, and T.F.H. Allen, an ecological theorist at the same university, wrote in *Hierarchy Theory*: "...line of reasoning forgets the value of the upper-level context...The lower-level mechanics, the "how" of the phenomenon, have nothing to say about "why.""[62]

To demonstrate understanding, the immunologist Irun Cohen suggested three ways: "we visualize *metaphors*, we make successful *predictions*, and we exercise know-how – *utility*."[63] Cohen explained that we show that we understand by grasping likeness, by foreseeing consequences, and by knowing use. Seeing an analogy between seemingly unrelated ideas is the key to making a metaphor. "Understanding amounts to translating one thing (the unknown) into the terms of another (a known). The unknown is *mapped* onto the known, and so becomes known," Cohen continued.[64]

When characterizing "understanding," Ginsburgh and Genuth stress that "...to invest yourself in it fully, you would come out with the full, mature, and complete understanding of that topic and how it relates to every other topic or issue in life."[65] And the same authors later declared: "...that it is at one and the same time complete, yet

---

[61] Russell Ackoff. *From Data to Wisdom:*
Presidential Address to ISGSR. Journal of Applied Systems Analysis, 1989, vol. 16, page 3.

[62] Valerie Ahl and T.F.H. Allen. *Hierarchy Theory: A Vision, Vocabulary, and Epistemology.* 1996, page 18.

[63] Irun Cohen, *Rain and Resurrection.* 2010, page 100.

[64] Ibid., page 100.

[65] Yitzchak Ginsburgh and Moshe Genuth. *Wisdom: Integrating Torah & Science.* 2018, page 117.

open to contributions from the outside!"[66]

Let us consider the difference between "wisdom" and the lower floors of the DIKW Pyramid. First, we must agree that there is no wisdom in "data," "information," or "knowledge." Instead, *each of the D-I-K components can be incorporated into wisdom!* That is not an automatic process, but an underlying goal that requires intensive creative efforts. Wisdom is a "totality" that incorporates all possible manifestations (presented by data) and all potential perspectives (used for obtaining "information"), and reflects a multi-dimensional approach to "knowledge." Clinical psychologist Sandra Carey advises: "Never mistake knowledge for wisdom. One helps you make a living, the other helps you make a life."

The DIKW Hierarchical Planes

Each domain in the D-I-K-W hierarchy is not just a staircase—it's a different world! Later, we'll learn about the Kabbalistic Worlds and reveal support for that viewpoint. Meanwhile, let us explore these worlds and get to know the entities that inhabit them.

Solitary, isolated, and non-communicative data inhabit the lowest of these worlds. Their sizes and configurations can vary, but since they are isolated, it doesn't matter. Knowledge is the world in which the most amazing creatures live. The forms of these creatures are whimsical and constantly changing because they are always evolving. However, they need some assistance in producing new forms on their own. Therefore, they must exploit the labor of people to do all the "dirty work" that needs to be done. Finally, inhabitants of the upper level know they are oriented to the center point (Figure 15).

---

[66] Ibid., page 121.

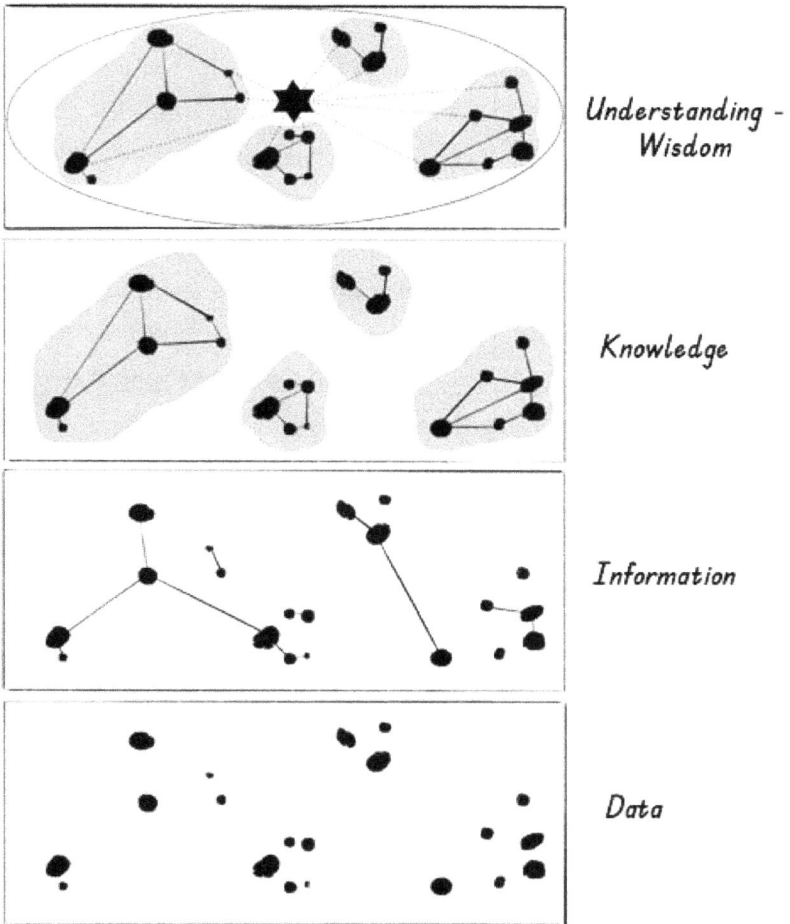

Figure 15. Each world of the Knowledge Hierarchy has its own inhabitants: Data as the Dots, Information as the Lines, Knowledge as the Areas, and Understanding-Wisdom as the Unifying Center connecting all Dots.

Our understanding of DIKW worlds can also be gained by comparing them with the levels of dimensionality we learned about in Chapter 2. While these worlds are real, they are presented as "point-line-area" visual models readily accepted by the human brain. Consider the following simple analogies: Point data is a one-dimensional world. When points are connected in figures, they create a two-dimensional world of information. Once the created figures achieve meaning, it leads to a three-dimensional world of knowledge.

Points are separate objects chaotically distributed below. They are dimensionless and are represented by a dot, the simplest geometric structure. They are at a potential stage and are not part of the Knowledge Pyramid until something happens... As Ginsburgh and Genuth nicely phrase it: "Something has to happen for them to move to the next stage and whatever this something is, it acts to invest life into the initial point."[67]

Data corresponds to single observations lined in the database. Information connects all points (a whole set) or selected points (sets of points) by lines of different intensity, creating a kind of network across the flat surface. Here is one more reference from Ginsburgh and Genuth: "The line-like structure is vulnerable to chaos, and therefore, it must continue its development. Stagnation in development can lead to a regression back to a point-like state. If the development does carry it forward, the structure can become area-like, a state in which each of its original elements is discovered to be a complex structure in itself."[68]

Examples of the DIKW Pyramid in Biology

Let's now take a few examples to illustrate the levels of the DIKW pyramid in biology. Here is a frequently used term: "gene." On the data level, genes are made up of linear sequences of DNA at a specific location on chromosomes in the nucleus of cells. On the "information" level, genes contain "instructions" for producing specific proteins that lead to the expression of a particular physical characteristic or trait. This could include a particular cell function or a particular morphological pattern, such as a hair color.

Conversely, on the "knowledge" level, a gene is the basic unit of heredity passed from parent to offspring. Geneticists routinely identify genes based on phenotypic differences in hybridization

---

[67] Yitzchak Ginsburgh and Moshe Genuth, *Wisdom: Integrating Torah & Science.* 2018, page 117.
[68] Ibid., page 116.

experiments or the analysis of family members. Initially, naturally occurring mutations were identified. Then, gene loss or inactivation had to be established by DNA sequencing or other methods. One of the standard techniques to study gene function is the so-called "gene knockout," in which one of an organism's genes is made inoperative ("knocked out" of the organism).

On the "understanding" level, the relationships between gene and biological processes appear to be more complex than initially anticipated. The gene is considered not just a linear sequence of DNA but a union of genomic structures that encode a coherent set of potentially overlapping functional products. In this view, several molecular products are the starting point from which elements in the genome are identified, giving rise to the set of multiple products.

On the "wisdom" level, all knowledge of biological processes leads to the holistic picture of biological individuality. Here, we prioritize the inseparable relationship between the entire constitution of genes belonging to an individual organism ("genotype"). In addition, the complete set of observable characteristics of an individual ("phenotype") are seen from the interaction of the genotype with the environment. "There is much more to heredity than the inheritance of nuclear DNA."[69]

Another application of the multidimensional approach is presented in an article my son, Roman, and I published in *Evolutionary Application* in 2017. In *Complexity and Biosemiotics in Evolutionary Ecology of Infectious Agents*, we discussed perceptional levels when describing infectious diseases. It's easy to see that the spectrum of descriptions is represented by hierarchical levels in Figure 16. It closely resembles the DIKW Pyramid. In fact, I was not even aware of this concept at that time.

---

[69] Paul Griffiths and Karola Stotz. *Genetics and Philosophy: An Introduction.* 2013, 278 pages.

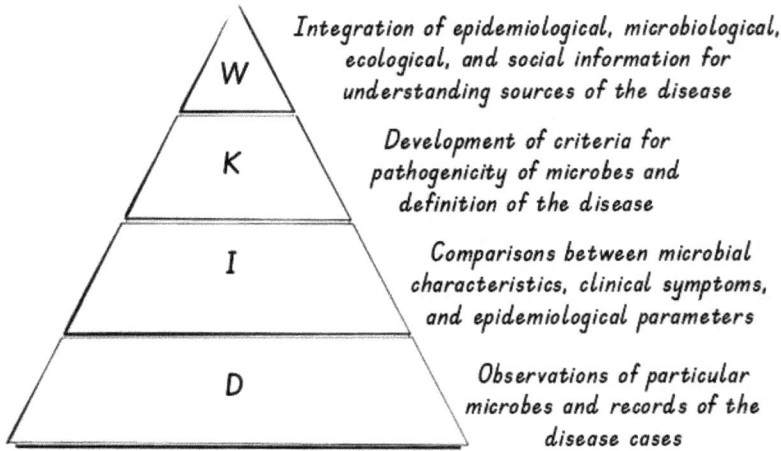

Figure 16. Perceived levels in descriptions of infectious diseases (modified from Kosoy and Kosoy, 2017).

At the lowest level, we detected a specific bacterium or a virus in the animal. At the second level, we hypothesized that this animal species was more likely to carry the pathogen. At the third level, we claimed that this animal species was the reservoir. Theoretically, a reduction of the disease could be expected by controlling the population of these animals. The third level describes concepts that explain the pathogen's origin, development, and complexity. At the fourth level is the evolutionary history of pathogen-host relations, microorganisms as an inevitable part of the animal or human body, the roles of pathogens in nature, and related ethical questions that can be considered.[70]

The Knowledge Hierarchy in Kabbalah

Knowledge transforms the flat network into a tree-like structure. The tree looks flat on a piece of paper, but a real tree exists in three-dimensional space. Understanding presents a level that "stands under" the Tree of Life ("Etz Hayim"). Here, I wish to emphasize the hierarchical structure of the knowledge-achieving process. Hierarchy

[70] Michael Kosoy and Roman Kosoy, *Complexity and Biosemiotics in Evolutionary Ecology of Infectious Agents.* Evolutionary Application, 2017, pages 1-10.

signifies the different roles an investigator plays in learning complex biological systems. A system is defined as "hierarchical" if it can be described as composed of stable, observable subunits unified by a superordinate relation.[71]

Remember that the DIKW Pyramid is called a "Knowledge Hierarchy." Each level in this pyramid represents a separate domain, or world. Within each world, complexity can increase and result in emergent patterns. However, whatever data is complex doesn't become information itself. Whatever information is complex doesn't become knowledge automatically. The same is true for higher levels.

From the perspective of Kabbalistic Biology, each level of the Knowledge Pyramid can be seen as a world in its own right:

- The world of data.
- The world of information.
- The world of knowledge.
- The world of wisdom.

When we see them defined in terms of the Kabbalistic principles, these worlds will have different names, and we will learn how to recognize them. While admitting to the problem of recognizing the concepts of "understanding" and "wisdom," it's important to note that there is no such a problem in Kabbalah, where "wisdom" (*Hokhmah*), "understanding" (*Binah*), and "knowledge" (Da'at) are three fundamental principles that cannot be misunderstood. It is not a problem with the translation of these words. In English and other languages, people use these words for the simple purpose of explaining what they mean. In Hebrew, the words "Hokhmah," "Binah," and "Da'at" are essential. They are rooted in the letters that create these words.

In Kabbalah, the terms *"Nistar"* and *"Nigleh"* refer to the two

---

[71] Valerie Ahl and T.F.H. Allen, *Hierarchy Theory: A Vision, Vocabulary, and Epistemology.* 1996, page 29.

distinct dimensions of knowledge and understanding. *Nistar* refers to the "hidden dimension" that explores the complex structure of the higher realms. This word comes from the Hebrew root that means "hidden" or "concealed." In contrast to the obscure nature of *Nistar*, *Nigleh* ("revealed") refers to the more accessible and literal interpretations of the texts and focuses on the practical application of the teachings in daily life. Together, these two dimensions offer a rich, multifaceted approach to thinking, allowing for a deeper engagement with the sacred teachings and a more profound appreciation of their complexity and nuance.

Contemplating hidden and revealed aspects of reality in Kabbalah, physicist Alexander Poltorak came to the following conclusions:

1. Everything has a hidden and a revealed aspect.

2. The hidden aspect is always more remarkable (or more significant in some sense) than the revealed aspect.

3. The bifurcation (division) of reality into hidden and revealed parts is illusory and exists only in our frame of reference—the view from below.

4. The hidden aspect will be revealed when the underlying unity of reality—including the unity of the hidden and the revealed—will be unveiled.[72]

I wish to return to the discussion of sorting out "knowns" and "unknowns." "Known knowns" is data; it can be collected in different ways, but it's what we know. "Known unknowns" is information that can be incomplete, but we know which data is missing. "Unknown knowns" is knowledge: we can formulate some concepts, but we may not know what is behind them. For example, although we can learn more about the mechanisms of evolution, the question of what evolution is remains unknown.

---

[72] Alexander Poltorak, *On Hidden and Revealed Aspects of Reality.*
https://quantumtorah.com/on-hidden-and-revealed-aspects-of-reality/

"Unknown unknowns" belong to wisdom. Can we achieve this level? If not, why talk about it? There are two main reasons. First, being aware of this level teaches us to be humble about our ability to know the laws of Nature. Such awareness is sadly missed in modern science. Second, this awareness provides an orientation into the invisible world. Nobody expects to *reach* the North Star, but it can certainly help orient us when we are lost.

In her beautifully written book, *The Wisdom of Not Knowing*, Estell Frankel calls us to embrace the unknown instead of ignoring and fearing it. From her professional experience as a psychotherapist, she argues that acceptance of the unknown is not only unavoidable, but productive. Frankel states, "In Jewish mysticism, divinity is the ultimate unknown and unknowable reality that, paradoxically, we are enjoined to "know." [73] She goes on to say the following, "All of our knowledge rests in a vast sea of not knowing ... Finding the balance between these energies is the key ... [Wisdom] involves a synergy of knowing and not knowing."[74]

How does this relate to the experience of a research biologist? First of all, scientists are people. It is not comfortable to acknowledge unknowns, especially in areas where you are recognized as an expert. I am not talking about limiting our knowledge, we know so little compared to what we don't know. Therefore, we look at the spots where we have light.

It is essential to accept the "Hierarchy of Knowledge" as separate worlds. We must describe biological life using different dimensions or aspects of reality. These can be grouped into four categories based on their visibility (apparentness) and invisibility (hiddenness). In the following chapters, we will examine some critical biological problems through the lens of visible and invisible spectra. The main lesson here is that we cannot fully comprehend biological life at only one level, in one domain, or one dimension.

The presented four worlds provide a framework for exploring the

[73] Estelle Frankel. *The Wisdom of Not Knowing: Discovering a Life of Wonder by Embracing Uncertainty.* 2017, page 2.
[74] Ibid., page 3.

complexities of existence and understanding the nature of knowledge of biological life:

1. The *visible world of what is visible* represents "known knowns" of data.

2. The *invisible world of what is visible* represents "known unknowns" of information.

3. The *visible world of what is invisible* represents "unknown knowns" of knowledge.

4. The *invisible world of what is invisible* represents "unknown unknowns" of wisdom.

The provided definitions may initially seem confusing, so it is time to learn some basic Kabbalistic concepts to comprehend them fully. We will discuss them in Part 3. Kabbalah warns about the dangers of "abstract -- discussions." Knowledge is intended to achieve results -- "to bring fruits." Why, then, do we talk about the unknown, invisible, hidden? *Because the biological world's most extensive and essential components are still unknown to us!* How can we ever claim to understand it if we lack any orientation in Nature's invisible world?

In the following chapters, I will demonstrate the significance of biological "unknowns," After that, I will present some orientation guidelines to help us explore this world.

Chapter 6

Levels of Visibility in Biological Objects

Biology is a complex and multifaceted field. Restricting our knowledge to only visible data would limit our ability to fully comprehend and appreciate the wonders of the natural world. Many aspects of biology are hidden from plain sight and require specialized knowledge, investigation techniques, and reasoning to uncover. "Visibility" in this context refers to the observable aspects of biological life, as well as the factors that underlie these observable phenomena. "Invisible" is not a scientific term. The word is exclusively used here as a tool and can be interchanged with the words "transparent" and "unapparent."

The central premise of Kabbalistic Biology is that despite the lack of visibility, information, and knowledge are still available to us. There may even be a variety of ways to achieve this objective. The issue of invisibility is not limited to a gradual transition from more visible to less visible states. What is more relevant and exciting is to distinguish between different levels (categories) of invisibility, each requiring specific actions. In this and the following chapters, we will explore how to identify different levels of transparency.

Introducing the Levels of Visibility
Embracing the various levels of invisibility will help us better understand biological life's complexities and how they function.

Domains represent different "planes" based on their levels of visibility. Each domain contributes to our understanding of the biological world at a different level of accessibility and has specific properties. Our understanding of biology can be enhanced by exploring these distinct domains, potentially leading to new discoveries and insights.

In the previous chapter, we discussed four aspects of knowledge that can be assigned to the planes of transparency. They distinguish objects and phenomena as 1) "known knowns," 2) "known unknowns," 3) "unknown knowns," and 4) "unknown unknowns." As you may recall, this classification led us to the D-I-K-W pyramid. Let's consider four planes/worlds of visibility representing all biological objects and processes. Understanding the reality of these worlds is the most important thing. Throughout the book, we will learn how to investigate these invisible worlds. This chapter will reveal how to distinguish between them. Table 1 represents some general characteristics of each level that we began to describe in the previous chapter.

Worlds	Dimensional representation	Knowledge hierarchy	D-I-K-W pyramid	Questions asked
The visible world of what is visible	1D- dot	Known knowns	Data	Is the data correct?
The invisible world of what is visible	2D -line	Known unknowns	Information	Who, where, when, and how many?
The visible world of what is invisible	3D-area	Unknown knowns	Knowledge	How can we use this information?
The invisible world of what is invisible	4D-hyperspace	Unknown unknowns	Understanding/ Wisdom	Why? What do I know?

Table 1. The Four Planes/Worlds of Visibility

140

1. *The visible world of what is visible* refers to the tangible, observable biological data that we can directly observe and interact with.

2. *The invisible world of what is visible* represents the foundation of information that connects visible biological pieces of data but is not directly observable.

3. *The visible world of what is invisible* involves hidden aspects of the knowledge of biology that can be indirectly observed or inferred through reasoning and conceptualizing.

4. *The invisible world of what is invisible* involves hidden aspects of understanding biological processes that can be indirectly observed or inferred through investigation, reasoning, and personal knowledge.

The World of Known Data, but Unknown Information

In biology, the visible world consists of separate biological objects. Each of these can be represented by their "known" (available) properties. There is no doubt about the following statement: "We know what we know!" This certainly applies to available data that we can extract and analyze once we receive it. Therefore, available data is "known knowns." Because each biological object interacts with the outside world, we can also receive data about visible environmental parameters.

For example, we can readily record measurements of biological objects (e.g., number of animals, size, color, etc.) or the occurrence of some events (e.g., birth, death, migration, etc.). We can record data on factors that could influence biological processes, such as climatic and geographic variables (temperature, precipitation, GIS coordinates, etc.). One could say metaphorically that the act of measuring and recording data is a way for us to make biological objects visible to us.

In many situations, we can see more objects with our physical eyes, but they remain inaccessible as data. I will demonstrate such a situation using an example from my experiences in bacteriology. The presentation demonstrates the difference between data ("what we know") and information ("what we don't know yet"). When we try to isolate and identify bacteria that invade the bloodstream, we place a certain amount of whole or diluted blood on a specific growth medium. This medium is placed in a Petri dish. After the incubation period specific to each bacterial species, we check for bacterial colonies that will appear on the surface of the medium.

We can note and record the appearance of bacterial colonies, their number, and growth conditions as DATA. As of this point, we have no information about these colonies. We don't know to which species they belong. However, if we are trained enough, careful, and accurate in observing such physical manifestations, such data will be reliable (Figure 17).

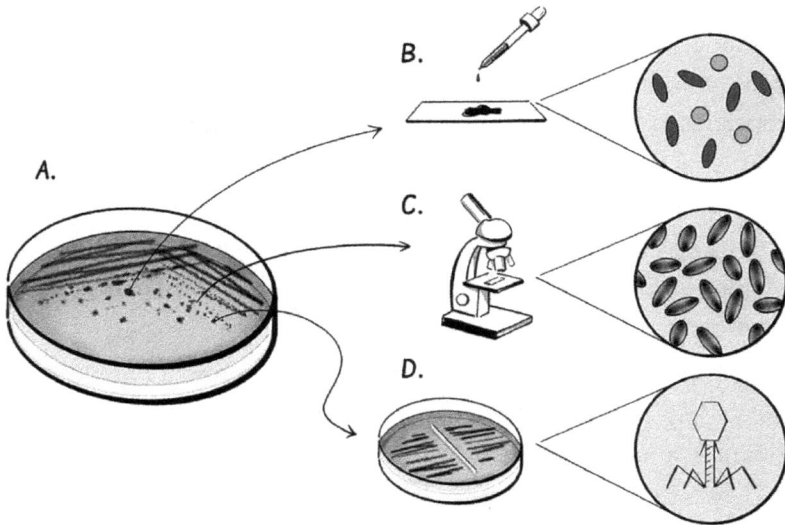

Figure 17. Bacterial colonies (Yersinia pestis, the agent of plague) on the surface of the medium in the Petri dish (A), color of the culture after special staining (B), shape of bacterial cells under a light microscope (C), and reaction to the specific bacterial virus – "bacteriophage" (D) represent "visible data." By connecting available data, it is possible to "visualize" information about these specific bacteria.

What's our next step? We can select one or a few colonies for further sub-culturing and characterization. The characterization process requires considerable effort, time, and resources. For example, let's say that we suspect that the observed colonies look similar to *Yersinia pestis*, the pathogen of bubonic plague. We must confirm or reject this speculation. Behold! Our speculations are *not* "data." These speculations could relate to our "knowledge," but they cannot replace "data." Data about colonies can be analyzed to obtain information, but at this point, we don't have enough data to generate meaningful "information." The only available data is about the physical appearance of the colonies!

For our next step, we can transfer one of the observed colonies onto a glass to characterize bacteria. Then, we can apply a specific stain, such as a Gram stain, on the bacterial culture to "see" whether the bacteria are either Gram-positive or Gram-negative. That gives us more data! We can examine the bacterial culture under a microscope to "see" the individual bacterial cells. We can note the shape of bacteria (round, rod-shaped, or a different shape).

Now, we have even more data. To check whether these bacteria match the plague pathogen, we can place a filter paper strip containing a specific "bacteriophage" (the virus that kills *Yersinia pestis* bacteria) over the bacterial culture. If it does not kill the bacteria, likely they belong to another related bacterial species. At this point, we might have enough data to generate information to identify the bacteria. Now, information about bacteria becomes "visible" (Figure 17, D).

The Invisible Information from Visible Data
We know when we don't have enough data to acquire the information we need. However, we can gain access to it if we are given enough time, modern technology, and resources. While we don't have this data yet, we are aware of it and know how to obtain it. The information we don't have yet but can receive from available data represents "the invisible world of what is visible" (or "known

unknowns"). We might know what *kind* of information is missing, but we don't know how little information we have when compared to all the information we need.

The invisible world of what is visible involves processes or mechanisms responsible for the *visible* aspects of biological life. While we can observe the results or effects of these processes, the processes themselves may not be directly observable. Remember, the amount of information we have versus the amount we do *not* have should not lead to discouragement; it should lead to awareness. We can only guess how much or how little we know. However, there are some practical steps we can take to handle such a problem.

The science of statistics has developed sophisticated tools to evaluate data that will produce more or less reliable, sufficient information. The price one pays, however, is dealing with probabilities, not "certainties," which can lead to the bane of scientific research: *uncertainty*. Even though we might have more information, can prepare available information better, and analyze the information available, *we still don't know how much we don't know*. For example, take the statement: "Every fox we observed is red; therefore, all foxes must be red." However, a rare red fox can be black, silver, or a cross between red and silver. Though the assumption about the red color of foxes can be statistically reasonable, it's not a foregone conclusion.

Modern science depends on an "inductive" approach: that is, any generalization based on particular observations. The inductive practice relies heavily on two premises. The first premise is: "What has happened in the past will continue to happen." The second premise is: "What happens while being observed happens even when unobserved."

In fact, neither of these premises can be proven correct! Rachel Gordon claims in the Introduction to Yitzchak Ginsburgh's book, *137: The Riddle of Creation*, "Induction has been called the mother of all problems."[1] Benjamin Fain, a world-renowned physicist wrote,

---

[1] Yitzchak Ginsburgh. *137: The Riddle of Creation*. 2018, page 4.

"When we have the results of a limited number of experiments, not only are we unable to determine the results of all subsequent experiments, but we cannot even determine that there is a particular order in the future reality... While there it was easy to make a guess at the future course of events, this is not the case when we know only fragments of the movements, paths, and solutions."[2]

For a certain conclusion to be based on specific evidence, it needs to be true. This relationship can be viewed as a math problem: Evidence + Assumption = Conclusion. Assumptions link things together. I wish to remind readers (and, above all, my colleagues) that the vast amount of obtained scientific data is still a minuscule part of potentially available data. In fact, most scientists realize this, but most non-specialists do not! They are often surprised that scientists cannot answer "simple" questions if they are supposedly so knowledgeable.

For example, why can't virologists give a straight answer about the next mutation of the coronavirus? After all, they are supposed to know so much about these viruses. Scientists can talk about some probable scenarios, but such conversations are speculative—they do not produce knowledge. Only actual "knowledge" can answer the questions of *why* and *how.*

Most scientists are aware of the incompleteness of information in their areas of expertise. I have my own criteria by which I can define the level of expertise of another scientist. It comes down to *how much one knows what is unknown* regarding a particular scientific problem. For example, I worked for years investigating bacteria belonging to the genus *Bartonella,* as well as diseases caused by such bacteria (Trench Fever, Cat Scratch Disease, Carrion's Disease). As the chief of the CDC's laboratory responsible for these bacteria, I knew leading experts in this area, followed related scientific publications, and participated in numerous international conferences where my colleagues and I shared information from our research. Even so, I

[2] Benjamine Fain, *Law and Providence.* 2011, pages 170 and 174.

145

could never claim to have more information about this topic than anyone else in the world. However, I felt confident that, at the time, I was one of the few people in the world who knew how much information we *didn't* have regarding specific questions about these bacteria!

Biologists have generated enormous data on numerous living creatures, their occurrences, their forms, genetic composition, etc. Modern technological advances have led to the exponential growth of new biological data. Today, two significant advances are producing and accumulating data related to biological research. First, sequencing and array technologies have fueled advances in life science research, translational and consumer genomics, and molecular diagnostics. The process of sequencing DNA and RNA nucleotides, ranging from gene fragments to whole genomes, has become much less expensive.

Second, high-performance computing has widely expanded the capacity of research institutions to process massive data generated from biological investigations. Computers and new software have introduced us to sequence assembly, genome annotation, simulation of cellular growth, system analyses of nervous and immune systems, preparation of databases for biological modeling, and many other activities related to biological research.

Examples of the World of Invisible Information in Biology

The exponential growth of data has been accompanied by scientists' humble awareness of incomplete information obtained after having analyzed new data. Below, I will provide a significant number of references to illustrate this point.

The magnitude of biological species on earth

Scientists constantly report new biological species. Lee Sweetlove, a Professor of Plant Science at the University of Oxford, counted around 8.7 million eukaryotic species on our planet.[3] Camilo Mora and his colleagues

---

[3] Lee Sweetlove, *Number of species on Earth tagged at 8.7 million.* Nature, 2011.

predicted that around *86% of land species and 91% of marine species remain undiscovered.* These authors conclude that some groups are much better known than others. They predict that only 12% of land animal species and 7% of land fungi species have been documented.[4]

The unseen myriads of microbes in the soil

It has been estimated that one gram of soil contains $>10^{10}$ bacterial organisms of 6,000 - 50,000 bacterial species. The impact of microbes on plant productivity is known chiefly for common nitrogen-fixing and root-associated bacteria.[5] An actual number of bacterial species in the soil is tough to predict, and any estimation depends very much on applied methods and acceptance criteria. Viral abundance in soils can range from below detection limits in hot deserts to over 1 billion per gram in wetlands.[6]

Diversity of viruses in mammals

The number of viruses on Earth is estimated to be around 10^{31}, which corresponds to roughly ten billion times the number of stars in the universe.[7] Recently published estimates reported the existence of between 40,000 and over one million virus species in mammals alone. The estimates represent actual data, but the absence of strict criteria leaves any statement open to further research.[8]

The number of genes in a human organism

It's hard to select an area of biological research where so much effort has been made as in human genomics. The achievements of this project are unquestionable. But here I cite Steven Salzberg, the Director of the Center for Computational Biology at Johns Hopkins University: "Scientists at the time believed that once we had the sequence in hand, we would fairly quickly be able to determine where all the genes were. Subsequent history

---

[4] Camilo Mora et al., *How Many Species Are There on Earth and in the Ocean?* PLOS Biology, 2011, vol.9, No.8, e1001127.

[5] van der Heijden et al., *The unseen majority: soil microbes as drivers of plant diversity and productivity in terrestrial ecosystems.* Ecology Letters, 2008, vol.11, no.3, pages 296-310.

[6] Kurt Williamson et al., *Viruses in Soil Ecosystems: An Unknown Quantity Within an Unexplored Territory.* Annual Review of Virology, 2017, vol. 4, no.1, pages 201-219.

[7] *Microbiology by numbers.* Editorial, Nature Reviews in Microbiology, 2011, vol. 9, no. 9, page 628.

[8] Colin Carlson et al., *Global estimates of mammalian viral diversity accounting for host sharing.* Nature Ecology & Evolution, 2019, vol.3, pages 1070–1075.

has proven otherwise: today there are several competing human gene databases, with many thousands of differences among them... The bottom line is that we don't yet know how many genes we have ... we cannot be sure that we have discovered all human genes and transcripts. For most other animal and plant species, we know even less about their gene catalogs."[9] There is increasing evidence suggesting that several proteins can carry out unrelated alternative functions, reported in the literature as "moonlighting."[10]

Obscure genetic variations

Cryptic variations in natural populations are invisible under normal conditions. In theory, cryptic genetic variations can fuel evolutionary ones when circumstances change. However, rare genetic mutations are not well characterized. In a recent paper, the team from the Broad Institute of MIT and Harvard found that rare mutations made a small, but important, contribution to the traits that they analyzed.[11]

Microorganisms as pathogens

In the study of epidemics, there is uncertainty regarding which microbes are pathogens that cause massive cases of the disease. Recent research has shown that many highly diverse bacteria operate along a functional continuum between pathogenicity and symbiosis ("living together") in natural hosts. Emergent properties of pathogens become evident only after diseases appear from the "invisible" side.[12]

Hotspots of emerging zoonotic diseases

Some results suggest that emerging infections are more likely to occur in regions with higher human population density and greater wildlife diversity. Conducted studies are mainly limited, given the lack of specificity of the predictors. Furthermore, this predictor set may not adequately represent a

---

[9] Steven Salzberg, *Open questions: How many genes do we have?* BMC Biology, 2018, vol. 16, page 94.

[10] George Kustatscher et al., *Understudied proteins: opportunities and challenges for functional proteomics.* Nature Methods, 2022, vol.19, No.7, pages 774-779.

[11] Daniel Weiner et al. *Polygenic architecture of rare coding variation across 394,783 exomes.* Nature, 2023, vol. 614, pages 492–499.

[12] Michael Kosoy, *Deepening the conception of functional information in the description of zoonotic infectious diseases.* Entropy, 2013, vol.15, page 1935.

range of potential mechanisms. A lack of an effect of rainfall, for example, does not discount the potential for other climatic factors to play a role.[13]

We can continue the list of examples forever; in fact, each biological discipline generates more questions than answers. This is a fact of scientific study. Thus, one can only recognize that we have access to a small portion of the information we must produce. However, we still need to obtain sufficient information when investigating a biological system.

Two trends characterize the current situation in biological research. The first trend is positive: great progress in developing new methods is resulting in a rapid accumulation of information about various biological systems; relevant new publications are emerging as a result. The second is the growing demand for additional information as we recognize the incompleteness of already available data and information. This situation is paradoxical because the expectation is that new discoveries should lead to an increase in available information and a productive way to use this information.[14]

There are a few reasons for the above paradox. First is the appearance of more sophisticated technology that generates previously unavailable data. The second factor is the demand for additional data discovered during the analysis. Third is the restriction of data by using a limited number of criteria. Fourth is the unknown temporal dynamic of various biological processes.

There are more technical problems that are hard to avoid: limited sample size, poor stratification of data collection, the design of experimental works with limited selected parameters, and others. Often, the process of analyzing data to receive information requires collecting missing data. Sometimes, "new gaps" appear as a side-effect of the attempts to fill gaps between point data.

One might compare the collecting of biological data to a connect-the-dots puzzle that reveals a hidden image. Collecting biological data involves gathering individual data points, or "observations," and organizing them in a meaningful way to reveal patterns or trends. In a dot-to-dot puzzle, each dot represents a specific point in space, and the challenge is to connect them

---

[13] Toph Allen et al., *Global hotspots and correlates of emerging zoonotic diseases.* Nature Communications, 2017, vol. 8, No.1, page 1124.

[14] Michael Kosoy, *Deepening the conception of functional information in the description of zoonotic infectious diseases.* Entropy, 2013, vol.15, page 1930.

in the correct sequence to reveal the intended image. Similarly, when collecting data, each data point represents a specific aspect of a larger phenomenon. The challenge is to collect and organize these points so they reveal underlying patterns or relationships (Figure 18).

Figure 18. As a metaphor, it can be said that collecting biological data is similar to putting together a dot-to-dot puzzle. The more data ("dots") are available, the more accurate information for an "image" can be obtained.

Like a dot-to-dot puzzle, collecting data can be a process that requires persistence and attention to detail. It often involves making careful observations, recording detailed information, and using specialized tools or methods to collect and analyze data. However, it's important to note that comparing collecting data with creating a dot-to-dot puzzle is not exact. While a dot-to-dot puzzle has a predetermined endpoint--a clear image that emerges once all the dots have been connected--collecting data is often an ongoing, iterative process. It may not have a clear endpoint or predetermined outcome. Collecting data may also involve more uncertainty and variability than a dot-to-dot puzzle.

"Visible" Data on the "Invisible" World of Bacteria

I will illustrate this point by going back to the question of how scientists identify plague caused by the bacterium of *Yersinia pestis* (page 17). When there is no active plague outbreak in an area, it can be challenging to detect this bacterium, even if it is present. Sometimes, we must examine hundreds and thousands of rodents before finding a positive one. While any test will reveal some data, it is mostly negative, with no evidence of actual plague.

Of course, negative data is also data! Many investigators of all ranks and ages need to remember this. Besides test results, the data includes geographic coordinates where the animal was collected, environmental variables, and animal characteristics such as species identification, sex, reproductive status, body measurements, and many others. On paper, every piece of data could correspond to a dot.

Let's imagine that we have a piece of data, such as finding a dead animal, that could indicate the presence of plague. It could be the discovery of antibodies in the blood or the culture of a bacterium resembling *Yersinia pestis*. Each piece of data adds a dot to the paper. One might find information about the plague when a few related data points are connected and analyzed. It might even be possible to represent the information indicating the existence of the plague by connecting isolated data points (dots).

As we continue the investigation and add more and more pieces of data, more animals are tested, antibodies are calculated, bacterial colonies are analyzed by observing their shape, getting their biochemical characteristics, and using genetic methods. Each piece of data adds more and more dots. Connections between dots result in more lines of information related to the identification of plague. Gradually, the accumulation of data represented by dots, and the information represented by the configuration of the "lines" connecting data points, made an image. This either supported or rejected the plague hypothesis. Doesn't this remind you of a "dot-to-dot puzzle?" (Figure 18).

Terra Incognita of Biology

Remember "The Streetlight Effect?" It's a fact that human beings strongly prefer to look at objects or phenomena where they are afforded a better view. Similarly, scientists usually select research areas where they can expect quick and productive results. This is a rational strategy when one intends to observe biological objects and phenomena. However, the analogy to the streetlight example is inexact.

Looking for a particular key in the wrong place can hardly provide helpful information! On the other hand, searching under a "streetlight" can still reveal helpful clues in the case of complex biological systems. The practical advice is to start under the "streetlight," and then move into the "shadows" ("a jump in the dark," in Waddington's words[15]). Avital used the Kabbalistic principle called "The Jumping of the Way," which refers to "transporting oneself from this limited timeline to another timeline, the next GREAT Shift from the darkness to Light, from what is "known" to that which is "Not Yet Known."[16]

The main concern should be avoiding reliance on the easily measurable proxy variable. It is important to recognize our tendency to explore selected places at the expense of neglecting vast territories!

Let's use an analogy with "Terra Incognita," a term used in cartography for regions that are unmapped or undocumented (from "unknown land" in Latin). On a map of the mid-17th century, the unknown south land was marked with the Latin term, "Terra Australis Incognita," roughly translated as the "Unknown (Incognita) Southern (Australis) Land (Terra)." Maps of the then-known world were later modified to include recently discovered lands and oceans. In the early 17th century, the top end of Australia was being discovered and charted and began appearing on contemporary maps. Some maps showed it correctly as a land mass unconnected to Terra Australis Incognita. However, other maps extended the coastline north from the Arctic Circle to incorporate the newly charted coasts to the south of New Guinea as part of Terra Australis Incognita (Figure 19). When a ship approached the coastal part, which was not reflected on the maps, it was a classic example of the emergence of an unfamiliar land in geography.

---

[15] C.H. Waddington, *The Scientific Attitude.* 1941, 128 pages.

[16] Samuel Avital, *Messages from "Nothingness,"* 2024, Kol-Emeth Publishers, page 5.

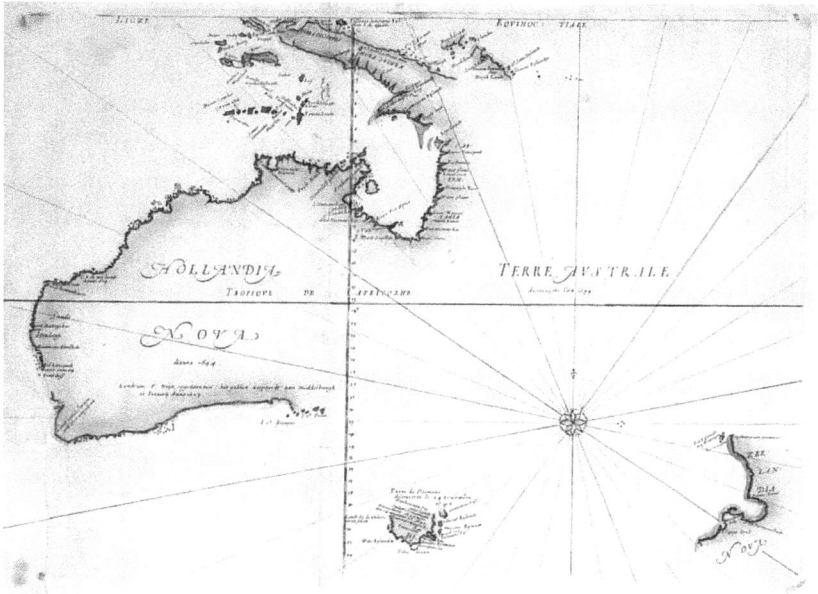

Figure 19. Terra Australis Incognita on the map of the mid-17th century represents pieces of land, rather than the whole continent of Australia.

We can draw an analogy between "Terra Incognita" and missed information with limited data in biology. Just as a mapmaker may have to rely on incomplete or limited information when creating a map of an unmapped area, biologists may have to work with incomplete or limited data when studying complex biological systems. In both cases, the lack of information or incomplete data can result in uncertainty, gaps in knowledge, and the potential for errors or inaccuracies. However, despite limited biological data, biologists may use various techniques to gather more information, such as experimental manipulations or statistical modeling.

People feel more comfortable dealing with what they can see. Whether these manifestations are appealing or frightening, human psychology has adapted to allow us to react to visible events.

We routinely acknowledge things we know, and we recognize events that we can explain in our daily lives. We also try to logically prioritize areas where we wish to learn more. But in some extreme, potentially dangerous situations, the unknown can produce extreme fear that makes it harder to understand what we know vs. what is rooted in baseless rumors.

Invisible elements cause fear. Fear of the unknown continues to exist among human beings. Epidemics of fatal diseases are a good example of fears implanted in our minds precisely because the cause of a particular epidemic is usually unknown. People are afraid of epidemics as invisible microbes move into their society. History's recounting of the Black Death has contributed significantly to fears affecting large populations. We may have the same feeling of dread after reading Boccaccio's *The Decameron* or Albert Camus' *The Plague* or looking at the art of Pieter Bruegel, the Elder. This fear is very much implanted in the human subconscious. Often, the news of an impending epidemic can arrive faster than the epidemic itself.

Scientists acknowledge what remains unknown and invisible. However, they believe it can eventually be identified as more information, more collective efforts, and more funds are provided. We proclaim that we know more about biological systems and processes, yet we misuse the word "know." Instead, we collect more information. I repeat: *knowledge is NOT limited to information received!*

The main message of this chapter is that "invisible" space has structure. When we are aware of such structure, we have the means to orient ourselves, even in dark spaces-- or when we are in the middle of Terra Incognita. To explore the realities of the "invisible" aspects of living processes, we must find new tools to help us navigate the remaining unknown territory.

Chapter 7

In Search of Knowledge of the Unseen in Biology

As we have pointed out in previous chapters, contemporary biologists can investigate only a small portion of biological natural phenomena. Biological life's invisible world, by the way, is *not* uniform. In the ever-accelerating process of data collection, biologists are often overwhelmed by how quickly information is accumulated; they are also daunted by how much inaccessible information is eluding them! It's not that the information doesn't exist; it's just not available yet. In the previous chapter, we called the lack of information we are aware of, "known unknowns." This chapter will expand on how we can explore the enormity of these invisible worlds.

In addition to discovering information not yet available to us ("known unknowns"), we will also learn about the world of "unknown knowns." This represents the level at which information about a particular biological system is available, but our knowledge of the *essence* of this system remains beyond immediate comprehension. We can label this realm as a world with "*known information, but unknown knowledge.*" At this level, even as we collect more information about biological phenomena, we may still be unable to fully explain our findings. This highlights the need for continued exploration and research into biological systems' complexity and depth at higher levels. Is this statement confusing? I'm not surprised!

The greatest problem we face is a strict adherence to traditional "logical reasoning," a binary thinking process based on "yes" or "no." To approach biological life's invisible reality, we must change our way of thinking. Although we cannot see it, the invisible world does have a structure. It is our task to learn how to gain access by revising our approach to it.

The World of Known Information but Unknown Knowledge

Using our known tools of investigation, reasoning, or specialized knowledge, we can infer aspects of biology from the visible world that might apply to the invisible world (or "unknown knowns"). While in this world, we cannot directly observe phenomena such as adaptation to the environment. We can, however, infer these from their effects. For example, evolutionary research requires us to look *beyond* the visible world, such as genetic variation, and explore the invisible world based on what we learn about this phenomenon's manifestations, both visible and potential.

A repeated motif you'll find in this book is the formula, "information ≠ knowledge." We discussed the difference between information and knowledge in previous chapters. The task now is to be able to discriminate between unavailable information ("known unknowns") and hidden knowledge ("unknown knowns") while we investigate the invisible side of all biological processes. Information *about* biological objects can contribute to attaining new knowledge, but it cannot replace it.

Before focusing on specific problems, let's consider examples where we could not blame our failure to receive information on inadequate knowledge of basic biological concepts. This chapter demonstrates how "unseen" biological reality is divided into different layers. For example, we can discover and describe more and more variants of coronavirus. We can investigate more and more bacterial species in a particular soil type. We can also measure more and more rabbits inhabiting a park. However, any missing information will always be "invisible" to us until we get it!

Suppose we had access to as much information as we wanted. There would still be a lack of clarity about the investigated biological phenomenon, despite the abundance of information available. Below are some examples of "invisible knowledge" in biology.

Natural selection

We can track changes in visible characteristics within a rabbit population over time, such as color. It is also possible to register the development of antibiotic resistance in bacteria, which is likely the result of the exchange and selection of bacterial plasmids. However, we cannot directly observe natural selection as a component of evolution.

Photosynthesis

We can't see photosynthesis, but we can see its effects. We can describe processes by which green plants produce oxygen and synthesize carbohydrate molecules, such as sugars and starches, through photosynthesis.

Digestion

It is not possible to directly observe digestion because it is a complex process that occurs within the body. However, digestion becomes partially "visible" through the observation of certain effects, like the breakdown of food by mechanical and enzymatic processes into substances needed by the body.

Nervous system

Scientists have learned a great deal about the structure and function of nervous systems. As with all biological systems, however, the nervous system of any animal (including man), as a whole system, itself remains "invisible."

Hormone regulation

While we cannot directly observe hormone production and transport, we can observe their visible effects, such as changes in behavior and physiology. For example, scientists could measure the receptors' functions to bind hormones and change the signal type to affect the recipient cell's metabolism. However, these observations individually do not accurately predict the full functional significance of the hormonal activity.

We Can See Dogs, but *not* "Dog-ness"

To illustrate the level of biology at which information exists ("visible") but knowledge is absent or limited ("invisible"), let's examine the concept of biological species. The concept of "species" is fundamental to the study of modern biology. It plays a crucial role in categorizing and classifying organisms based on their shared characteristics and evolutionary relationships. Species provide a framework for organizing and interpreting biological diversity and allow us to predict the behavior, ecology, and evolution of different organisms. It also provides a common language and set of standards so we can communicate about biological research and knowledge.

In 1753, Carl Linnaeus, the great Swedish scientist, introduced the system of naming species with a *binominal* (two-part) name-- now informally called a "Latin name."[1] The first part of the name – the generic name – identifies the taxonomic "genus" to which the species belongs. The second part – the specific "name" – distinguishes the species within the genus. For example, domestic dogs belong to the genus *Canis,* and within this genus, we find the species *Canis familiaris.*

If I were to ask, "Have you seen the dog species?" the answer may seem obvious because everybody has seen dogs. Suppose I ask similar questions, such as whether people have seen the red fox species (Vulpes vulpes), the gray squirrel species (Sciurus carolinensis), or the bobcat species (Lynx rufus). In that case, people may have heard of them, even if they never saw them alive but did see them in pictures. These questions, however, are not as obvious as you might think. *None of us can see a biological species with our physical eyes.* We can only see *representatives* of the species as individual animals.

Thus, experts can only know about the *existence* of this species. Zoologists can use the term, "dog-like" mammals, but they cannot

---

[1] Carl Linnaeus. *Species Plantarum.* 1753.

identify a dog's "essence" or "dog-ness" (Figure 20). To attempt to do so would be like trying to define grass as "green" because of its color. Seeing something green in color is not the same as seeing the essence of "greenness."

Figure 20. We can observe animals of a particular species, e.g., dogs, but we cannot directly see the species itself as a "dog-ness."

Does a Species Exist?

The task of describing a biological species is challenging. In systematics, the *place* of a biological species is of utmost importance. The science of systematics, one of the oldest branches of biology, develops classification systems and assigns scientific names to organisms (nomenclature). By classifying organisms, scientists can expect the presence of particular characteristics while working with them.

Initially, Carl Linnaeus relied only on the size, shape, color, and many other external features of various species. Sometimes, the parameters could be extended to include more sophisticated indicators. For example, zoologists use various shapes of teeth to identify some species of rodents. One common indicator is the number of chromosomes considered typical for a given species. For example, humans typically have 46 (23 pairs) of chromosomes, while gorillas have 48 (24 pairs), cats have 38 (19 pairs), dogs have 78 (39 pairs), and so on. Zoologists can also assess the animal's behavior, ecology, nutritional requirements, and many other parameters.

Over time, biologists have expanded their use of genetic methods to identify species. One method, "barcoding," is used to utilize a specific region of DNA called the "DNA barcode." Such a genetic fragment can be queried in a DNA library, which contains information about other species linked to their barcodes. Confusion might arise if DNA barcoding is used when genetic information suggests that a biological organism is "close" to a species to which it does not belong. Regardless of how many listed methods exist, not one of them can actually define the essence of a biological species. Consequently, people cannot "see" a biological species beyond their individual members.

There is a 124-year-old optical illusion known as "Duck or Rabbit?" American psychologist Joseph Jastrow first used the duck-rabbit drawing in 1899 to point out that perception is not only what one sees, it is also a mental activity (Figure 21).

Figure 21. The duck-rabbit image highlights a curious phenomenon called "aspect perception."

When presented as a duck, those unfamiliar with the image may initially see only a duck. However, if alerted to the possibility of another embedded image, suddenly, a rabbit may pop out.[2] Depending on how much attention one pays to the right or left side of the image, one will perceive different things. Paradoxically, the

[2] I.C. McManus et al., *Science in the Making: Right Hand, Left Hand. II: The duck-rabbit figure.* Laterality, 2010, vol.15, pages 166–185.

image is both one thing and another at the same time! You can see either a duck or a rabbit, one after the other, but not both at the same time!

Ludwig Wittgenstein, the great philosopher of the 20th century, distinguishes two types of *seeing*. First, we have the standard, direct act of seeing — interpreting the stuff that is broken down into sense data for the eyes. For example, we see a body shape and an ear shape specific to a rabbit. But we can also "generalize." For example, we can assign the animal with a characteristic of cottontail rabbits to the genus *Sylvilagus*.

The duck-rabbit image is just a metaphor. However, in my professional activity, I often have to make a choice when I assign an organism to a taxonomic category. If different taxonomic criteria lead to alternative variants, we must choose and then define the criteria to use. For example, desert cottontails have longer ears compared to mountain cottontails. In comparison, mountain cottontails' hind legs are covered with denser fur. Since morphological identification is not always reliable, genetic markers can help. Although the distribution of both cottontail species overlaps, there are extensive geographical areas where only one of these species has been registered. You cannot expect to find a desert cottontail in Oregon, or a mountain cottontail to appear in Texas. However, in western Colorado, where I live, both species have been known to appear.

A cottontail's habitat could help identify the animal. Mountain cottontails occupy thickets, sagebrush, loose rocks, and cliffs. In contrast, desert cottontails are found in valleys and low deserts. Our preference for some morphological, genetic, geographic, or ecological characteristics might influence the identification of the species.

Experts can change the indicators used to identify some species, and thus affect the assignment of individual animals to a particular species. Does this mean that the species identification process depends on the investigator and that a "real" species does not exist in nature? I would separate these two questions and answer "yes" to the first and "no" to the second.

Although the appearance of organisms and other apparent parameters can help identify a species, these identifiers do not necessarily define it. A biological species does not have a specific appearance. It is strictly defined by the combination of characteristics attributed to all representatives of that particular species.

There is growing interest in the concept of "biological species" to help us understand the existence of species in nature. This concept defines a species as "a member of populations that actually or potentially interbreed in nature." The biological species concept is closer to its core identity than to the methods of identification, which are limited to separate indicators. However, it still doesn't allow us to *see* species. In addition, this concept has certain limitations that do not apply to identifying fossil species, organisms that reproduce asexually, or species that freely hybridize.

I cannot count how many times I warned my students and young scientists not to jump to conclusions when they were ready to announce the discovery of a particular bacterial species. For example, one day, a student excitedly told me that he had found evidence of plague based on the fact that he had detected *Yersinia pestis* in a rodent collected in New Mexico. Not wishing to dampen his enthusiasm, I suggested that we slow down a bit. First of all, we had no evidence of the presence of plague, as a disease. Instead, the young scientist had detected a fragment of a gene typical of the plague pathogen. A more reliable report would have included culturing bacteria with properties specific to this species. Based on some indicators, we could claim that this particular bacterial isolate likely belonged to the *Yersinia pestis* species. However, we still could not "see" the species; we could observe only some indirect indicators.

To summarize, we must use our "knowledge" to define biological species. Knowledge and preferences can be determined differently by experts. As a result, there could be conflicting opinions about the taxonomical identity of many species. Members of some expert communities often try to standardize taxonomic criteria, but biologists find themselves constantly revising these groups. To further complicate the issue, at times, the scientists' revisions might be accompanied by mutually exclusive arguments.

As a result, many biologists evade questions about the essence of biological species. Some of them claim, instead, that all species are simply mental constructs and products of formal agreements between scientists. Such statements are patently untrue: of course, biological species exist in nature! In fact, species are one of the fundamental units of comparison in virtually all subfields of biology, from anatomy to behavior, development, ecology, evolution, genetics, molecular biology, paleontology, physiology, and systematics.

Biological species are actually *"units of evolution."* In nature, biological species evolve, not just our perception of them. Charles Darwin titled his most important book, *On the Origin of Species*.[3] The essential role of species in evolutionary processes has continued into the more recent history of the discipline, including the period of the Modern Evolutionary Synthesis. Two of the most important publications describing the central role of species in the 20th century are Theodosius Dobzhansky's *Genetics and the Origin of Species*[4] and Ernst Mayr's *Systematics and the Origin of Species*.[5]

Biological species are also units of both ecosystems and the entire biosphere. Most biologists describe biodiversity by counting the number of species. Often, when we are concerned about threats to biodiversity, we talk about the extinction of certain species due to the impact of human activities on them, as well as the spread of non-native species and species hosting animal diseases.

The recognized role of biological species is not limited to theoretical biology; it's crucial for scientists if the need arises for practical actions. We must know which species present danger and

---

[3] Charles Darwin, *On the Origin of Species by Means of Natural Selection*. 1859.

[4] Theodosius Dobzhansky, *Genetics and the Origin of Species*. 1937.

[5] Ernst Mayr, *Systematics and the Origin of Species*. 1942.

which species can be helpful, which species should be controlled, and which species should be protected.

Many Biological Phenomena Remain Invisible to Us

Another important concept in biology is that of "natural populations" of animals and plants. A population is the number of organisms of the same species living in a particular geographic area at the same time. This makes it sound as if a "population" is more tangible than a "biological species." Not true! Populations can occur on various scales. A local population can be confined to a small area, like fish in one pond. However, this locality of populations can also operate on a regional, countrywide, island, or continental scale; it may even make up an entire species! If individual members of local populations can interact with other local populations, they are considered a "metapopulation."

We can study populations by using many particular parameters that include, such as:

- Population size.

- Range of distribution.

- Population density.

- Demographics (birth and death rates, sex ratio and age distribution, fluctuation in numbers over time, many others).

- Migratory patterns.

- Population genetics (genetic variants frequencies within and between populations).

- Morphological features.

- Adaptation to specific local environment.

- Life history traits.

While having so much information, we still can never actually "see" biological populations! It is not enough to try counting how many individuals belong to a particular population. In addition, biologists use a variety of approaches to define and identify populations in natural settings. We know that populations of animals and plants exist in nature, and we can also effectively study them. We can endlessly continue to collect additional data. We can improve our ability to obtain information from available data. However, we don't know the essential characteristics of practically any natural population, and we cannot even agree on what a "population" is. Some experts might focus on geographic habitation, others emphasize dynamic synchronicity, and others might argue about genetic similarity.

We can say the same about "biological communities." In biological terminology, a biological community is an interacting group of various species in a common location. For example, a forest of trees and undergrowth plants rooted in soil containing bacteria and fungi constitutes a "biological community." We know that biological communities exist in nature and we can study them in detail. We can collect more and more information about various components of biological communities. However, knowledge of the essence of any biological community remains an "unknown known." This will depend on our decisions and the resources available to select a scale for the investigation.

The composition of biological communities changes over time. Communities evolve through a process called "ecological succession." Successions create a continually changing mix of plant species within communities. At every stage, certain species have evolved life histories so they can exploit the particular conditions of the community. We can identify and estimate the density of plants belonging to some species. In other words, we can "see" these plants. However, we cannot "see" stages of succession. We can only define them by the presence of lichens and some plants, such as small annual plants, grasses, and perennials, or shade-tolerant trees like oak or hickory. Importantly, we can learn succession stages and use them if needing to take conservation actions (Figure 22).

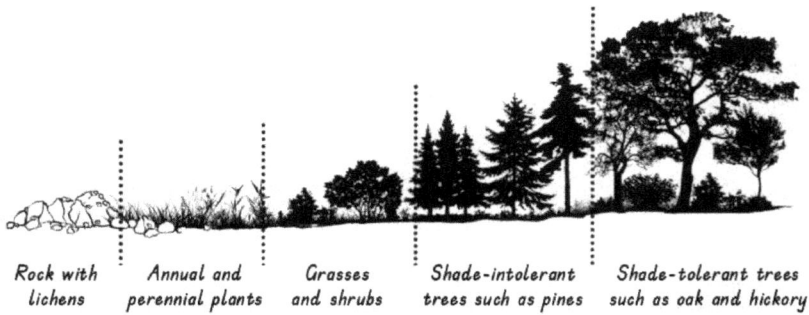

Rock with   Annual and   Grasses   Shade-intolerant   Shade-tolerant trees
lichens   perennial plants   and shrubs   trees such as pines   such as oak and hickory

Figure 22. Ecological successions create a constantly changing mix of plant species within communities. While we can "see" the plants, we cannot "see" stages of succession.

The same reasoning applies to ecosystems. An "ecosystem" describes a single environment that includes every living (biotic) organism and non-living (abiotic) component contained within it. An "ecosystem" embodies every aspect of a single habitat, including every interaction between its variety of elements. Therefore, no matter how much we learn about a particular ecosystem, it will always remain an "unknown known."

Any biological system that includes many elements includes "unknown knowns." We know much about systems, such as the immune system, the blood circulatory system, the lymphatic system, the respiratory system, and the digestive system. Still, they all remain "unknown knowns" when we consider them as "whole" systems. Models of species, communities, ecosystems, nervous systems, and other biological systems can help us make biological processes within such systems "visible."

The Invisible Aspects of Epidemics

Here's another example. Do we understand epidemics? Of course, we know that epidemics occur, and we are somewhat familiar with elements relating to the epidemic process: infectious agents, animal reservoirs, vectors transmitting pathogens, and susceptible victims of epidemics. But can we know "epidemics" as a unique phenomenon? Epidemics as a whole will always remain "unknown knowns" because of numerous biological, environmental, and social factors that influence the trajectory of developing any epidemic. The interactions between these factors are numerous, and they are not proportional, nonlinear, or multi-scaled. At each moment, we can consider a unique but invisible underlying background that identifies critical points. This will reveal how the infectious process will be expressed, whether in animal populations or human societies.

The advent of the COVID-19 pandemic served as a stark reminder that the most pressing issue at hand extended far beyond data limitations and information gaps around viral variants and population immunity. This crisis transcended scientific inquiry and exposed the profound complexity of a natural phenomenon. It defied prediction or prevention by experts and non-specialists alike.

The pandemic unfolded swiftly and ferociously, taking the world by surprise. As it swept across continents, it laid bare the limits of our knowledge and capacity to control nature's forces. In the face of such a formidable adversary, it became evident that the crisis demanded more than just data and information; it required a deep, holistic knowledge of the intricate web of factors in play. Scientists grappled with viral variants, racing to gather data and develop countermeasures. Yet, even as they worked tirelessly to unravel COVID-19 mysteries, it became increasingly clear that handling the pandemic required not only the accumulation of more and more data and information. It was a global event that tested our resilience, adaptability, and solidarity capacity.

For non-specialists, too, the pandemic prompted a profound reckoning. It forced individuals from all walks of life to confront the fragility of our existence, as well as the interconnectedness of our world. It underscored the importance of collective action, following expert guidance, and recognizing that our actions have consequences that ripple throughout society. In this crucible of uncertainty, the COVID-19 pandemic underscored a fundamental truth: that the challenges posed by the natural world are not mere puzzles to be solved through data and information alone. They are profound questions that call for a deeper knowledge of the complex and ever-changing dynamics of life on this planet.

Dark Biodiversity

The importance of biological diversity has become the center of many environmental health and preservation problems. Biological diversity has been studied for centuries, and its patterns have been linked to ecological processes. Ecologists refer to the observed community's diversity; they can also calculate a set of species present in a particular region that can potentially inhabit a particular observed community because of suitable local ecological conditions. However, the observable patterns provide only a portion of the information needed to better understand the processes affecting a particular population or community.

Recently, ecologists proposed measuring what's known as "dark diversity." The idea was to measure a portion of the species pool *not* observed in the focal community.[6] It constitutes species in the surrounding landscape or region that could potentially colonize the site, but are currently absent or not detected. Dark diversity invites parallels with dark matter. Dark diversity, like dark matter, cannot be seen or measured directly. *Dark diversity, however, can be accurately estimated using readily available data.* A key benefit of dark diversity concepts is that they can be applied to a wide range of conservation applications, such as managing invasive species, recovering ecology, and estimating health risks.[7]

The Junk DNA

An excellent illustration of the difference between the worlds of "invisible information" vs. "invisible knowledge" can be found in the results of the Human Genome Project. The Human Genome Project was a landmark global scientific effort whose signature goal was to generate the first sequence of the human genome. Scientists from

---

[6] Rob Lewis et al., *Estimating dark diversity and species pools: an empirical assessment of two methods.* Methods in Ecology and Evolution, 2016, Vol.7, Issue 1, pages 104-113.

[7] Camilla Fløjgaard et al. *Dark diversity reveals the importance of biotic resources and competition for plant diversity across habitats.* Ecology and Evolution, 2020, vol.10, No.12, pages 6078-6088.

many countries participated in this project, thanks to massive allocated resources. In 2003, the Human Genome Project produced a genome sequence that accounted for over 90% of the human genome.

However, the project also demonstrated that only less than two percent of the human genome includes DNA sequences directly encoding proteins. The rest, more than 98% of DNA sequences, were commonly called "junk DNA," large portions of the genome that do not code for proteins. Many mysteries still surround the issue of noncoding DNA and whether it is worthless junk or something more. Portions of it, at least, have turned out to be vitally important biologically. A vast majority of the thousands of genetic loci associated with human disease are located outside of coding regions. This may be because we have not examined gene expression in disease-relevant cell types or conditions. Many studies of gene expression across tissues are based on human samples obtained opportunistically or post-mortem, mostly from adults.[8]

Scientists learned that some non-coding DNA is used for RNA components such as transfer, regulatory, and ribosomal RNA. New information about non-coding parts of genome DNA increasingly helps us understand their "hidden" presence in gene expression, cellular signaling, and other biological processes (Figure 23). Still, many other DNA regions are not transcribed into proteins or used to produce RNA molecules, and their functions remain unknown. Although we can observe the visible effects of genetic mutations and variations, such as the development of diseases or changes in physical traits, we still do not understand the function and evolutionary significance of non-coding DNA.[9] Even beyond the question of its functionality (or lack of it), researchers have only recently begun to appreciate how noncoding DNA can be a genetic resource for cells, particularly for the control of gene activity.

---

[8] Benjamin Umans et al. *Where Are the Disease-Associated eQTLs?* Trends in Genetics, 2021, vol.37, no.2, page 109.

[9] John Parrington, *The Deeper Genome: Why There Is More to the Human Genome Than Meets the Eye.* 2017, 400 pages.

Figure 23. The saying goes, "One man's trash is another man's treasure." More than 98% of our genome was considered "junk DNA," but it might still be able to play a key role in disease or evolution.

Today, a heightened focus on "junk" DNA is radically transforming our understanding of the genome, and hinges on metaphor and imagery. In her book, *Junk DNA: A Journey Through the Dark Matter of the Genome*, Nessa Carey provides a compelling introduction to junk DNA and its critical involvement in such diverse phenomena as genetic diseases, viral infections, sex determination in mammals, and evolution.[10]

Let's summarize the results of the Human Genome Project. The project has produced an enormous amount of data and generated extensive information about human DNA. It was even able to identify many genes responsible for various bodily functions, therefore increasing our knowledge of the human genome. However, the project also demonstrated how little we know about the function of the whole genome. Scientists still don't know the purpose of most of the genetic material stored in human chromosomes. Overall, the experience from the Human Genome Project highlights our need to

---

[10] Nessa Carey, *Junk DNA: A Journey Through the Dark Matter of the Genome*. 2017, 360 pages.

recognize the limits of the "invisible" knowledge yet to be discovered, despite the accumulation of available ("visible") information.

Entering the Invisible World of Biology Requires Knowledge

I wish to conclude the list of "unknown known" examples by mentioning the most essential one of all "the essence of biological life." At the beginning of this book, I mentioned a limited number of books and other publications entitled "What is Life?" Usually, such books are written by physicists or philosophers but not by biologists. Why is this so? The answer is simple: "Information" is inadequate to answer such a question. The answer belongs in the realm of "wisdom." Unfortunately, biologists learn how to get information about biological *organisms*, but fail to receive knowledge about the essence of biological life.

Data with measurements of individual animals and plants can either remain the same or be improved by additional observations. Data analysis from measuring animals and plants will result in information shared with colleagues. As a result, we gain information related to animals and plants. Knowledge of biological species, populations, and communities belongs to different levels than information about some parameters.

We need to realize that all listed phenomena represent "unknown knowns." They constitute the structure of biological knowledge. Yes, we are confident that species, populations, ecosystems, nervous systems, and other biological systems exist. *But what they really mean is not known.* Therefore, we must define them according to how much knowledge we have about them. *Knowledge depends on defining what we understand under those categories.*

The definition of "unknown knowns" requires knowledge. *Knowledge is always based on personal or collective experience.* The general public does not need to study "unknown knowns." More important is that the scientists who understand this concept use such knowledge to produce more knowledge. Using the flashlight metaphor, we must bring our "flashlight" to illuminate dark places in biological research. The Zohar, the most fundamental book of

Kabbalah, suggests that "light might be revealed through darkness."[11]

The invisible "unknown knowns" are entirely different from the invisible "known unknowns." They do exist in different worlds. We cannot *see*, let's say, species at the level of data or information; we cannot *see* species from the level of individual animals or plants. *Knowledge cannot be expressed at the level of information.* Metaphorically, it is like a sphere or another 3D object that can be seen as a 2D image while passing through it.

Species are, in a way, the reality of a higher-dimensional world! To illustrate this point, let's use the metaphor from the world of Flatland. We can visualize a "dog as a biological species" as a Sphere (the concept of a perfect figure). Inhabitants of the Flatland ("local biologists") will see an image of a particular dog when they encounter such a conceptual entity passing the level of an individual organism. The image of the organism can be more or less detailed, depending on the available information.

We must learn to recognize various domains of the *invisible* world of biology for our orientation. Still, the mentioned two *invisible* realms ("incomplete information" and "hidden knowledge") do not include the total "invisible" world of biology. In the next chapter, an even more mysterious level of reality is awaiting us.

---

[11] Estelle Frankel, *Sacred Therapy.* 2005, page 37.

Chapter 8

Transcending the Hidden Landscape of Biological Life

When exploring new dimensions of biology in the *visible* world, the data available to us is in the realm of "known knowns." To navigate in the *invisible* worlds of biology, we must acquire some new navigational tools. In Chapter 6, we explored a world where information is incomplete due to a lack of data. Such a world remains partially invisible to us because we are unable to access data that would give us insights into that realm of existence.

In Chapter 7, we examined the world about which we do have available information regarding cells, microbes, plants, animals, etc. However, the entire universe of biological knowledge (species, populations, natural selection, adaptation, etc.) remains hidden from us. The guidance that follows will enable us to explore the next level of "invisibility."

What's Invisible in the Invisible World?

The "invisible world of what is invisible" represents the most elusive and mysterious domain of biological life. This term covers aspects that are both concealed and difficult or impossible to understand. This domain may include yet-undiscovered laws of biology, unknown dimensions, or other biological phenomena beyond our current understanding.

For this knowledge, we must look beyond the surface to fully grasp these hidden dimensions. We also need to examine the deeper underlying mechanisms and relationships between the variables responsible for these hidden dimensions. Developments in quantum physics have led to a profound shift in our understanding of subatomic particles. Rather than considering them "tiny bodies," we now describe them as "condensed energy." Does this new perspective affect biologists, as well? Because we deal with biological cells, organisms, and ecosystems, all of which are macroscopic objects. Most biologists argue that the problems of quantum physics should not affect our investigations. However, this is not accurate when exploring Kabbalistic Biology.

In Chapter 2, I mentioned a famous thought experiment widely known as "Schrodinger's Cat." A cat is undoubtedly a biological object! The conditions of the "experimental" cat created by Schrodinger were "both alive and dead at the same time." The conditions of "alive" or "dead" are often described in biological terms. "Why can't we see an actual cat be alive and dead at the same time?" Rephrasing this question using quantum physics terminology, "Why can't we observe a cat in a quantum superposition of possibilities?" Theoretical physicist Sky Nelson-Isaacs has given us an insightful answer:

> "...because we are not observing the cat... Any interaction we have with the cat ... will always provide a definite result. We will always see a "classical" world, even if that world is purely quantum because that's exactly what "seeing" does... The mainstream view makes a proposition about the macroscopic world: it is both defined (the cat is in a single state) and objective (the "world out there" exists without defining who is observing it)."[1]

---

[1] Sky Nelson-Isaacs, *Living in Flow:*
The Science of Synchronicity and How Your Choices Shape Your World. 2019, 320 pages.

Here is the point: the "invisible cat" is a cat that contains *invisible aspects* of the "biological" cat. We see the body of the "visible" biological cat, patterns of its movement, and specific reactions of the cat's organism to a particular stimulus, and we can microscopically investigate the cat's cells and tissues. We can also culture microbes from the cat's tissues and investigate them. However, this is *not* a "biological" cat! We cannot observe aspects such as its evolutionary origin, embryonic development, connection with other cats and microbes, behavior, or physiology.

When we are at the intersection of quantum physics and biology, we are challenged to consider the hidden aspects of life that lie beyond our current comprehension. By acknowledging the limits of our understanding and embracing the complexity and interconnectedness of living systems, we are positioned to uncover new insights into the nature of Life itself. "Quantum Biology," as an emerging field, has the potential to deepen our understanding of various biological processes. While we can now observe the visible effects of these processes, the underlying quantum mechanisms and their implications for biological explorations remain largely unknown.

The "Inside or Outside" Dilemma in Biology

The "inside or outside dilemma" in biology refers to the challenge of distinguishing which factors or variables are internal or external to a biological system. It sounds like a simple choice. After all, whatever is inside a cell membrane, e.g., the gel-like cytoplasm, cytoplasmic organelles suspended in the cytoplasm, the nucleus, etc. is clearly internal. It can be said that anything outside the cell would be considered "external."

While I agree that such a view is valid at other levels, it is not so at the level of the "Invisible World of What is Invisible." This "inside" or "outside" dilemma becomes even more complex when we consider the broader context in which biological systems are situated. There is no such thing as an isolated biological system. It is constantly interacting with its environment and with other biological systems. Thus, since many factors can affect biological processes, it can be difficult to categorize them as completely "inside" or "outside" the system. Lawrence Kushner, a Reform rabbi, and the scholar-in-residence in San Francisco, apparently agrees, saying, "In the light of a closer biological observation ... the line between external and internal begins to blur."[2]

Here's a useful example: To study the hidden dimensions of epidemics, we know we must go beyond the linear relationship between a specific pathogen and its susceptible host. This may involve the complex, often obscure relationships between different factors and variables that contribute to epidemics. We already know that many environmental and social factors play a role in epidemics, such as one's access to health care. Many other biological variables have a significant impact on the epidemic process, such as immune status, the presence of other infections, physiology, mood, stress level, nutrition, etc. "Are these variables part of the epidemic process? Or are they simply outside influences affecting the system?" This is the question at hand. Right now, the answers are unknown.

Take another example: the immune system, a critical part of the internal workings of the body. It is also closely linked to environmental factors, such as stress. Past exposure to a particular pathogen can either enhance the immune response (e.g., vaccination) or impair it (e.g., Dengue virus). The behavior of individuals within a population can have a significant impact on the spread of disease, but it can be difficult to perceive this behavior as entirely "inside" or "outside" the biological system.

Through the simple act of breathing, we encounter the "inside-outside dilemma" in our daily lives. When we inhale, air enters the body through the nose or mouth. It embarks on a journey through the intricate pathways of the respiratory system until it finally reaches the lungs. At this point, the air could be considered "inside" the body,

---

[2] Lawrence Kushner, *The River of Light.* 1993, 153 pages.

having crossed the boundary between external and internal environments. Similarly, when we exhale, we release air from our lungs, which travels through our respiratory system before leaving our bodies through the nose or mouth. During this process, the air is "inside" the body until it leaves the body through the nasal passages or the mouth. Then, it becomes "outside" the body.

We need to remember that the air we breathe, as well as our exhalations, are not simply external aspects of our bodies that interact with them. In a certain way, the air actually becomes a part of the organism. Taking in oxygen is crucial to producing energy within the body, and exhaling carbon dioxide is essential to removing waste products from the body. Furthermore, the air we breathe is not a static entity. It is constantly changing, thanks to environmental factors such as pollution or allergens, which constantly alter the air we breathe. These external factors can adversely affect the quality of the air we breathe, which may significantly affect our bodies' internal functioning.

When we consider that viruses and bacteria are exchanged during the breathing process, the inside-outside dilemma becomes even more complicated. There is more to breathing than just inhaling and exhaling air; we are also inhaling and exhaling any viruses and bacteria that may be present in the air. Under some circumstances, the virus and microorganisms become a part of the organism. We will discuss this further in the following paragraphs.

Breathing is a vital process in biological life. Many biological organisms cannot exist without breathing. This activity is a fundamental biological phenomenon essential to the survival and functioning of most living organisms, including all those relying on aerobic respiration. This process includes oxygen intake, carbon dioxide removal, pH regulation, olfactory communication (detecting volatile chemical compounds such as smells), and even stress reduction. Although breathing is an integral part of our lives, we cannot claim to understand the process as a whole. Breathing is both internal and external to organisms. While we cannot *see* breathing, we can observe and measure its results. Similarly, while we cannot *see* the wind, we can observe and estimate its effects.

Habitual exposure makes things invisible. As Neil Theise, professor of pathology at the NYU Grossman School of Medicine, noted, "Fish are unaware of the water they move through, just as we humans typically fail to notice the air around us unless there is a breeze."[3]

The Hidden World of Microbiota

Joshua Lederberg coined the term "microbiota," emphasizing the importance of microorganisms inhabiting the human body in health and disease. *A microbiota is an ecological community of microorganisms that literally share our body space.* Some calculations indicate that the number of bacterial cells occupying a human body is much greater than the number of our own cells!

Readers may ask why they encounter the word "microbiota" in some places and the word "microbiome" in others. What is the difference? Though we use these two terms interchangeably, there *is* a subtle difference, at least for those of us who work in this discipline. "Microbiota" usually refers to the totality of microorganisms found within an organism or with a specific part of the organism. The "microbiome," on the other hand, refers to the complete set of microbial genes present in an organism or in particular tissues like the gut or skin.

Microbiota can be vital for both human and animal metabolism, bringing recipients embryonic development, physiology, immunity, and behavior patterns. Experimental studies have revealed that the microbiota of animals can dramatically alter the population dynamics of animals. The "inside-outside" dilemma is relevant when considering the potential impact of microbiota on health. While microbiota is generally considered beneficial to our health, certain changes in the microbiota can contribute to the development of

---

[3] Neil Theise, *Notes on Complexity.* 2023, page 119.

diseases such as inflammatory bowel disease or allergies in human beings.

The microbiome is a complex, dynamic system that consists of hundreds of trillions of microbial cells. While physicians often talk about a "healthy" or "poor" microbiome, these qualities are hard to measure. Simply counting gut bacteria numbers is insufficient for both research and treatment. Instead, investigators need to use a variety of parameters to describe microbiomes. These include the microbial taxa composition, their heterogeneity and dynamics, the stability and resilience of microbial networks, the definition of core microbiomes, and other parameters. Although the microbiome is vast and intricate, we can still develop some indicators of microbiota health that can be used to treat and prevent illness. However, we must recognize that these are simply markers within the entire system.

From one perspective, the microbiota can be considered an internal factor since it is part of the body's internal workings. There is a constant exchange between microorganisms that are part of the microbiota and microorganisms common to the environment outside biological organisms. A microbiota is not separate from the microbial communities of the environment, such as water, soil, and air. Furthermore, exposure to new microorganisms in the environment can alter the composition of the microbiota and potentially affect its role in keeping the body healthy. Microbiota significantly influences the state of the immune system, which in turn regulates the composition of microbial communities inside organisms and the arrival of new microorganisms from the surrounding environment.

We now know that many pathogens live and interact with other microorganisms, including other bacteria, protists, and fungi in animal microbiomes. The interaction between pathogenic and commensal microbes (those living inside an organism without causing harm) may substantially influence disease processes. Bacteriophages, in particular, shape the bacterial community inside the body.

I was honored to participate in the first Pathobiome Conference dedicated to "Pathogens in microbiotas in hosts," organized by the French National Veterinary School of Maisons Alfort, Paris, in 2015. While discussing these questions, we entered an area of knowledge that could hardly be comprehended, despite having massive information about components of the complex pathobiome system.

The pathobiome defines the pathogenic agent integrated within its biotic environment, such as the microbiome. In order to understand the pathobionts, we require:

1. Clear information on any direct or indirect effect the microbiome has on pathogenesis.
2. Accurate knowledge of microbiomes.
3. Knowledge of environmental factors that may disrupt the pathobiome and lead to the onset of pathogenesis.
4. An understanding of the impact of the microbiome on the persistence, transmission, and evolution of pathogenic agents.[4]

The microbiome may also be seen as the connection between people and their environments. Amber Benezra, a sociocultural anthropologist from the Stevens Institute of Technology, said, " ... what I love about the microbiome is that it brings together social intimacies of life with our biological selves in ways that show us that those two things are inextricably entangled."[5] Despite their similar manifestations in physiological functions, human microbiomes are composed very differently, due to differences in the colonization

---

[4] Muriel Vayssier-Taussat et al. *Shifting the paradigm from pathogens to pathobiome: new concepts in the light of meta-omics.* Frontiers in Cellular and Infect Microbiology, 2014, vol.4, 29.
[5] Cited from Suzanne Ashaq et al. *Introducing the Microbes and Social Equity Working Group: Considering the Microbial Components of Social, Environmental, and Health Justice.* 2021, vol.6, issue 3.

process.[6]

Thus, it is not clear whether microbiota is an integrated part of an animal or a human organism, or a combination of miniature organisms that invaded the host organism. Fernando Baquero and Florentino Nombela from Ramon y Cajal University Hospital in Madrid claimed that the human microbiome is now considered to be our "last organ."[7] Is it rational for one to kill microorganisms with a high dose of antibiotics, knowing that this might also kill a part of one's own body at the same time? I am not advocating abandoning antibiotics. The key is to keep the dynamic balance between maintaining our internal microbial world and protecting the organism from invasion by pathogenic microorganisms. The balance of such a system is very sensitive to many factors acting both internally and externally.

The main point of this chapter is that you cannot expect "yes" or "no" answers in the "invisible world of what is invisible." We can generate tons of sequence data and produce a lot of useful information that presents numerous microorganisms inhabiting various tissues of the human or animal body. However, animals as ecosystems that include entire microbial communities will remain invisible, regardless of how much information we accumulate about the components, structures, and functions.

A Hidden Holobiont

While the microbiome represents a hidden world that we cannot understand fully, the concept of *hologenome* leads us to an even deeper level of knowledge. The term "hologenome," includes the Greek element "hólos," meaning "all, whole, total," and encompasses not only the genome of a multicellular organism but all of the

---

[6] Luke Ursel et al. *The Interpersonal and Intrapersonal Diversity of Human-Associated Microbiota in Key Body Sites.* Journal of Allergy and Clinical Immunology, 2012, 129, pages 1204-1208.

[7] Baquero and Nombela, *The microbiome as a human organ.* Clinical Microbiology and Infection, 2012, vol. 18, suppl 4, pages 2–4.

microbes and viruses that inhabit it.[8] The hologenome is the sum of all the genes of the host and all its symbionts, including microbes that constitute the microbiome.

Microorganisms and viruses are everywhere! They occupy diverse niches and ecosystems. They associate stably or transiently with all animals and plants on Earth. The interactions between hosts and microbes can lead to very intimate connections. In fact, some microorganisms and viruses carry genes that encode critically important functions for their animal and plant hosts. These functions include nutrition, development, protection against parasitoids, and thermal tolerance, among others.

The intriguing aspect of human genetics unveils an unexpected connection to our distant evolutionary past: the presence of genes originally inherited from viruses, known as recombination-activating genes (RAGs). These genes serve as a testament to the complex history of life on Earth. They shed light on the fascinating interplay between viruses and our genetic makeup. RAGs are a prime example of how genetic material can be repurposed and integrated into an organism's genome over millions of years. These genes, which are crucial for the immune system's function, allow us to generate a diverse array of antibodies and immune receptors. Long ago, an ancestral virus integrated its genetic material into the genome of a distant ancestor of modern-day vertebrates. This viral DNA eventually gave rise to RAGs. These genes became essential tools to support the vertebrate immune system, enabling it to adapt and respond to a wide range of threats.

The hologenome concept considers the host and its microbial partners to be a single, integrated unit, or *holobion*. This unit functions and evolves together. Although each holobiont could be a large organism, such as an animal, fungus, or plant, together with its microbiome, the holobiont itself cannot be physically seen.

---

[8] Eugene Rosenberg and Ilana Zilber-Rosenberg, *The hologenome concept of evolution after 10 years.* Microbiome, 2018, vol. 6, page 78.

An organism is not whole without its symbionts. For example, without its rumen microbiome, a cow is unable to digest plant material and, therefore, would not be the herbivore it is. Joan Roughgarden and her colleagues argued that holobionts are actual units of selection.[9] These are essential physiological and metabolic processes, often required to provide nutrients. For example: corals rely on microalgae for photosynthesis and energy production, and plants in the family Fabaceae require bacteria that infect their roots for nitrogen fixation.

Another example is the light-producing organ of the Hawaiian bobtail squid, which requires the symbiotic bacteria *Vibrio fischeri* to mature fully. After an adult squid releases these bacteria into the water, the bacteria can colonize the immature organs of newly hatched larvae. The squid's immune system specifically tolerates these bioluminescent bacteria, allowing a stable association between bacteria and the host by producing a functional light organ.

Mendoza and her colleagues demonstrated that genes in the genome of vampire bats, as well as genes in the bat microbiome, had been shaped in specific ways to cope with the dietary challenges posed by a blood-sucking diet.[10] These investigators observed that the genes in the vampire bat's genome were insufficient to explain some of the bats' adaptations to hematophagy. This included the capacity to deal with blood-borne pathogens or to adapt to a protein-based diet. Mendoza and her collaborators concluded that some of the genes of the microbiome, despite not being physically integrated into the bat genome, had evolved as a part of an "extended bat genome."

Although holobiont remains and will remain invisible, the concept has become a powerful tool to enable scientists to face the most pressing ecological and evolutionary problems of our time. This perspective challenges traditional notions of individuality. It also

---

[9] Joan Roughgarden et al., *Holobionts as Units of Selection and a Model of Their Population Dynamics and Evolution.* Biological Theory, 2018, vol. 13, pages 44–65.

[10] Zepeda Mendoza et al. *Hologenomic adaptations underlying the evolution of sanguivory in the common vampire bat.* Nature Ecology and Evolution, 2018, vol. 2, no.4, pages 659-668.

highlights the importance of considering the intricate relationships between hosts and their microbiota when we study biological systems. By considering the holobionts as an independent selection level, we can better understand the underappreciated modes of genetic variation and adaptation to rapid environmental change in evolution.[11] The holobiont perspective also reveals why we must consider the microbiome as an integral part of the host organism, rather than as a separate entity. The diverse aspects of holobionts, such as anatomy, metabolism, immune function, and development, all function together as a whole.

The Extended Organism

"Extended organisms" are not only considered as individuals. They are also an interconnected system that includes the individual and other organisms, such as the microbiota. Furthermore, the extended organism concept illustrates the existence of an invisible world in biology. For example, a living organism's physical body may be observable to the naked eye, but its complex interactions with its microbiota and environment cannot.

J. Scott Turner, Emeritus Professor of Biology at the SUNY College of Environmental Science and Forestry in Syracuse, New York, wrote the book, *The Extended Organism: The Physiology of Animal-Built Structures*.[12] As a physiologist, Turner sees an external organism in terms of its mechanisms. He asks how it works and how it alters the flow of matter and energy. He seeks information via the organism and between the organism and its environment. At the beginning of his book, Turner asks the question: "Are animal-built structures considered independent objects, or are they part of the animals themselves?" The answer he gives is that animals build structures, not as frozen behaviors but rather as external organs of physiology and even extensions of their phenotypes.

---

[11] Ricardo Guerrero, Lynn Margulis, and Mercedes Berlanga, *Symbiogenesis: the holobiont as a unit of evolution. International microbiology.* 2013, vol.16, No.3, pages 133–143.

[12] J. Scott Turner, *The Extended Organism: The Physiology of Animal-Built Structures.* 2000, 235 pages.

In support of his claim, Turner provided several examples:

- An indivisible link can be drawn between sponges on coral reefs and the flow of energy and matter in the environment as a whole.

- The tunnels dug by invertebrates in marine muds are devices for tapping into one of the largest potential energy gradients on the planet.

- Earthworms manipulate the physical properties of the soil to make it an "accessory kidney" that enables them to survive in an otherwise forbidding environment. The silken webs of diving spiders and some types of aquatic cocoons function as accessory lungs and gills for these creatures.

- The mating songs of male crickets are amplified by a constructed "calling burrow."

The large mound nests constructed by certain species of African termites are not simply houses for the colony. They are also accessory gas exchange systems that allow termites to adapt to a wide range of environmental conditions and grow mushrooms.

The concept of extended organisms highlights the interdependence and specialization of different organisms within ecosystems. By studying these complex and highly organized biological systems, we can recognize previously unknown connections between individual organisms and the environment.

Turner was influenced by another book, *The Extended Phenotype*, written by evolutionary biologist Richard Dawkins.[13] Keep in mind that a phenotype is a set of observable characteristics of an individual after its genotype has interacted with its environment. Dawkins' main idea is that the designation "phenotype" should not be limited to biological processes such as protein biosynthesis or tissue growth. Instead, it should be extended

---

[13] Richard Dawkins, *The Extended Phenotype*. 1999, 313 pages.

to include every effect a gene has on its environment, inside or outside the body of the individual organism. Dawkins sees the extended phenotype as the extension of the effect of genes beyond the outermost boundaries of the organism. He wonders whether these extended phenotypes assist in transmitting genes from one generation to the next.

One of the manifestations of the extended phenotype is the ability of animals to adjust their environments by building structures similar to the idea of the extended organism. Dawkins provides two specific examples to illustrate his point: caddisfly houses and beaver dams. Caddisflies are insects with aquatic larvae and terrestrial adults. The most remarkable thing about them is their ability to build a wide variety of tiny, portable houses. The larvae exude silk fibers from their salivary glands, which they use to attach to surfaces. The caddisfly larvae use their silk to fasten pieces of spruce needles, leaf fragments, and stones to form tubes in which they can live while roaming the river for food. Many of their homes come in fascinating shapes, and they often provide protection from predators, as well as camouflage from the outside world.

The beaver dams are an even more popular example. Beavers build them to create a pond to protect beavers from predators and to hold their food during the winter months. As a result of these structures, the natural environment is altered to enable the overall ecosystem to build upon the change. This makes beavers a keystone species that has provided a tremendous impact on the ecosystem. As defined by zoologist Robert T. Paine, a "keystone species" is one that has a great effect on the natural environment, disproportional to the abundance of this species. The world's largest beaver dam was described in Wood Buffalo National Park, Canada. From end to end, it measures approximately 800 meters long!

A beaver dam and a caddisfly house are just selected examples. In reality, every organism has a visible or invisible extension. Prairie dogs are incomplete without their colonies and their sophisticated burrow system. Predators are neither complete without their prey, nor are their prey complete without their predators.

Another fascinating natural phenomenon that illustrates the concept of an extended organism is the way animals leave traces of their life activity within the territory they have inhabited for generations. This is how the system of "biological attractors" is created. Objects that attract the attention of animals and influence how they relate to their territory and its resources. The scent lingers in the environment, enabling olfactory communication and allowing rivals to identify the territory's owner even in their absence. During communications between animals, scent marking serves several purposes, including individual or group recognition, age and social status recognition, site familiarity, and sexual attraction.

This example of scent marking demonstrates how animals interact with their environment, leaving lasting traces that influence their behavior and that of others. The ability to interconnect with other organisms via their surroundings exemplifies the concept of an "extended organism." This occurs when the boundaries between an individual and its environment become blurred.

In the same way, parasites are seen as extensions of the hosts they occupy. One form of the extended phenotype is the manipulation of host organisms by parasites. "Parasite manipulation" refers to how, in some parasite-host interactions, the parasite can modify the host's behavior to enhance the parasite's fitness.

Another form of extended phenotype is the way a parasite acts at a distance to affect its host. A common example is how cuckoo chicks manipulate their host's behavior, which elicits intensive feeding by the host birds. Here, the cuckoo does not interact directly with the host (which could be meadow pipits, dunnocks, or reed warblers). The relevant adaptation occurs when the cuckoo produces eggs and chicks that adequately resemble those of the host species. This enables them to avoid immediate rejection from the invaded nest!

Extended organisms are not visible, nor can they be fully described. Therefore, no organism can be fully described if its extended phenotype is ignored. Using the perspective of extended organisms is not just an intriguing concept. It could be an effective tool for planning tasks relating to environmental protection and healthcare issues.

Extended Spider Cognition

Here is yet another perfect example of an extended organism: the spider with its web, which is produced within the body of the spider itself. "Where is the end of the body of a spider sitting in the spider net?" is a thought-provoking question from philosopher Elena Knyazeva. It invites us to re-examine our perception of the boundaries between an organism and its environment (Figure 24).

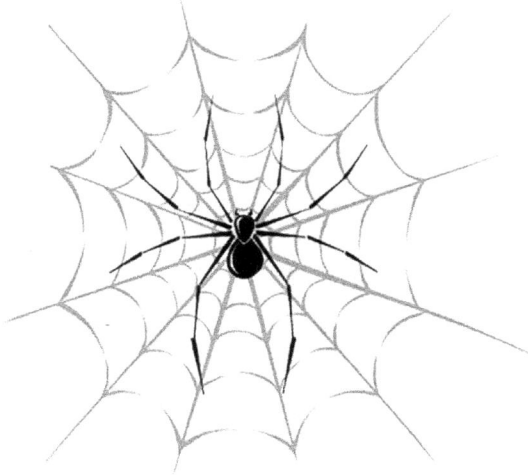

Figure 24. Where is the end of the body of a spider sitting in the spider net?

It could be said that when looking at a spider from a strictly anatomical perspective, the spider's body ends where its physical structures, such as its legs and abdomen, end. However, the spider web is not just a physical extension of the spider's body. It is a functional extension of its biological and behavioral traits. In some ways, it could be said that the spider's body *includes* its web, as the web is essential to the spider's survival and success. Thus, the web's function as an extension of the spider's sensory system blurs the line between the body of the spider and the environment in which it lives.

Many spiders build webs specifically to trap and catch insects to feed on. A spider's ability to construct a web is an adaptation that allows it to capture prey and perform other vital functions. The web itself plays an active role in the spider's interactions with other organisms. Spider webs are extremely intricate. The design could go beyond being purely functional, as seen in how spiders actively decorate their webs.[14]

Moreover, a spider web is not a static structure but an adaptive one that responds to environmental changes. The spider can modify its web's structure and placement in response to changes in prey availability, wind patterns, or other environmental factors. It can repair and maintain the web to keep it functional over time. Some spiders replace their entire web every day! Larger webs cost the spider more energy to produce, which really adds up with each rebuild. Larger catches will more than offset the increased energy output. Consequently, the spider and the web it weaves are part of a broader system that encompasses the surrounding environment. With this in mind, it becomes less relevant to distinguish between the body of the spider and that of its web, as both contribute to their ecological role within its habitat.

According to Hilton Japyassu and Kevin Laland, spiders are also an important model for a system that proposes "extended cognition."[15] There are two main viewpoints on biological cognition: one that sees cognition as being in the brain ("central cognition") and the other as extending beyond the brain into the body and environment ("extended cognition"). The first view holds that the brain is the primary location of cognition, while Japyassu and Kevin Laland argue that cognition is distributed throughout the body and the environment. The central cognition view predicts that animals living in complex environments should have more extensive or

---

[14] Marc Théry and Jérôme Casas, *The Multiple Disguises of Spiders: Web Colour and Decorations, Body Colour and Movement*. Philosophical Transactions: Biological Sciences, 2009, vol.364, no.1516, pages 471-480.

[15] Hilton Japyassu and Kevin Laland, *Extended Spider Cognition*. Animal Cognition, 2017, vol.20, pages 375-395.

complex brains. In contrast, the extended cognition view described above suggests that animals may use tools or their environment to assist with information processing.

Japyassu and Laland suggest that spiders, as small predators with complex webs, are prime candidates for extended cognition. They found that spiders could outsource their thinking to their webs! Spiders actually use their webs to sense and process information about their environment, such as where prey is located or changes in wind patterns. Spiders also adapt to their environment and find food in different ways. Their brain size or its complexity does not limit them; instead, they can use their webs to extend their thinking process! This means that their thinking happens in the brain *and* the spider web! Scientists believe other animals may also use niche-constructed structures like webs to extend their thinking and adapt to their environments.

The Invisible World of Biological Networks

The study of networks that govern cellular, organismal, and ecological processes is one of the areas currently being explored in biology's invisible world. Christopher Kilgore of the University of Texas stated that the "network" has become a dominant metaphor in contemporary thought.[16] Analyzing biological networks requires a sophisticated mathematical apparatus, from molecular interactions to ecosystems, to demonstrate how life is interconnected. The main players in networks are points (or "nodes") connected by lines (or "edges").

Frank Emmert-Streib and Matthias Dehmer compared networks to microscopes; both allow us to visualize an underlying phenomenon to be examined with the eyes. Let me be clear: all elements of networks, such as "points," "edges," "sets of layers," "topological lattice," and "hidden metric space," are invisible and abstract concepts. However, these concepts have proven to be

---

[16] Christopher Kilgore, *Rhetoric of the Network: Toward a New Metaphor.* Mosaic, 2013, vol.46, no.4, pages 37-58.

invaluable in modern biological research. Network analysis has become an essential factor in helping us understand the complex relationships and interactions that drive biological processes and systems. These interconnections include gene regulation, metabolic pathways, immunological landscapes, and ecological dynamics.

The network is instrumental for representing complex systems with unknown components and unknown connections between partially known and unknown components. One such example is the investigation of viruses' extraordinary abundance in ecosystems. Despite the growing interest in viruses as sources of diseases, viral dynamics among various animals, plants, and bacteria are highly complicated to examine. Since the advent of molecular techniques and high-throughput sequencing, methods such as co-occurrence, nucleotide composition, and advanced statistical frameworks have been widely used to infer interactions between viruses and their natural hosts. These have generated a great amount of data and information. However, *understanding* the enormous biological effect of viruses is still in the rudimentary stage.[17]

We do know that viruses can control the abundance and function of microbes and animals. The recent COVID-19 pandemic provides a compelling example. Viruses regulate *all* forms of life; they are also genetic repositories that exchange genes between organisms. We can integrate diverse data about the viral world by applying a network framework. A multilevel approach could integrate previously disconnected data sets to explain why viruses are abundant in an ecological context (Figure 25).

---

[17] Robert Edwards and Forest Rohwer, *Viral Metagenomics*. Nature Reviews Microbiology, 2005, no.3, pages 504-510.

Figure 25. A network of "invisible" connections exists between viruses and animals in the ecosystem.

Network reconstruction is naturally confined to what we have observed. Little is known of how our incomplete observations can impact our interpretation of the available data. Mohammad Sadeh, with his colleagues from the University of Regensburg, Germany, demonstrated that if there were missing observations or hidden factors, a reliable reconstruction of the full network would not be feasible.

By using network-based models and mathematical tools, scientists can extract meaningful insights from large and complex data sets, identify critical components or pathways, and make predictions about system behavior under various conditions. These models can also assist in the development of targeted therapeutic interventions, ecosystem management strategies, and other practical applications. Despite their metaphoric and invisible nature, biological networks are indeed real and functional. They serve as a powerful metaphor for capturing the complexity and interconnectedness of living systems, allowing researchers to address pressing challenges in biology and related fields. As applications of these networks continue to grow, they will, however, remain invisible forever!

The Puzzle of the "Hidden Obvious"

We have now established the conformity of invisibility over a visible segment for any biological object or phenomenon. It is true, however, that some essential parameters can help differentiate categories of invisibles in biology from each other. Such differentiation creates an opportunity for our orientation in hidden realms of biological life.

In the preceding chapters, we highlighted examples of domains with incomplete information despite having abundant data about a particular biological system. Furthermore, we demonstrated the existence of a level of reality (or "world") that gives us access to considerable amounts of information about a system. However, the knowledge about it remains mostly invisible or only partially visible. Finally, we have described in this chapter several significant concepts where we work on the assumption that we cannot know some of the fundamental laws of biology.

Despite their omnipresence and profound impact, the essential characteristics of biological life remain concealed from our immediate perception. They belong to the world of "invisible knowledge." For example, consider the interconnectedness of species within ecosystems. The complex web of relationships between species in an ecosystem is a prime example of this level of invisibility. While it would be impossible to observe these interactions directly, their effects on the overall health and balance of the ecosystem are undeniable.

A species' remarkable ability to adapt to its environment is another example of hidden knowledge in biology. This knowledge encourages us to interact with the natural world with respect, humility, and commitment to its preservation. Rabbi Nilton Bonder wrote: "When you travel the hidden territory of what is hidden, you know that your knowledge does not come from the "world of confusion," precisely because you are aware of its existence."[18]

By acknowledging and exploring the "hidden obvious" of biological life, as expressed by Kabbalah Master Samuel Avital, we can cultivate a more profound appreciation for the intricacies and mysteries that underpin our existence. Avital explained that the puzzle of the "hidden obvious" is crucial for exploring the amazing "invisible" links of "relationship." He wrote: "These invisible links reveal the connection to each other…, yet they are so obvious that we tend to overlook them. We take these things so completely for granted that they disappear from our window of awareness, and as a result, they become a "puzzle" to us that we do not know how to solve."[19]

There has never been a more critical time to explore the hidden aspects of biological life than it is now. As we face unprecedented threats on a global scale, such as the loss of biodiversity, pandemics, and food insecurity, understanding the underlying processes and mechanisms of life is crucial for our survival and that of our planet.

This exploration of invisible aspects of biological life goes beyond a mere metaphor. It represents a necessary step towards devising innovative solutions and taking decisive action to address these pressing issues. Kabbalah offers an exceptional perspective on the hidden realms. By transcending the hierarchy of invisible realms, Kabbalah provides a framework for us to delve into the deeper aspects of existence and understand the interconnectedness of all things.

---

[18] Nilton Bonder, *Yiddishe Kop.* 1999, page 93.
[19] Samuel Avital, *From Ecstasy to Lunch.* 2020, page 266.

PART THREE:
Toolbox of Kabbalah for Traveling through the Hidden Worlds

Chapter 9

Entering the
Pardes Garden

As we have previously discussed, various levels of visibility exist in the field of biology. Within each level are examples of unknown, hidden biological objects and phenomena. Knowing this, the question arises: How can we recognize and penetrate these levels of transparency? Specifically, how can we embrace biological realities within the following three realms:

The visible world of the invisible.

The invisible world of the visible.

The invisible world of the invisible.

These tasks present significant challenges, as traditional university courses have yet to equip us with the tools to address such questions. This chapter will explore metaphors and principles derived from Kabbalah to gain insight into intangible realities. While many ancient traditions acknowledge the existence of invisible worlds, Kabbalah is unique in its active pursuit of exploring worlds that offer varying levels of transparency.

As we approach Kabbalah as a source of untapped knowledge, it might be more useful to describe what Kabbalah is *not*, rather than refer to what others believe it is. Kabbalah is neither a science, nor a religion, nor a philosophy. Kabbalah comes from the Hebrew word, *L'Kabel*, meaning "to receive" or "something that is received."

Because Kabbalah has become an increasingly popular word in modern culture, I wish to make a disclaimer. I am using the word "Kabbalah" to identify it as a continuation of the authentic tradition. This tradition originated from the ancient texts of *Sefer Yetzirah* and *Sefer haBahir*, among other sources. These were fundamentally outlined in all of the books of *Zohar* and were further developed by the great Ari (Yitzhak Ben Shlomo Luria Ashkenazi) in the 16th century.

The teaching of Kabbalah has remained accessible to subjected circles of followers for more than a thousand years. The process of revealing the teaching of Kabbalah was carefully restricted; however, in the 20th century, Kabbalah found its way into many different circles. Even so, to date, a significant number of important texts of Kabbalah have still not been translated. Finally, in 2018, Daniel Matt, an acknowledged scholar of Kabbalah, completed his multi-volume annotated translation of the Zohar after almost twenty years of work. It was published by Stanford University Press in 12 volumes under the title, *The Zohar: Pritzker Edition*. I will also refer to several modern sources that honor this tradition and help me interpret the classical sources more easily.

The following chapters will introduce principles of Kabbalah that can offer researchers new tools that can widen their exploration of the natural unity of biological life. Let me introduce these principles by beginning with a short story from *Hagigah 14b*, one of the 63 tractates of the Mishnah, the first major written collection of the Oral Torah. The story describes four Sages entering an orchard known as "Pardes." The word "Pardes" comes from an Old Iranian word meaning "orchard or walled enclosure." Interestingly, the word "paradise" entered English from the same form.

However, in this story, the Pardes is not an ordinary orchard but a place that signifies the presence of the Source of all knowledge. The sages who entered the area were Simon ben Azzai, Simon ben Zoma, Acher (the "Other," a name given to Elisha ben Avuya), and Rabbi

Akiva. As a result of their adventures in Pardes, Ben Azzai died, Ben Zoma went mad, Acher lost his faith, and only the great sage Akiva was able to enter the orchard and depart in peace.

Four Planes to Read Texts

The varieties of interpretation of this simple short story have had a major role in developing the tradition of reading texts of the scriptures. In one of his lectures, Moshe Idel, Emeritus Max Cooper Professor in Jewish Thought at the Hebrew University, stated, "...the Pardes is an unexplained parable for an unrevealed secret. There is a crucial vagueness here, and one must make the assumption that this sort of vagueness **does not represent a defeat but an opportunity** [bold is mine] to introduce new meaning to an open text, as in Umberto Eco's account of reading texts as open texts. The Pardes comes as a generalized metaphor for the danger zones of experience, seen as something good for the few, but pernicious for others."[1]

Torah can be studied on different levels or planes. The word *Pardes* (in Hebrew, *Pey-Reish-Dalet-Sod*), is an acronym formed from the initial letters of four Hebrew words: *Pshat, Remez, Drash,* and *Sod*. In the story, the varying experiences of the four scholars represent each of the four escalating levels of reading and understanding the Torah.

The word *Pshat* means "simple" and refers to the literal meaning of the Torah's text. Passages of the scriptures discuss events, list names, and offer certain instructions. The act of reading the text requires a clear, direct understanding of its simple meaning, providing details that leave no room for confusion or doubt. However, Life demands that our interpretations move beyond the literal so that we may better understand and deal with the world around us. Rabbi Nilton Bonder wrote: "To understand the full significance of the literal, we must recognize that it does not exist in

---

[1] Felicia Waldman, *Edenic Paradise and Paradisal Eden: Moshe Idel' Reading of the Four Sages Who Entered the Pardes.* Journal for the Study of Religions and Ideologies, 2007, vol.6, no.18, pages 79-87.

isolation.It is merely a component of reality formulated within a void that renders its meaning comfortable to us."[2] Bonder called for a "total fidelity to literality" as the fundamental prerequisite for any interpretations.

Remez, or "hint," refers to the allegorical interpretation of the Torah. Remez deals with allusion as opposed to explicitness. It provides a symbolic meaning to each passage in the text of the Torah. However, for such a reading, one must learn how to decipher (decode) the text. In the process, we have to be very careful-- remember the fate of Ben Zoma, who lost his mind! Bonder wrote: *"Kabbalists protect themselves from madness (from the void) by remaining in steady contact with the* [literal meaning -- Pshat]."[3]

Drash means to resort to, to seek, to seek carefully, to enquire, to require. This reading can offer more insight into material that isn't clear from the text. The interpretation could be accompanied by a more personal understanding of the messages contained in the text. The *Drash* level classifies things by their types and ignores differences among individuals. A key rule of *Drash* is that there *"can be no Drash without a prior Pshat."* It means that any interpretation, even a very deep interpretation, should not ignore any word of the Torah text and avoid contradictions with it.

The word, *Sod*, which means "secret," refers to the hidden meaning of the Torah text. The search for hidden meaning can take a great deal of time, effort, and concentration. However, even these conditions cannot suffice without a direct mystical experience, or the disciple's receipt of secret knowledge from an honorable teacher. The *Sod* level is specific to Kabbalah. However, the other three modes of reading, analysis, and interpretation are required steps on the way toward achieving the level of Kabbalah.

Why are three modes of interpretation so important to those who study Kabbalah? First, they acknowledge that the knowledge is at a level of interpretation (*Sod*) that could not be achieved simply by reading certain books.

---

[2] Nilton Bonder, *Yiddishe Kop*. 1999, page 8.
[3] Ibid., page 10.

The second explanation is more important in our introduction to Kabbalistic Biology. As per Kabbalah, every level of *Pardes* (*Pshat, Remez, Drash,* and *Sod*) offers a unique way to receive knowledge in concrete, practical ways. All are equally important, although the levels range from lower to higher. One cannot understand the whole text until every level has been accessed. Rabbi Akiva could only emerge with his enhanced wisdom and experience after he possessed all four modes of interpretation. These studies related not only to a specific text of the Torah but to *any* text, including the text of biological life.

"Do Not Say, Water, Water"

There are certain words at the beginning of the Pardes story that may confuse readers. These words, however, also sparked many thoughtful comments among Talmudists regarding the story. Rabbi Akiva declared, "When you reach pure crystal stones, or *shayish* (commonly translated as marble), do not say: "water, water." After proclaiming this, Rabbi Akiva referred to Psalm 101:7: "He who speaks falsehood shall not be established before My eyes."

What, then, is the mysterious warning: "Don't say water, water?" Hananel ben Hushiel, an 11th-century Talmudist in Kairouan, Tunisia, commented upon Rabbi Akiva's words, saying, "It appears like water that has waves of water within it, but there is not even a single drop there but only the radiant reflection of the pure stones that pave the palace that appear to reflect like water. The one who asks: "What is the nature of this water?" is cast away."[4]

The warning to *not* say, "water, water" is equally important and challenging. In Tikunei Zohar, there is a separate appendix to the Zohar consisting of seventy commentaries on the opening word of the Torah, *Bereishit* ("in the beginning"), written in the Midrashic style. It reports the following incident regarding the four sages who entered the *Pardes.* In Tikun 40, the ancient Saba [an old man] asked

---

[4] Sefaria, https://www.sefaria.org/sheets/202494.80?lang=bi&with=all&lang2=en

the question referring to Genesis 1:16, "There shall be Raqia [firmament] between the waters and it shall separate between water [above the firmament] and water [below the firmament]. Since the Torah describes the division of the waters into upper and lower, why should it be problematic to mention this division?"

The famous Kabbalist Moshe ben Jacob Cordovero (known as RAMAK) lived in 16th-century Tzfat. He explained in *Pardes Rimonim* ("Orchard of Pomegranates"), "There is no separation between one water and the other; they form a single unity from the aspect of the Tree of Life... This is the meaning of "do not say that there are two types of water, lest you endanger yourself because of the sin of separation."[5] Cordovero describes the nature of the connection between the two glasses of water through the use of a hidden meaning of the Hebrew letter, "*yud*" within another letter, A*lef.* It would be untimely to provide his explanation in this chapter.

So, why is this message so relevant for the Kabbalah student and specifically for those at the threshold of learning Kabbalistic Biology principles? Water exists in various states, such as liquid, vapor, or ice. No matter what its state may be, it remains fundamentally the same substance, water, no matter what its form may be. Here, the word "water" is not limited to its physical meaning. Water is considered the source of Life or the flow of Life. The relationship between the Upper Water and the Lower Water is the guiding thread through the Book of Zohar and other authentic Kabbalah sources. Although the two Waters exist in separate worlds-- the upper and lower--they have the same nature, essence, and source.

Let us rephrase the warning. Do not say "life, life." When investigating biological life on different planes, from its elementary manifestations to the ineffable secrets, remember that it is the same Life! We look at biological life from different perspectives, but ultimately, all "Life" is "One!" As we investigate living creatures and biological processes at different levels using different processes, we

---

[5] Moshe ben Jacob Cordovero, *Pardes Rimonim, Orchard of Pomegranates.* 2007, 340 pages.

must remember that we are investigating the same organism or process. Losing this awareness is the source of every mistake in biology!

Four Olamot

Let us look back at *Pardes'* four levels of learning. When we pass through the four levels of the *PaRDes*, we are at the threshold of yet another discovery. Each level of learning opens the door to one of the separate worlds. In Kabbalah, these worlds (*olamot*) are *Asiyah, Yetzirah, Briah,* and *Atzilut,* first implied in *Sefer ha-Bahir.*[6] Thus, *Pshat* relates to the world of *Asiyah; Remez* relates to the world of *Yetzirah, Drash* relates to the world of *Briah,* and *Sod* relates to the world of *Atzilut.*

The word *olam* (*olamot* is plural) has many meanings, including "world," "existence," and "permanence within the limits of time." The etymology of this word suggests the link between *olam* and the verb, *alam,* meaning "to hide, conceal, and be secret." The four worlds (*olamot*) are distinguished from one another via their relative proximity to their source. The light, *Ohr* in Hebrew, is present in *all* worlds. These worlds are illuminated by removing vestments and screens.[7]

Throughout the whole text, Zohar emphasizes that the source of everything is "both concealed and revealed."[8] However, Zohar's phrase, "similar to the upper world," means that our world is only *in the image* of the upper world, not identical to it.

In 1937, Yosef Yitzchak Schneersohn, the sixth Rebbe of the Chabad Lubavitch Hasidic movement, was asked two questions about the nature of *Olamot* discussed in Kabbalah. His response systematically defined and quantified four primary worlds: *Asiyah, Yetzirah, Briah,* and *Atzilut.* In his book titled, *The Four Worlds,*

---

[6] *Sefer ha-Bahir* is one of the oldest and most important pf all ancient texts. See Aryeh Kaplan, *The Bahir Illuminated.* 1979, 244 pages.

[7] Bruce Friedmam, *Mystery of Black Fire, White Fire.* 2016, page 232.

[8] *The Zohar,* 1:39b, 234b; 2:95a, 178a; 3:65b, 71b, 72a-b, 146b, 289a

Schneersohn wrote: "Although the worlds are without limit and number, they have been divided into two categories – revealed worlds, and hidden worlds that are not revealed."[9]

Each world receives its essence, or "Life Force," from the world above it as they penetrate and interact with one another. The worlds coexist at different levels of reality. The terms "higher" and "lower" do not refer to spatial dimensions. They refer only to differences in the degree of concealment and dimming of the original Life Force. Samuel Avital uses the image of the "spiritual funnel" (*mashpeh* in Hebrew) to illustrate the passage of the Life Force from the upper *Olam* to the lower one (Figure 26).

---

[9] *The Four Worlds: A Letter by Rabbi Yosef Yitzchak Schneersohn of Lubavitch*, 2003, page 24.

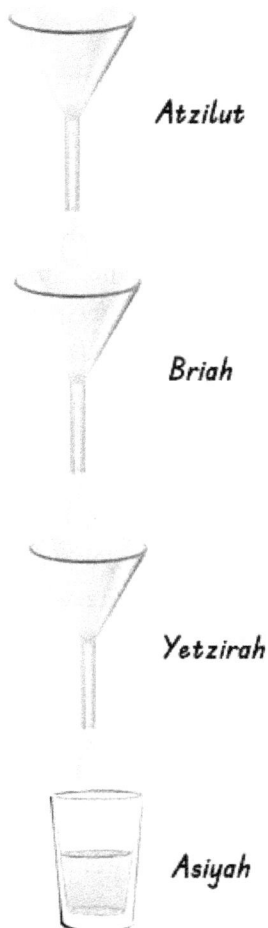

Figure 26. Each olam (world) receives its essential force ("Life Force") from the world above: Briah from Atzilut, Yetzirah from Briah, and Asiyah from Yetzirah.

There are multiple definitions of each world (*olam*). Authors tend to use different analogies to describe the worlds' uniqueness. According to Rabbi Nissan Mindel, the four worlds are analogous to the stages of building a house: 1) the initial desire, 2) the mental concept, 3) the detailed design, and 4) the final construction.[10] This analogy embodies the idea that the two upper worlds are undifferentiated and

---

[10] Nissan Mindel, *The Philosophy of Chabad*. 1973, page 82.

conceptual. In contrast, the two lower worlds are differentiated and manifest in specific forms.

The questions and answers that make sense in one world or level of reality may not make sense in another world.[11] According to Jason Shulman, the "four universes" (his version of "four worlds") have unique dimensions representing a "topographic map of the levels of reality."[12] These worlds represent the increasing levels of integrated energy, matter, and information. The four universes are points of view that give us new insights and perspectives on every aspect of reality. From these points, everyone can perceive the realities, from fragmented parts to whole systems.

From *Asiyah* to *Atzilut*, the four universes expand in transparency as the quality of reception becomes clearer. An interpenetration of these worlds can occur from the "bottom-up" or the "top-down." Each world is a replica of the world immediately above it that is "projected" into the world below. Here is how Shulman presented the relationships between *olamot* in their ability to become transparent:

"Asiyah (not Transparent to);
   Yetzirah (not Transparent to):
      Beriah (not Transparent to):
         Atzilut (Completely Transparent to):
      Beriah (Transparent to Yetzirah & Asiyah):
   Yetzirah (Transparent only to Asiyah):
Asiyah (not Transparent; sees only itself)."[13]

The World of Asiyah

The world of *Asiyah* is the lowest of all worlds. "*It is therefore called Asiyah, which connotes the lowest capacity, deed, or actuality,*" wrote

---

[11] Estelle Frankel, *The Wisdom of Not Knowing*. 2017, page 132.

[12] Jason Shulman, *Kabbalistic Healing*. 2004, page 122.

[13] Ibid., page 14.

Rabbi Yosef Yitzchak Schneersohn.[14] The essential nature of *Asiyah*, he added, is "where matter dominates form." The *Asiyah* (or Asiyatic) world manifests "physical" or "materialistic" forms (*Asiah Gashm*, or "bodily Asiyah"). "*Asiyah* is where tangible reality – matter – exists," wrote clinical psychologist Edward Hoffman in his book, *The Way of Splendor*.[15] Everything in *Asiyah* is concrete and obvious.

Asiyah also emphasizes simple actions. The word *Asiyah* comes from the root *asa*, which means "to do, to make." This is the world of action. It represents everything that occurs. Because the physical realm is so specific and locatable, all the differences between things are relatively easy to recognize and to be measured. Here, the subject and object are delineated and separated. Any personal attitude toward an investigated object is neither relevant nor pursued. *Asiyah* is dedicated to seeing the world in a purely objective way. Zalman Schachter-Shalomi noted that "the way that we know what is true in the realm of function is, 'if it works, it's true.'"[16]

Here, I must add that all observations in this world are reliable only by their limits. Everyone sees something, depending on the reason for their searches. The danger is in perceiving reality through a lens of conditions that could lead to predispositions, prejudices, or distortions.

The World of Yetzirah

In the world of *Yetzirah*, all objects acquire shapes, forms, and connections. Yetzirah represents the plane of "formation." The word *Yetzirah* is from the Hebrew word *yatzar*, meaning "to form." Rabbi Yitzchak Ginsburgh illustrated this world by using a fetus as an example: "The moment of formation, when the fetus' form becomes apparent, is called 'the formation of the child'' and corresponds to the

---

[14] *The Four World: A Letter by Rabbi Yosef Yitzchak Schneersohn of Lubavitch*, page 30.

[15] Edward Hoffman. *The Way of Splendor: Jewish Mysticism and Modern Psychology*. 1993, 264 pages.

[16] Zalman Schachter-Shalomi, *Credo of a Modern Kabbalist*. 2005, 399 pages.

world of Yetzirah."[17] In the words of Edward Hoffman: "Yetzirah contains the patterns or blueprints, but not physicality itself."[18]

The process of formation requires a connection between the system's elements. According to Aryeh Kaplan, best known for his Living Torah, "Yetzirah is called 'the universe of Malakh," which means "messenger," and denotes the concept of transmission. The realm through which the message is transmitted is Yetzirah, the dimension of space Ruah [meaning "wind"] corresponds to the universe of Yetzirah, implies motion and transference through space."[19] Zalman Schachter-Shalomi, commonly called "Reb Zalman," noted that "Yetzirah is naming the Between-ness."[20]

The Yetziratic world is subjective, the opposite of the Asiyatic world. Let's suppose that in Asiyah, everybody who used the same "tools" should receive the same results. On the Yetziratic plane, the interpretation of the results will be influenced by the interpreter's school, colleagues, accepted terms, and popular concepts.

One of the significant pitfalls of Yetzirah that interferes with our understanding is the idea that there are limits to our ability to understand. Alas, this dilutes our sense of confidence that we will receive enlightenment. According to Shulman, a fundamental quality of *Yetzirah* is "completely self-reflective: it knows itself only by looking at itself... In Yetzirah, reflecting upon life becomes as important as participating in it."[21] Shulman has also noted that in *Yetzirah*: "we enter the world of the unseen."[22]

The World of Briah

The world of *Briah* closely relates to the realm of concepts. Its name comes from the Hebrew *bara* meaning "to create." *Briah*

---

[17] Rabbi Yitzchak Ginsburgh, *The Mystery of Marriage*. 1999, page 327.

[18] Edward Hoffman. *The Way of Splendor: Jewish Mysticism and Modern Psychology*. 1989, 247 pages.

[19] Aryeh Kaplan, *Inner Space*. 1990, page 31.

[20] Zalman Schachter-Shalomi, *Credo of a Modern Kabbalist*. 2005, page 74.

[21] Jason Shulman. *Kabbalistic Healing*, 2004, page 20.

[22] Ibid., page 129.

encompasses ideas but lacks specific forms or structures. Adin Steinsaltz once said that this world contains different spheres of matter, each with its own rhythms relating to the past, present, and future.

Yitzchak Ginsburgh commented that in the world of *Briah*, there are no longer things, only raw materials. "The mind has to divorce itself from thinking about 'things.'... In the world of creation, uncertainty is a statement about the mind... The mind has to reform itself completely around this idea of uncertainty, which takes away our notion of things." For example, a biological species is not a "thing," nor is individuality; nevertheless, this is a term used by biologists, not philosophers.

In this world, the observer affects the world and thus affects the picture that the observer perceives. According to Michael Laitman, the founder of Bnei Baruch – Kabbalah Institute, "the perception of reality is like an average picture between the attributes of the observer and the attribute of the observed object."[23] Here is the difference between the worlds of *Yetzirah* and *Briah*. In *Yetzirah*, we do not affect the world; the picture of the world changes in our eyes because *we* change. In *Briah*, we can affect the world, and our perception of the world is a combination of the individual's qualities and those of the world.[24]

The World of Atzilut

The highest world is called the *Atzilut*. This name is derived from the root, *etzel*, meaning "nearness." Rabbi Yosef Yitzchak Schneersohn explained this world in such a way: "Atzilut connotes emanation and separation. For Atzilut is not like a new being that comes into existence through the creation of something from nothing. Rather, it is only revelation of the concealed."[25] As stated in *Sefer Yetzirah*:

---

[23] Michael Laitman, *Kabbalah Science and the Meaning of Life*. 2006, page 134.

[24] Ibid., page 135.

[25] *The Four World: A Letter by Rabbi Yosef Yitzchak Schneersohn of Lubavitch*, page 26.

"There is a flame that exists within the coal and one that exists outside it."[26] This is *Atzilut*: "It is not more than a revelation of the concealed[27]."

Edward Hoffman commented that "*Atzilut* ...involves the endless, undifferentiated energy." This is the world of perception beyond conception. This world is totally abstract; it assumes no form whatsoever. The distinction between the World *Atzilut* and other worlds is so great that it is stated there is a *massach* ("curtain") or *parsa* ("covering") to separate them. The essence of this world marks the emergence from potentiality into actuality from the concealed power to the actualization and effectuation of change in the lower worlds.[28]

So, why do we need to learn about the *Atzilut*? Reb Zalman (Schachter-Shalomi) provides the following answer:

> "The answer that makes the most sense to me is that this fourth level of Atzilut was conceived precisely because it felt like something important was missing; to describe the experience of [life] on only three levels was incomplete ... How does one know that [this] exists? ... That's the basic question of epistemology about knowing anything. I couldn't have invented that by myself... It can only be Atzilut because it cannot be proven."[29]

However, this world is very important: it enables us to have different views of the big picture and to develop a higher tolerance for the positions of other scientists. The famous saying that "many questions are meant only to be asked – never answered" characterizes the level of *Atzilut.*

The Four Realms Corresponding to the *Olamot*
Reb Schneur Zalman of Liadi (known as the Alter Rebbe), the

---

[26] *Sefer Yetzirah,* 1:7.

[27] *The Four World: A Letter by Rabbi Yosef Yitzchak Schneersohn of Lubavitch,* page 26

[28] Gershom Sholem, *Kabbalah.* 1974, pages 96-105.

[29] Zalman and Siegel, *Credo of a Modern Kabbalist.* 2005, page 74.

founder of Chabad, has left us a very insightful framework to study all the possible worlds we can live in. Brazilian rabbi Nilton Bonder has carefully represented these words (or realms) in the following way:

The Apparent Realm (World) of what is Apparent.

The Hidden Realm (World) of what is Apparent.

The Apparent Realm (World) of what is Hidden.

The Hidden Realm (World) of what is Hidden.

Bonder proposed such wording as "a way of breaking out of structures of ignorance that fail to take into account the various aspects of reality."[30] In Chapter 6, we introduced the planes of transparency that nicely correspond to the realms, according to Schneur Zalman. In fact, this classification leads us also to the D-I-K-W pyramid (Table 2)

Olamot	Levels of reading a text	Realms, according to Schneur Zalman	Planes of transparency	D-I-K-W pyramid
Asiyah	Peshat	The Apparent Realm of What is Apparent	The visible world of what is visible	Data
Yetzirah	Remez	The Hidden Realm of What is Apparent	The visible world of what is invisible	Information
Beriah	Derash	The Apparent Realm of What is Hidden	The invisible world of what is visible	Knowledge
Atzilut	Sod	The Hidden Realm of What is Hidden	The invisible world of what is invisible	Understanding Wisdom

Table 2. The realms according to Schneur Zalman.

Bonder wrote: "Since ancient times, the kabbalistic tradition has maintained that reality is layered, like an onion. By peeling off layers one by one, we can dissect reality much more effectively than if we perceive only one facet of it."[31] That is why the concept of "olamot"

---

[30] Nilton Bonder, *Yiddishe Kop.* 1999, page 5.

[31] Ibid., page 4.

provides a unique opportunity to revolutionize biological science. That is why the concept of *olamot* provides a unique opportunity to revolutionize biological science. We have talked about biological data, biological information, biological knowledge, and biologists' wisdom. It's crucial to understand that these domains represent completely different realms (*olamot*). The expression, "information is not knowledge," becomes a famous slogan. Still, it takes thoughts and actions to consciously accept that there is a barrier between the following spheres of activity: obtaining information and reaching knowledge.

There is no question about which world is superior; *Asiyah, Yetzirah,* or *Briah*? That would be like asking, "What would be better data or information, or what knowledge would be superior." We cannot get information without data, just as we cannot get knowledge without information. The difference is that we can collect data without processing it to obtain information. A similar way of looking at the question of superiority is that we can specialize in gathering and processing information without going deeper into receiving knowledge. We must be aware of the existence of separate worlds (*olamot*) or planes in the knowledge hierarchy. We begin with data and then move towards information, followed by knowledge, and conclude with wisdom.

We will explore *Olamot's* worlds further in Chapter 16, showing how they are essential to our ability to explore biological objects and processes. Only one world (*Asiyah*) represents visible forms, while the other three worlds represent the invisible connection between objects and functions. Whether visible or invisible, these worlds exist. They are real and unique. They can lead us to the practice of assigning different properties to each world, aiding our ability to orient ourselves in the invisible worlds.

The final important lesson of Kabbalah is that all these worlds have a similar structure, referred to as *Etz Hayim* (*"Tree of Life."*). Rabbi Shlomo Eliyashiv, a famous kabbalist in the early 20th century, known as the Ba'al HaLeshem, wrote: "...all the layers are exact duplicates of each other each being true according to its particular essence." Thus, learning the fundamental structure of a lower world can lead us to better knowledge and understanding in the higher worlds. We will begin to explore common structures of the *Olamot* "worlds" in the next chapter. We will begin to explore this structure in the next chapter.

Chapter 10

The Tree of Life and the Tree of Knowledge

In the previous chapter, four sages in the Pardes story viewed the "orchard" differently. Within the orchard (forest) are trees. However, there are different levels of exploration when trees are the subject. For some reason, Rabbi Akiva was the only one of the four sages who could perceive the orchard in its entirety. In this chapter, we will explore the powerful image, the "Tree of Life" associated with Kabbalah. The Tree of Life is considered a roadmap for those investigating the life of all creatures, from tiny microbes to endless ecosystems.

Trees as a Metaphor for Knowledge

The image of a tree as the organizing structure of the whole world is prevalent in many cultures. The tree is a popular visual archetype that acts as a compelling metaphor for the organization of knowledge, including concepts such as immutability, hierarchy, centralization, and a recurrent branching structure. Aristotle believed that the world could be seen as a "ladder of perfection," and that one could rank biological organisms accordingly.

The great English philosopher, Francis Bacon, wrote in *The Advancement of Learning* in 1605: "The distribution and partitions of knowledge are not like several lines that meet in one angle, and so touch but in a point, but are like branches of a tree that meet in a stem, which hath a dimension and quantity of entireness and continuance, before it comes to discontinue and break itself into arms and boughs."[1]

Ben Shneiderman, a professor in the Department of Computer Science at the University of Maryland, noted in the foreword to Manuel Lima's *The Book of Trees*: "Trees can inspire a Newton to notice the proverbial falling apple or a Darwin to see the Shema for all life-forms."[2] In this context, it is very unexpected to see the use of the word, "Shema," the Hebrew word that declares the absolute unity of the Creator.

David Bohm, one of the most original scientists and thinkers of the 20th century, was particularly attracted to trees and admired their complexity and the dappled light that filtered through the branches. He found trees "wild and unconstrained," yet they betrayed a subtle underlying order that he could see as he looked from trunk to branch and from branch to twig.[3]

Yet, both in Kabbalah and biology, the concept of "the Tree of Life" was granted a pivotal and fundamental place. In biology, the image of a tree has played an important role in supporting the idea of evolutionary progress. In 1809, Jean-Baptiste Lamarck presented a figure of a tree "serving to show the origin of the different animals."[4] In Darwin's *On the Origin of Species*, the tree is an essential metaphor demonstrating the author's vision of evolutionary history. The tree diagram that illustrated Darwin's thinking significantly contributed to its wide acceptance among rather reluctant readers.[5]

The Tree of Life has become one of the most important organizing principles in modern biology. Laura Hug, with her colleagues, wrote in *Nature Microbiology*, "The tree highlights major

[1] *Francis Bacon: The Major Works*, edited by Brian Vickers, 2008, page 175.
[2] Manuel Lima, *The Book of Trees: Visualizing Branches of Knowledge*. 2008, page 9.
[3] F. David Peat, *Infinite Potential: The Life and Times of David Bohm*. 1997, page 29.
[4] Jean-Baptiste Lamark, *Philosophie Zoologique*. Quoted from J. David Archibald *Aristotle's Ladder, Darwin's Tree: The Evolution of Visual Metaphors for Biological Order*, 2014, page 61.
[5] Theodore Pietsch, *Trees of Life: A Visual History of Evolution*. 2012, page 87.

lineages currently underrepresented in biogeochemical models and identifies radiations that are probably important for future evolutionary analyses."[6]

In his "tree model," Bruce Friedman, a physicist at Syracuse University and deeply familiar with Kabbalah, illustrates possibilities and temporality. As opposed to an emphasis on present or immediate gratification, Friedman emphasizes that the future remains a potential to be realized, and thus able to be discussed only in terms of possibility. He wrote the following in his book, *Mystery of Black Fire, White Fire:* "The depiction as a 'tree model' arises by conceiving of the future as a structure comprised of ever-proliferating branches, each of which represents a possibility. The past and the present are the trunk of the tree. The trunk flows through the process of actualization of the possibilities in the present."[7] The "now" of the temporal location depends upon the perspective of the individual considering it.

"Marble" and "Wood"

In order to contemplate the relationship between the metaphorical meaning of the Tree of Knowledge and practical ways to conduct scientific activities, I refer to "marble" and "wood," the paradox formulated by Albert Einstein. Specifically, Einstein felt that the left side of General Relativity's equation was "made of fine marble," but the right side was "built of low-grade wood."

When explaining the symbolism of this word pair, Michio Kaku, in his book *Hyperspace*, emphasizes how Einstein was frustrated by such a paradox.[8] While developing the unified field theory for 30 years, Einstein was disturbed by a fundamental flaw in his central equation. He had linked the curvature of space-time to "marble" because of its clean and elegant geometric structure. He also considered the laws of matter and energy as a combination of

---

[6] Laura Hug et al. *A new view of the tree of life.* Nature Microbiology, 2016, vol. 1.

[7] Bruce Friedman, *Mystery of Black Fire, White Fire.* 2016, page 178.

[8] Michio Kaku, *Hyperspace.* 1995, page 98.

confused, seemingly random forms manifested in all physical objects. Therefore, he compared these forms to "wood." Einstein had believed that "marble" alone would one day explain all the properties of "wood."

Addressing Einstein's dilemma, Kaku recalls that the laws of nature simplify and unify in higher dimensions. He later stated, "The laws of nature simplify when self-consistently expressed in high dimensions.... Which fixes the dimension of space-time to be ten."[9] Why ten? Many authors have noted the remarkable resemblance between ten Sefirot and ten dimensions implied by super-string theory. We will learn more about the Sefirot in Chapter 11, but let's talk first about the Tree of Life.

The "marble/wood" paradox can be seen as an effective tool to explain the difference between the physical sciences, such as physics and mathematics, and the biological sciences. While physicists and mathematicians often search for an elegant formula or theory to explain all physical manifestations, biologists must contend with the richness, variability, and unpredictability of biological life. While crystallography, the science of crystals, marble is a subject in solid-state physics and dendrology, the science of woody plants like trees ("wood") is a part of biology. Marble might sometimes *feel* alive. The architect Christopher Alexander wrote: "The quarries at Carrara, in Italy, are famous because the marble from that place feels intensely alive. Another marble may feel more ornate but less alive. Artificial marble ... feels much less alive. Yet none of the three is actually alive, biologically."[10]

Biologists may be excited when they find regularities in new information or discover novel concepts. However, they must remain devoted to observations made under natural or laboratory conditions. There is no way hypotheses and concepts can contradict

---

[9] Ibid., page 177.

[10] Christopher Alexander, *The Nature of Order. Book One: The Phenomenon of Life.* 2002, page 32; Christopher Alexander is an architect, scientist, and builder who was the winner of the first medal for research ever awarded by the American Institute of Architects.

the apparent richness and unpredictability of biological life.

"Wood" refers to the real and concrete material biologists work with. Trees are made of "wood." Wooden trees make up a forest. This could include not only trees' wood but also the flesh of animals, the protoplasm of cells, or components of ecosystems. The higher the biological hierarchy we reach, the more complex and unpredictable the system becomes.

Biologists cannot dream of transforming such "wood" into "marble." The choice is between two other options. The first choice is to patiently continue studying "wood." The second is to search for ways to explore the worlds of "wood" and "marble" without creating contradictions between them. The latter requires finding ways to explore the hidden, invisible aspects of biological systems without losing sight of the natural world's richness and diversity. Taking this path is not easy, but Kabbalah is here to guide us.

The Tree of Life and the Tree of Knowledge: One Tree, *Not* Two

The written Torah tells us about the Tree of Life ("Etz Hayim") in Genesis 2:9. But there was also another tree--"the Tree of Life in the garden, and the Tree of Knowledge of Good and Bad."[11] Both trees have been the subject of numerous interpretations and insinuations. However, in human history, the concept of the Tree of Knowledge is unique. One commonly referenced statement claims that consuming fruit from the Tree of Knowledge has led to all the troubles human beings have experienced! Trees of Life and Trees of Knowledge are often seen as symbols of dualities due to their contradictory relationship.

Kabbalist Moshe Chaim Luzatto, known as "Ramchal," who lived in early 18th century Italy, had the following comments about the Tree of Knowledge and the Tree of Life:

> "Certainly, no verse can disregard its plain meaning. The
> trees were trees, the fruit was fruit, and the act of eating was

---

[11] *Tanach*, The Stone Edition, Mezsorah Publications, 1996, 2:9, page 5.

eating. However, the fruit was so ethereal, and the act of eating was so ethereal, that our thoughts cannot picture it, as our thoughts are capable only of picturing things of a corporeal nature."[12]

According to Martin Buber, an influential philosopher and writer, the expression "knowledge of good and evil" refers not to an acquired consciousness of moral categories but to an acquired *access* to the tension between the contradictory propositions that exist in all Creation.[13]

Though the dualistic view of the relationship between the Tree of Life and the Tree of Knowledge is valid, there is yet another view. Thirteenth-century Rabbi Ezra ben Shlomo of Gerona made the critical point that both trees are the same tree![14]

Reading Zohar, I came to think about these Trees as the manifestation of *one Tree*. In the portion of the Haqdamat Sefer ha-Zohar [1:8a], the introductory section to the Zohar where the Tree of Life and the Tree of Knowledge are discussed, we read, "Two joined together as one. There are two points: one hidden and concealed, another existing overtly. Since they are inseparable, they are called ... beginning – one, not two. Whoever attains one attains the other... Because it is the Tree of Good and Evil. If a person is deserving, it is good, if not, evil."[15]

The commentary to this text provided by Daniel Matt indicated that in Kabbalah, the Tree of Life symbolizes Tiferet, while the Tree of Good and Evil symbolizes Malkhut (or Shekhinah), the two sefirot, which we will discuss later.[16] I would not exaggerate if I said that Tiferet and Shekhinah's relationship is the central narrative in Zohar. The Zohar also contains the following passage: "Come and

---

[12] *Da'at Tevunot*, page 114, Siman 126, quoted from Joel David Bakst *Beyond Kabbalah*, 2012, page 283.

[13] Quoted from Nilton Bonder, Yiddish Lop. 1999, page 60.

[14] Quoted from Gershon Winkler, *Kabbalah 365: Daily Fruit from the Tree of Life*. 2004.

[15] *The Zohar*, Pritzker Edition, Vol. One, page 50.

[16] Ibid., page 51.

see when this *tree bearing fruit* joins the *fruit tree*: all these supernal species."[17]

The identity of both trees was stressed by Rabbi Joel David Bakst in The Secret Doctrine of the Gaon of Vilna: "... the Tree of Knowledge and the Tree of Life are one and the same. The Tree of Knowledge is only that part of the Tree of Life that is branching out into this lower and limiting dimension that then produces the phenomenon of multiplicity."[18]

Eating Fruits from Both Trees

The interplay between the Tree of Life and the Tree of Knowledge became a central theme in Kabbalistic teachings. However, as I will demonstrate, it is valid as a powerful metaphor for science in general and biological science in particular. Understanding life as something unseen and exploring the physical manifestations of biological organisms and their relationship, is the central mystery we face when trying to explore the biological world. Since the beginning of biology as a science, scientists have picked "fruits" from trees and other biological organisms for analysis. Making a choice is symbolized by the Tree of Knowledge. We cannot know in advance the consequences of making our choices; we only discover them after picking the fruits and eating them!

Is a Tree of Knowledge limited to be the Tree of Knowledge of Good and Bad? The root of this problem is in the dualistic thinking that was typical of the primitive stage of humankind, long before Aristotle formally documented it. A choice between good and bad is grounded in the emotional nature of every human being. However, we humans are not restricted to the emotional realm! We actually have the mental capacity to overcome dualistic perceptions.

According to Martin Buber, best known for his philosophy of dialogue, the expression "knowledge of good and evil" refers not to an acquired consciousness of moral categories, but to an acquired

---

[17] Ibid., page 362.

[18] Joel David Bakst, *The Secret Doctrine of the Gaon of Vilna*. Volume II. 2009, page 107.

access to the tension between the contradictory propositions.[19] In Kabbalah, the right side represents "good," while the left side represents "bad." In the Genesis story, a woman ate fruit from the Tree of Knowledge of Good and Bad. Her choice was restricted to selecting fruit from the right or left side of the tree. That is the world of two dimensions – to move right or left.

In modern science, we must always make similar selections. Will you study animals or plants? Will you investigate selected organisms at the level of cells or organisms? Will you apply methods designed for studying nuclear acids or proteins? In contrast to The Tree of Knowledge, the Tree of Life provides us with opportunities beyond dualistic restrictions. We pay attention to the central trunk, the root system, the branching system, leaves, and seeds, among others. The most critical task is accepting a tree as a whole system that has not been reduced to the sum of the parts. The Tree of Life symbolizes the unity of all beings.

Life experience is derived from the Tree of Life, which draws energy from nature, yet offers itself in service of knowledge. Nature teaches us that Life cannot be equated with reason alone. Notably, a vertical tree trunk is a tool for climbing up and exploring a new dimension from up to down. Rachel Pollack, an award-winning novelist from New York, wrote, "People climb trees. They find their way, branch by branch, into the higher regions. Before the invention of machines of flight, balloons, gliders, airplanes, giant trees where the only way people could leave the surface of the earth."[20]

As we learned from the Pardes story, the Tree of Life provides more than three dimensions by penetrating the four worlds of *Olamot: Asiyah, Yetzirah, Beriah,* and *Atzilut.* The Tree of Life becomes a prototype for a ladder, like the stairways to heaven that Jacob saw in his dream in Genesis 28:12.

Is there any prohibition for eating from both trees, the Tree of

---

[19] Nilton Bonder, *Yiddishe Kop.* 1999, page 60.
[20] Rachel Pollack, *The Kabbalah Tree.* 2005, page 2.

Life and the Tree of Knowledge? A hundred years ago, Aaron David Gordon, a Utopian mystic and original thinker, made an exciting point: "The trouble with man is not that he ate from the tree of knowledge, but that he did not eat from the tree of life."[21] Rachel Pollack noted, "When they [Adam and Eve] ate from knowledge – without wisdom or understanding – they became conscious of the devastating power [of knowledge] for the first time."[22]

According to Jewish tradition, people may eat from the Tree of Life on the Shabbat. It can be seen as the achievement of the status won by experiencing life from the perspective of oneness after the completion of the weekly cycle. The Hasidic masters say that creation is restored to its root in oneness by refraining from "doing" on the Shabbat and returning to the ground of "simply being." Estelle Frankel referred to the midrash that says Adam and Eve could have eaten from the fruit of the two trees if only they had waited till Shabbat. She wrote: "Indeed, it is a message to all of us that by linking our need for self-assertion and separation (the tree of knowledge) with the awareness of our inseparability from all being (the tree of life), we reach … unity and duality dance harmoniously together."[23]

Michael Rosenak, a Professor of Jewish Education at the Hebrew University of Jerusalem, wrote: "we may learn from the imagery of the kabbalists that there is a tension between the revelation of restriction (of commandment and prohibition), which is symbolized by the Tree of Knowledge, and the revelation of creativity, harmony and the absence of boundaries – symbolized by the Tree of Life." [24]

Here are two more important quotes from Rosenak's *Tree of Life, Tree of Knowledge,* which will be helpful in our search. The first, "this interlinking suggests that one should avoid seeing the Tree of Knowledge – with its clear, fixed, and "particularistic" categories of holy and profane, with its intimation of the world as divided into

---

[21] Cited from Margaret Chatterjee, *Studies in Modern Jewish and Hindu Thought.* 1997, page 31.

[22] Rachel Pollack, *The Kabbalah Tree.* 2005, page 28.

[23] Estelle Frankel, *Sacred Therapy.* 2005, page 232.

[24] Michael Rosenak, *Tree of Life, Tree of Knowledge.* 2001, page 334.

"inside" and "outside."[25] And the second, " ... the Tree of Knowledge comes first, and the distinctions it makes and the specific commitments it demands illuminate the path to the Tree of Life."[26]

Gershom Scholem, the preeminent scholar of Kabbalah, wrote in 1971:

> "The trees in Paradise are not merely physical trees; beyond that, they point to a state of things which they represent symbolically. In the opinion of the Jewish mystics, both trees are, in essence one. They grow out in two directions from a common trunk. Genesis tells us that the Tree of Life stood in the center of Paradise, but it does not indicate the exact position of the Tree of Knowledge. The Kabbalists took this to mean that it had no special place of its own but sprouted together with the Tree of Life out of the common matrix of the divine world... **they were models for two possible forms of life.** "[27]

Rachel Pollack expressed this idea more poetically, " ... **life lies inside knowledge, embedded in it**, the way the four levels of PRDES lie embedded in each other. If we can reach true knowledge, knowledge joined to understanding and wisdom, not separated from them, then indeed the full Tree of Life will stand revealed before us, with no barriers or abyss to separate the higher and lower principles."[28] Later in the same book, Pollack wrote more explicitly, "we come to the idea of the Tree of Knowledge hidden inside the Tree of Life."[29]

Science is the activity of gaining knowledge. As scientists, we approach the Tree of Knowledge of Good and Bad. What is good? Life! What is bad? Death! However, from the point of view of the Tree of Life, death does not exist. From the point of view of the Tree

---

[25] Ibid., page 335.

[26] Ibid., page 336.

[27] Gershom Scholem, *The Messianic Idea in Judaism.* 1971, pages 69 and 70 (bold is mine).

[28] Rachel Pollack, *The Kabbalah Tree: A Journey of Balance & Growth.* 2005, pages 31-32 (bold is mine).

[29] Ibid., page 43.

of Knowledge, there is a *choice* between life and death.

The Zohar tells not only of a Tree of Life but also of a Tree of Death. "Observe that there are two Trees, one higher and one lower, in the one of which is life and in the other death, *and he who confuses them brings death upon himself...* "[30] The two trees are seen as opposing forces that are constantly struggling for dominance in the world. Choosing life is the foundational statement in Jewish tradition, "I have set before your life and death, ... therefore choose life, that you and your descendants may live" (Deuteronomy 30:19).

Death is well-known to those in biological science! This does not only apply to the death of animals and plants; cells constantly die, populations crash, biological species disappear, and all cellular processes come to an end. It can be said that biological death is the opposite of biological life. Kabbalistic Biology views death as a natural part of life. Without death, a biological cycle would not be complete. Such a statement cannot be regarded as a pure declaration. As part of the self-replication process, it provides a transitional direction for life. Furthermore, it provides practical tools for measuring any biological object during the process of "becoming."

The Crystal Structure of the Tree of Life
Let's return to Einstein's Marble-Wood paradox. Another translation of the word *shayish*, in the Pardes story is the term "crystal stone." Marble is a crystal stone. There are a number of ways in which the scientific definition of a "crystal" may be defined. However, the first one is based on the microscopic organization of the atoms (elementary components) inside a crystal in a regular pattern that reflects its internal symmetry. The Encyclopedia Britannica defines crystal structure as the arrangement of its unit cells, "It is the smallest unit of volume that permits identical cells to be stacked together to fill all space. By repeating the pattern of the unit cell over and over in

---

[30] *The Zohar*, vol. 3, Harry Sperling et al. (translators) Soncino Press, London, 1978, page 410. Quoted from Robert Haralick, *God Consciousness, The Exercises: Working with the Sefirot and Netivot*, Torah Books, Pomona, New York, 2014, page 43.

all directions, the entire crystal lattice can be constructed."[31]

There is no doubt about the crystal structure of the Tree of Life when viewed from the perspective of its most familiar images. For comparison, I have placed two popular images of the Tree of Life together with an image of the crystal structure of Titanium (Ti) to illustrate how they have a similar crystal structure (Figure 27). Titanium is the ninth-most abundant element in Earth's crust and the most biocompatible metal due to its resistance to corrosion from many sources. These include bodily fluids, bio-inertness, capacity for the formation of a direct interface between an implant and bone without intervening soft tissue, and high fatigue limit.

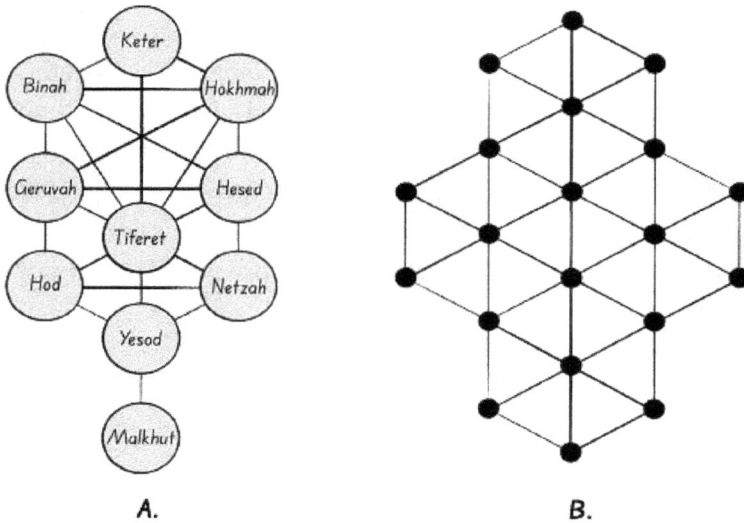

A. B.

Figure 27. According to Isaac Luria, the crystal structure of the Tree of Life (A) looks like the hexagonal crystal system of Titanium, the ninth-most abundant element in Earth's crust (B).

The Kabbalah teaches us that things are not always as they appear to be on the surface. Estelle Frankel, who has taught Kabbalah for over forty years in both academic and spiritual settings, noted that reality is always multilayered, and the boundaries between good and bad are

---

[31] https://www.britannica.com/science/crystal/Structure

not always clear-cut.[32] What may seem bad on the surface may paradoxically turn out to be good, and vice versa.

Claude Levi-Strauss, the French anthropologist credited with developing "structuralism" in cultural anthropology, distinguished two levels of reality comprising a "surface structure" and a "deep structure." Surface diversity conceals underlying unity. Levi-Strauss wrote: "Behind this diversity, there lies something deeper, something common to all its aspects."[33]

Choosing between the "crystal-marble" and "wood" perspectives presents a significant challenge for biologists, even when they try to disregard it. Biological research aims to obtain reliable data and information on the structure of all biological entities. The formulation of overarching natural laws that explain the complexity and dynamism of most biological processes is often reduced to a specific principle.

It might also be meant to describe patterns observed in living organisms. Discovering more general laws in biology, akin to those in physics and chemistry, has remained an elusive goal for many generations of biologists. Pursuing generalized statements, while maintaining fidelity to observed facts, necessitates good intentions, years of experience, and a deep understanding of the fundamental laws governing all life on Earth.

Another deep insight gained from the interconnectedness between the Tree of Life and the Tree of Knowledge comes from the similarity in their structures. Both trees transcend the world hierarchy (*Olamot*). Despite their differences, both trees have roots, grow out of a common trunk, have branches on their right and left sides, and produce fruits.

Samuel Avital writes in *The Invisible Stairway*: "The root of the tree which serves as a foundation upon which the trunk is built, and the branches grow into a complete state of being we call the Tree of Life.

---

[32] Estelle Frankel, *Sacred Therapy*. 2005, page 233.

[33] Claude Levi-Strauss, quoted in Jerome Levi *Structuralism and Kabbalah: Science of Mysticism or Mystification of Science?* Anthropological Quarterly, 2009, vol. 82, page 936.

So, metaphorically speaking, we need the root, the trunk, the branches, the leaves, etc. And water, of course, to nurture the tree to live and prosper, in order to build, create, and form the existence of the whole Tree of Life, knowledge, practical wisdom, and existence."[34] The structures of the Tree of Life and the Tree of Knowledge have more profound and more productive similarities, which we will discuss in the next chapter.

---

[34] Samuel Avital, *The Invisible Stairway.* 2005, page 48.

Chapter 11
Thirty-two Pathways of Knowledge

The *Sefer Yetzirah* (The Book of Formation) is the oldest and most mysterious of all Kabbalistic texts. The first stanza of the book begins with the words, "Thirty-two mystical paths of Wisdom..." In fact, most of the *Sefer Yetzirah* text describes the 32 paths. As the second stanza explains, these paths manifest as the 10 digits and the 22 letters of the Hebrew alphabet.

Aryeh Kaplan, a physicist who became a prominent rabbi, gave us the most complete and accessible description of *Sefer Yetzirah*. According to Kaplan, the ten digits constitute the inner structure and makeup of the Four *Olamot* "Worlds." [35] It was said in the Mishna that "the world was created through ten utterances." [36]

The Ten Ultimate Sefirot
In the Tree of Life, the ten sefirot represent the ten digits. The word *sefirot* is a plural form of *sefirah*. According to the Hebrew tradition, first, the root letters are defined. Then, other words using these letters are examined to gain a deeper understanding of those words. In this case, the word sefirah has three root letters: "*samekh – pey -reish.*" In the first verse, the *Sefer Yetzirah* lists other keywords with the same root letters: *Sphar* – number, *Sippour* – story, and *Sepher* – book,

---

[35] Rabbi Aryeh Kaplan, *Inner Space.* 1990, page 37.
[36] Pirke Avot, 5:1.

which, in reality, "is one, different, and the same."[37] All these words relate to the meaning of sefirah. The word *sefirah* is also connected with the word "sapphire," the gemstone with a hexagonal crystal structure. You may recall that we noted the symbolism of the crystal structure of the Tree of Life in the previous chapter.

There are ten sefirot. The book *Sefer Yetzirah* makes the statement about this number very clear: "Ten ineffable sefirot, ten and not nine, ten and not eleven... Examine inside them and probe them." Not surprisingly, the number ten represents completeness! The names of sefirot from up to down are *Keter, Hokhmah, Binah, Hesed, Gevurah, Tiferet, Netzah, Hod, Yesod,* and *Malkhut.* In Hebrew literature, you can see different spellings of the names of some sefirot.[38] Hebrew letters have precise positions; all variations relate to English equivalents.

The interaction between the various sefirot is depicted through a network of connecting *tzinorot* ("channels" or "pipes"), which provide an exchange of energy and information. These connections suggest various subgroupings of the sefirot, each reflecting a common dynamic among the sefirot that they include. There are two basic descriptions of the concept of sefirot. According to the first, the sefirot is a system of vessels to receive the primordial force. According to the second, the sefirot are particular ways of perceiving and understanding objects and processes. The greatest Kabbalist, Isaac ben Solomon Luria (or Ari), stressed that both definitions are proper, depending on the functions of sefirot.

Understanding the functions of sefirot requires learning the meaning of each *sefirah* and becoming aware of their interrelations. According to Moses ben Jacob Cordovero (or Ramak), a central figure in the historical development of Kabbalah who lived in 16th-century Safed, "the nature of any created things depends upon the manner in which the various sefirot have been combined in its formation."[39]

---

[37] Samuel Avital, *The Invisible Stairway.* 2005, page 267.

[38] I selected the provided spelling to keep the names simple.

[39] Sholem, *Kabbalah.* 1974, page 115.

According to the book *Zohar*, sefirot are compared to the organs of the human body:

"For there is not a member in the human body but has its counterpart in the world as a whole. For as a man's body consists of members and parts of various ranks all acting and reacting upon each other so as to form one organism, so does the world at large consist of a hierarchy of created things, which, when they properly act and react upon each other together form literally one organic body." [40]

During the last decades, the images of sefirot became familiar to many people because of their frequent presentation in popular culture. Unfortunately, as the concept of sefirot received more public exposure, simplified visions emerged to describe it. The problem is that any common, conventional language is limited in its ability to describe the essence of sefirot. Thus, English "equivalents" of sefirot names should be approached with caution. Being able to apply the concept of sefirot requires deep knowledge, which is only possible with many years of dedicated studies of written and oral sources.

Below, I will introduce some selected properties of the sefirot. By doing so, I am using some concepts from Kabbalah, including the "sefirotic structure" of the Tree of Life, to formulate specific dimensions. These will be able to measure unseen realities of biological life that we will discuss in the following chapters.

The Three High *Sefirot*

Keter is the first and highest of the ten sefirot and represents absolute dominance over all forces of reality. The word *keter* means "the Crown," indicating its position at the top of the Tree of Life. As a crown is separate from one who wears it, Keter is separated from the sefirot that comprises the body of the Tree of Life.[41] For this reason, Keter is hidden.

However, the existence of Keter is essential in life because of its connection to the manifested world. Aryeh Kaplan describes the

---

[40] *Zohar* 1, 134a, cited from Sanford Drob, *Symbols of the Kabbalah*. 2000, pages 167-168.

[41] Sanford Drob, *Symbols of the Kabbalah*. 2000, page 209.

central paradox of Keter when he says, "Keter is *Ain* -- Nothingness as well as *Ani* --Selfhood."[42] According to Kaplan, only Malkhut, the lowest of sefirot, manifests and actualizes Keter's abstract potential. Keter and Malkhut share the *Ani-Ain* identity. *Ani* (the Hebrew word for "I") refers to something that acts as a cause. It is called *"Ain"* (the Hebrew word for "nothingnes") when the same cause is seen as an effect of a higher cause.[43] For us, the relationship between *Keter* and *Malkhut* is significant, and we will consider the impact of such a relationship in Chapter 13. Between these two poles of existence are all the other sefirot.

Hokhmah (or Chokhmah) is the second sefirah from the top. The word *Hokhmah* means "wisdom." Kaplan describes Hokhmah as the sefirah that "contains in potential all the laws of creation. Moreover, it constitutes the basic set of axioms that determine how these laws will function in actuality."[44] While Hokhmah cannot be seen, it animates everything. In the Zohar, Hokhmah is named "inner thought" and is considered to be void of all individuality, instantiation, and separateness.[45]

According to Moses Cordovero, Hokhmah is "being from nothingness." The status of "the beginning and not being itself" requires a third point (Binah) to reveal existence.[46] If the ineffable nature of Keter symbolizes "nothingness," Hokhmah is "being" or an "arena in which things and ideas begin to be formulated."[47]

Kabbalah teaches that certain kinds of questions have the power to inspire wisdom. To explain the essence of Hokhmah, I will provide a passage from the Introduction to the Zohar that distinguishes between questions that ask *mi* (who) and *mah* (what). It says, "The first, concealed one – called Who – can be questioned. Once a human

---

[42] Rabbi Aryeh Kaplan, *Inner Space.* 1990, page 49.

[43] Ibid., page 49.

[44] Ibid., page 58

[45] *Zohar,* 1, 2a, cited from Drob, 2000, page 214.

[46] Moses Cordovero, *Or Ne'erav.* VI:1, 35a, cited from Drob, 2000, page 214.

[47] Y. David Shulman, *The Sefirot.* 2000, page 219.

being questions and searches, contemplating and knowing rung after rung to the very last rung – once one reaches there - What? What do you know? What have you contemplated? For what have you searched?"[48] The question "what" (*mah*) is a component of the word *Hokhmah*. When the letters of this word are reversed and broken down into two words, it spells *koach mah*, meaning "the power of what."[49] True wisdom derives from the humble awareness that we are not separate from our Source. Without this kind of humility, all our knowledge remains conceptual, and we cannot acquire true wisdom.

Binah is the third sefirah, the final of the three upper sefirot. The word means "understanding." Binah represents the fulfillment of concealed Hokhmah. The relationship between Hokhmah and Binah is referred to in the Zohar as "two companions that never separate." According to Rabbi Shneur Zalman of Liadi, Binah represents the "dimensions" of explanation, understanding, and manifestation.[50]

Binah is portrayed in Kabbalistic literature as the "cosmic mother" giving birth to all seven lower sefirot.[51] Another meaning of Binah in Hebrew comes from the word "womb." If Hokhmah is viewed as a seed, Binah nurtures it. According to the mathematic metaphor, Binah represents the beginning of a substantial existence as a circle.[52] The Talmud explains Binah as "understanding one thing from something dissimilar."[53] Binah transforms the blueprint into reality. Joel Primack, a professor of physics at the University of California, Santa Cruz, and Nancy Ellen Abrams, a former Fulbright scholar, used the Big Bang metaphor to describe the process of unification of Hokhmah and Binah: "There is probably no name more appropriate for the Big Bang ... than "Hokhmah-Binah."[54]

Aryeh Kaplan compared Hokhmah (Wisdom) with the

---

[48] *The Zohar*, Pritzker Edition, vol. 1, page 6.

[49] Estelle Franklin, *The Wisdom of Not Knowing*. 2005, page 51.

[50] Shneur Zalman, *Likutei Amarim-Tanya*. cited from Drob, 2000.

[51] Isaiah Tishby and David Goldstein, *The Wisdom of the Zohar*. 1991, vol. 1, page 282.

[52] Sanford Drob,2000, page 216.

[53] *Sanhedrin* 93b, cited from Kaplan, 1990, page 57.

[54] Joel Primack and Nancy Ellen Abrams, *The View from the Center of the Universe*. 2006, page 201.

nonverbal right hemisphere of the brain, while Binah (Understanding) is associated with the verbal left hemisphere. [55] As the Kabbalists explain, wisdom is normally only experienced when it is "clothed" in understanding. The *Sefer Yetzirah* formulated the connection of these sefirot very clearly: " ...understand with wisdom and be wise with understanding."[56]

The Seven Low Sefirot

The seven lower sefirot differ drastically from the three upper sefirot. In the Lurianic Kabbalah, the teaching of Ari Luria, who lived in the 16th century, the seven lower sefirot were shuttered by access to the primordial force. A turning point in the development of our reality has resulted in incomprehensible manifestations and a multitude of living organisms.

Hesed (or Chesed) is the first of the lower sefirot. The Hebrew word *tiferet* means "lovingkindness" or "mercy." The main property of Hesed is the tendency to achieve unlimited expansion and growth. Natural processes and systems are interdependent, with organisms and components relying on one another for survival and function. In a sense, Hesed represents this interconnectedness and the kindness and generosity that sustain it. As a final point, Hesed can inspire us to seek new knowledge and push the boundaries of scientific discovery. In the context of scientific research, this could be interpreted as the importance of collaboration and cooperation between scientists from different disciplines, backgrounds, and cultures.

The sefirah of Gevurah is the counterpart to Hesed and shapes its endless expansion. This sefirah represents the principle of measure, limit, and restraint. The word *gervurah* means "strength." New forms are created through the destruction of the old ones. The main property of Gevurah relates to barriers. A key characteristic of Gevurah is its ability to control and contract. By contraction, withdrawal, and concealment, Gevurah can generate usefulness,

---

[55] Aryeh Kaplan, 1990, page 39.
[56] *Sefer Yetzirah*, Chapter 1, Mishna 4.

activity, progress, and decomposition.

One way to think about the role of Gevurah in the natural world is the principle of cause and effect. Many processes and systems in the natural world are governed by cause-and-effect relationships, with various factors and conditions leading to specific outcomes. Gevurah represents causality and the power and strength required for specific outcomes. In a scientific context, Gevurah can also be interpreted as the need for an objective evaluation and assessment of scientific findings. As a result, it encourages researchers to submit their work for scrutiny and evaluation and to revise their ideas in light of new evidence.

Tiferet, the sixth sefirah, is the center of the Tree of Life. Tiferet embodies the idea of wholeness, balance, and harmony. It coordinates and reconciles the opposing forces of Hesed and Gevurah. Many natural systems, from ecosystems to the human body, operate within a delicate balance, with various components working together dynamically and interdependently. Tiferet can be seen as representing this balance and harmony, as well as the interconnectedness of all biological objects.

Traditionally, Tiferet is associated with the heart, which is associated with emotions and feelings. Thus, Tiferet could be related to the importance of empathy and compassion in scientific research and discovery. Finally, Tiferet is also associated with the concept of beauty, which in a scientific context could be interpreted as the beauty of Nature and the wonder of the natural world. Scientific research often uncovers the natural world's mysteries, gaining us a deeper understanding of its intricacies and complexities.

The sefirah Netzah signifies firmness, endurance, persistence, and the determination to continue initiated processes and overcome internal and external obstacles. The word *netzah* relates to the word *menatzerah* meaning to "conquer" or "overcome." *Netzah* also means "to conduct" or "orchestrate." In addition to determination, tenacity, and persistence, *Netzach* combines patience, perseverance, and guts. Reliability and accountability establish security and commitment, which are essential to endurance. To make progress and achieve breakthroughs, scientists must be willing to work hard, overcome obstacles, and persist in the face of challenges. Netzah can be seen as representing these qualities of determination and persistence.

The sefirah of Hod keeps Netzah focused. It shapes conditions to support the endurance of Netzah. The sefirot of Netzah and Hod are called "two halves of a single body." If endurance is the engine of life, humility is its fuel. Hod is commonly translated as splendor or "gloria." However, this sefirah has another meaning: it is related to the word *hoda'ah*, which means "submission," "message," and "acknowledgment."

Knowing strength is coming from a higher place allows one to endure far beyond one's usual capacity. One way to understand the role of Hod in natural processes is by recognizing the presence of humility and receptivity. Natural systems are often highly complex and dynamic, and they operate according to a set of principles and laws that are not always obvious to us. To gain a deeper understanding of these systems, we must approach them with a sense of humility and openness, recognizing that there is much we don't know and much we can learn.

Additionally, Hod is associated with communication and the exchange of ideas. When exploring natural processes, we must understand the various components and interactions that make up a system. This is best achieved if scientists and researchers communicate and collaborate to gain a deeper understanding of natural systems and to develop effective strategies for managing them. Finally, *Hod* is also associated with gratitude, which could be interpreted as a deep appreciation for the natural world and its many wonders. It represents the wonder that drives scientific exploration and the recognition that we are all part of a vast, interconnected web of life.

The sefirah Yesod serves as a container and funnel for the other sefirot. The word *yesod* means "foundation." According to Aryeh

Kaplan, a foundation doesn't simply support a building, "it establishes the proper relationship between the building and the ground."[57] Yesod represents the principle that underlies the Upper World. It could be compared to physics laws explaining natural phenomena. Yesod could be viewed as a gateway between the higher and the physical realms, just as scientific models and theories can bridge the gap between abstract and concrete aspects of reality. As Tiferet mediates between Hesed and Gevurah, so does Yesod unite the opposing sefirot, Netzah, and Hod.

Malkhut is the lowest of the ten sefirot. This sefirah can be viewed as manifestation and materialization. In the natural world, scientific theories and models are commonly tested through observation and experimentation to manifest or materialize the ideas and concepts they represent. While Malkhut is closely associated with the physical world, it is also seen as a bridge between the lowest (physical) and highest realms. Malkhut is considered the point where the Life energy (*shefa*) that flows through the other sefirot enters into the physical realm. It is through Malkhut that we can experience and interact with the physical world. Moshe Cordovero refers to Malkhut as the architect who brings about creation and declares that nothing reaches the lower world except via its portals.[58]

All of the other nine sefirot are described as activities. However, Malkhut cannot be depicted as an activity, but is instead a state of being or existence.[59] Schneur Zalman noted that Malkhut is the origin of time and space, "the attribute Malkhut from which space and time are derived and come into existence."[60] In addition, Malkhut refers to sovereignty, which includes ethical considerations and the responsible use of scientific knowledge. It is essential for researchers to be aware of their work's impact on society and the environment, as well as to use their expertise and knowledge ethically and responsibly.

---

[57] Arieh Kaplan, 1990, page 64.

[58] Moses Cordovero, *Pardes Rimmonin*, 11:2. Cited from Schochet, *Mystical Concepts*.

[59] Bruce Friedman, *Mystery of Black Fire, White Fire*. 2016, page 229.

[60] Rabbi Shneur Zalman, *Likutei Amarim-Tanya*. Quoted from Sanford Drob, 2000, page 227.

Network of Sefirot

In the words of Estelle Frankel, "the sefirot provide an imaginative pathway to approach the unknowable."[61] The sefirot exist not only as individual entities but also as configurations of them. There are two primary models to represent a relationship between sefirot: the circular model (*igulim* –circles, waves) and the linear model (*yashar* – straightness, uprightness).

In the first model, in which the circle represents the female principle, there is hidden potential. As per Lurianic Kabbalah, the *tzimtzum*, the contraction of an infinite light within itself, created otherness and apparent separation. The result of this contraction was an empty (vacant) round space - *khalal panui*.[62] The ten circular sefirot encircle the round vacant space, one on the other. As ten concentric "circles," the sephirot act sequentially and independently from each other, from Keter in the closest proximity to the source to Malkhut at the center. Keter encircles Hokhmah, which encircles Binah, and so on, until Yesod encircles the last sefirah -- Malkhut (Figure 28).

All potential reality was contained within this place without distinction, without beginning or end. Rabbi DovBer Pinson compares the *igul* (a singular form of *igulim*) to a seed that will eventually grow into a full-blown tree. As the full tree exists undifferentiated within the seed, so too, the *igul* is one seamless whole.[63]

---

[61] Estelle Frankel, *The Wisdom of Not Knowing*. 2017, page 29.

[62] DovBer Pinson, *Thirty-Two Gates of Wisdom*. 2008, page 16.

[63] Ibid., page 16.

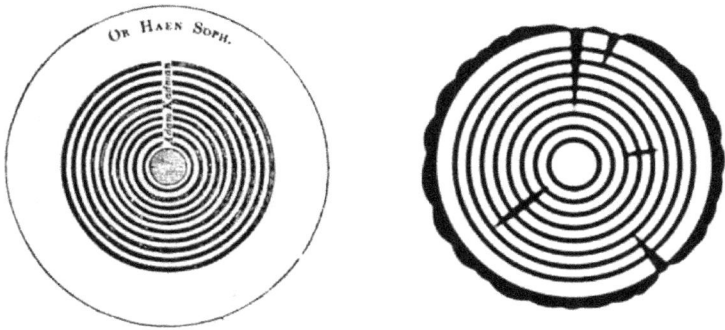

Figure 28. According to the circular (igulim) model, ten Sefirot are arranged in ten concentric circles that resemble tree cross-sections with growth rings.

The linear model (*yashar*) represents the relationships between them in an organic way. Furthermore, this model is much more popular. If the *igulim* model signifies possibility, the *yashar* model represents actual reality. This model has distinct points and an up-down sequential structure with a definite beginning and end. The *yashar* points are the sefirot.

According to the *Yashar* model, the ten sefirot are arranged in three vertical pillars (or columns): the right, the left, and the central (or middle). The right one is the pillar of "expansion," or "masculine." The left one is the pillar of "contraction," or "feminine." The middle is the pillar of "balance." As you move from top to bottom, you will find Hokhmah, Hesed, and Netzah on the right pillar. On the left pillar are Binah, Gevurah, and Hod. On the middle pillar are Keter, Tiferet, Yesod, and Malkhut (Figure 29).

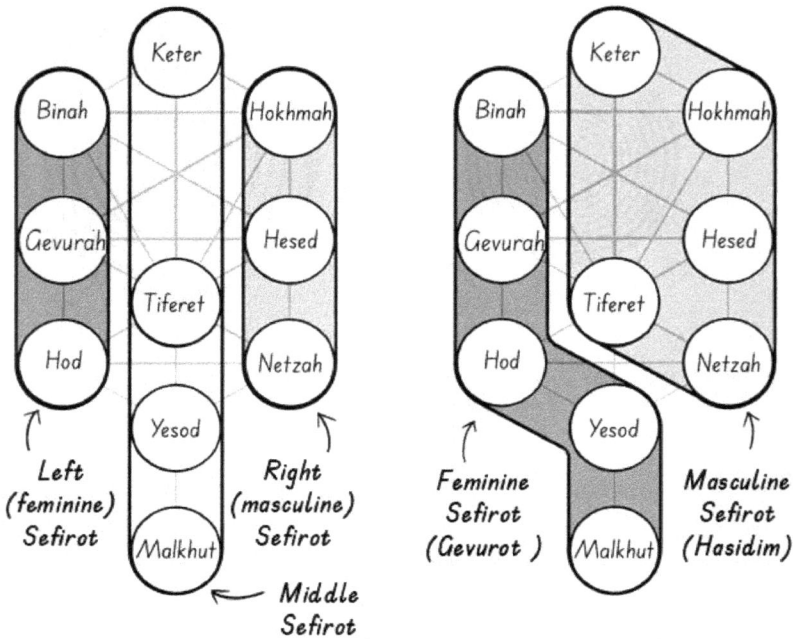

Figure 29. The Ten Sefirot can be arranged in two or three vertical pillars.

Although the sefirot are usually divided into three pillars, they can also be arranged in two arrays, one on the right and the other to the left.[64] All the sefirot on the right side and the two upper center ones would be considered "masculine." Feminine" sefirot on the left includes the three generally on the left and the two lower sefirot in the center. Since the five masculine sefirot are on the Hesed side, they are called Hasadim ("Hasadim" is plural of Hesed). Similarly, the five feminine sefirot are called the five Gevurot since they are on the side of Gevurah ("Gevurot" is plural of Gevurah).

Comparing the arrangement of sefirot in two or three pillars, Aryeh Kaplan noted that when the sefirot are in their "normal" state, arrayed in three columns, they are in a state of equilibrium. However, when the sefirot of the central column moved to the right and left to divide the sefirot into two arrays, a powerful tension is provided. In

---

[64] Aryeh Kaplan, *Sefer Yetzirah.* page 33.

such a mode, powerful forces can be directed and channeled.[65]

Nine of the sefirot are arranged in triads. Keter, Hokhmah, and Binah create the upper triad, called "Ha-Ba-D." The sefirot of the upper triad acts together as the "guiding force" for a lower configuration and is called the Mohin of growth.[66] Hesed, Gevurah, and Tiferet form the central triad, called "Ha-Ga-T." They act together as the second level of the guiding force for a lower configuration. The lower triad is made of Netzah, Hod, and Yesod, called "Ne-H-Y." The Sefirah of Malkhut stays separate from the triads (Figure 30).

---

[65] Ibid., page 33.
[66] Raphael Afilalo, *Kabbalah Concepts.* 2006, page 40.

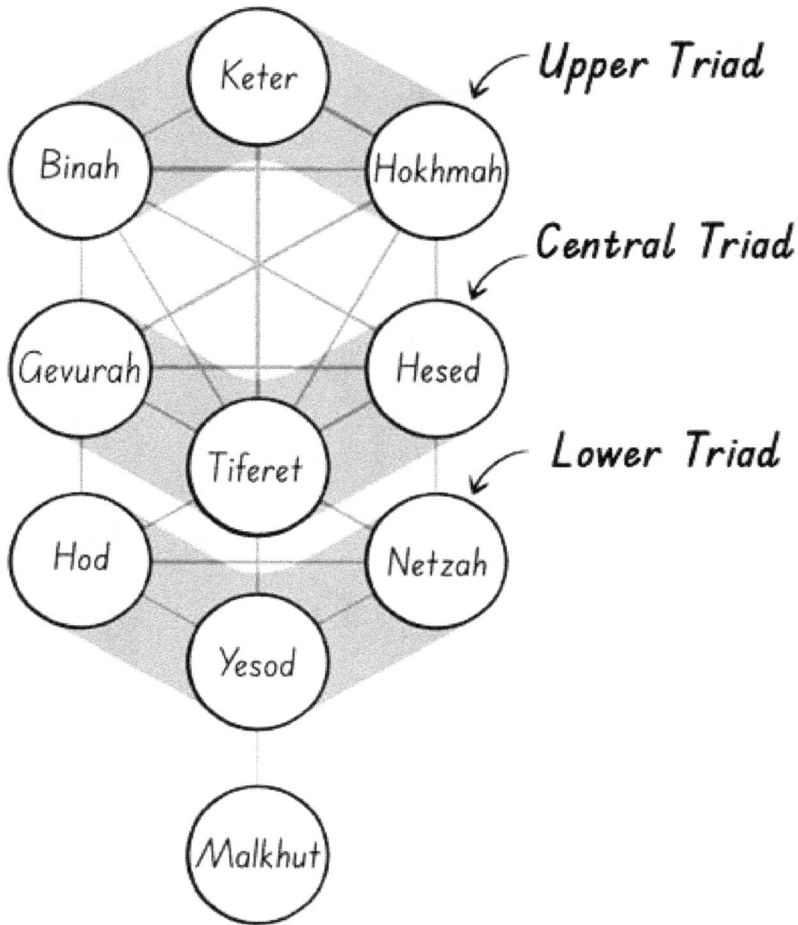

Figure 30. The Ten Sefirot are arranged in three triads.

Configuration of Partzufim

The ten sefirot are also arranged in particular configurations named *partzufim*, a plural form of *partzuf* (Pey-Reish-Tzadi-pai) meaning "personae/faces/forms/configurations." The basic purpose of the individual *partzuf* is to integrate disparate elements into a more coherent whole. Each partzuf is a harmonious unified whole of entities expressed by sefirot. The partzufim configurations are in a constant state of action, illumination, and interrelations between sefirot.

The names of five main partzufim are *Arich Anpin* - "Long Face," *Abba* - "Father," *Imma* - "Mother," *Zeir Anpin* - "Small Face," and *Nukvah* – "Daughter." Each sefirah is associated with a particular partzuf. Keter is associated with *Arich Anpin*; Hochmah and Binah are associated with *Abba* and *Imma*, respectively; six sefirot from Hesed to Yesod from Zeir Anpin; and Malkhut is associated with *Nukvah* (Figure 31). *Arich Anpin* is different from the other configurations. All the other configurations are his "branches."

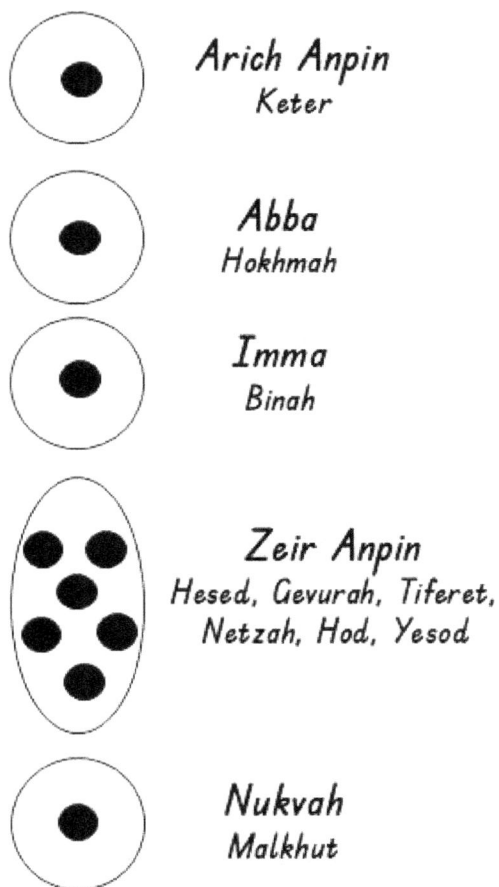

Figure 31. The Sefirot are arranged in five Partzufim.

The concept of *partzufim* is derived from the Zohar, but its full doctrinal significance emerged in the 16th-century Lurianic Kabbalah. Based on the Lurianic teaching, the four worlds (*olamot*) reach stable forms through the reconfiguration of the original sephirot into partzufim. This rearrangement leads to the transformation from *Tohu* -- "Chaos" to *Tikkun* – "Rectification." Thus, partzufim corrects two flaws within the world of *Tohu:* the immature vessels and their lack of inter-relationship.

The guidance of the world is dependent on the different positions and interactions of the masculine and feminine partzufim. The construction of a partzuf is achieved by way of *zivug* (union, coupling) of two higher masculine and feminine partzufim. Figuratively, each *zivug* is followed by a period called *ibbur,* gestation (pregnancy in Hebrew) inside the higher feminine configuration, and "birth," when it is revealed.[67]

Partzuf *Zeir Anpin*, often called by the initials, Z'A, is a dynamic configuration in a constant process of moving from a weaker state to one of growth to renew the forces and influence. To act, *Zeir Anpin* needs to get the directive forces, or *mohin*, literally meaning "brains." The reception of *mohin* is divided into four phases, called the first and second stages of *katnut* (smallness) and the first and second stages of *gadlut* (greatness). Partzufim are characterized by their fully expanded vessels and their relationships. Each element of position within the universal hierarchy is constantly affected by higher strata of reality.[68]

The Alternative Sefirah — Da'at

In many images of the Tree of Life, you can see one more circle, along with ten others representing the sefirot mentioned above. This circle is called Da'at, meaning "knowledge." Unsurprisingly, there has been a lot of confusion and disagreement regarding Da'at's status. The earliest Kabbalistic source, *Sefer ha-Yetzirah*, clearly states: "Ten

---

[67] Ibid., page 62.
[68] Yitzchak Ginsburgh, *What You Need to Know about Kabbalah. 2006.*

Sefirot of Nothingness – ten and not nine -- ten and not eleven."[69]

Arguments on this point go back to 16th-century Tzfat (or Safed, the center of Kabbalistic thought in Galilee, at that time Ottoman Empire). Moses Cordovero (known as RaMaK, a central figure in the historical development of Kabbalah) excluded Da'at as a separate Sephirah while counting Keter as part of the ten sefirot.[70] In the opposing view, Isaac Luria, known as Ari or Arizal, thought that Da'at was one of the sefirot, while Keter was excluded.[71]

Aryeh Kaplan, in his commentaries to *Sefer Yetzirah*, called Da'at "the quasi-sefirah."[72] Kaplan explains that "Da'at can only manifest in the absence of Keter. Keter and Da'at are then said to be mutually exclusive."[73] Kaplan referred to Da'at as "applied logic" -- the ability to combine the basic information given (Hokhmah) and make it logical (Binah). Compared to Hokhmah and Binah, Da'at represents externalization and an ability to express intelligence.

While associating remembrance of the past to Hokhmah and anticipation of the future to Binah, Kaplan stressed that "the present is the only place where we really have Da'at – Knowledge."[74]

In *The Sefirot*, Y. David Shulman wrote: "The Da'at is an intermediate phase before the undivided energy descends into Sefirot more directly linked to our reality." [75] In addition to being called "knowledge," Da'at could also be "perception" or "connection." Da'at is the first balanced fullness of Keter after uniting the complementary energies of Hokhmah and Binah. Da'at contributes to the blending of Hokhmah and Binah.

There are two aspects of Da'at, according to the Kabbalists. First, there is an "upper" aspect (*Da'at Elyon*) directly derived from Keter, which brings about a dialectical union between Hokhmah and Binah.

---

[69] Aryeh Kaplan, *Sefer Yetzirah*. page 38.

[70] Moses Cordovero, *Pardes Rimonim*. 3:1.

[71] Chaim Vital, *Etz Chaim*. 23:8; 25:6; 42:1.

[72] Aryeh Kaplan, *Sefer Yetzirah,* page 51.

[73] Ibid., page 51.

[74] Ibid., page 59.

[75] Y. David Shulman, *The Sefirot*. 1996, page 73.

Moreover, a lower aspect of the sefirot (*Daʿat Tachton*) is the vehicle through which the intellectual activity of the upper three sefirot is channeled into the lower seven sefirot.[76]

Keter and Daʿat have an alternative relationship. Thus, no matter which relationship is used, the number of sefirot will always be ten, regardless of how Keter and Daʿat are related. There are ten sefirot in the Tree of Life, and Keter is one of them, while Daʿat is hidden. We have ten Sefirot in the Tree of Knowledge, including Daʿat as one of the Sefirot, while Keter is hidden (Figure 32).

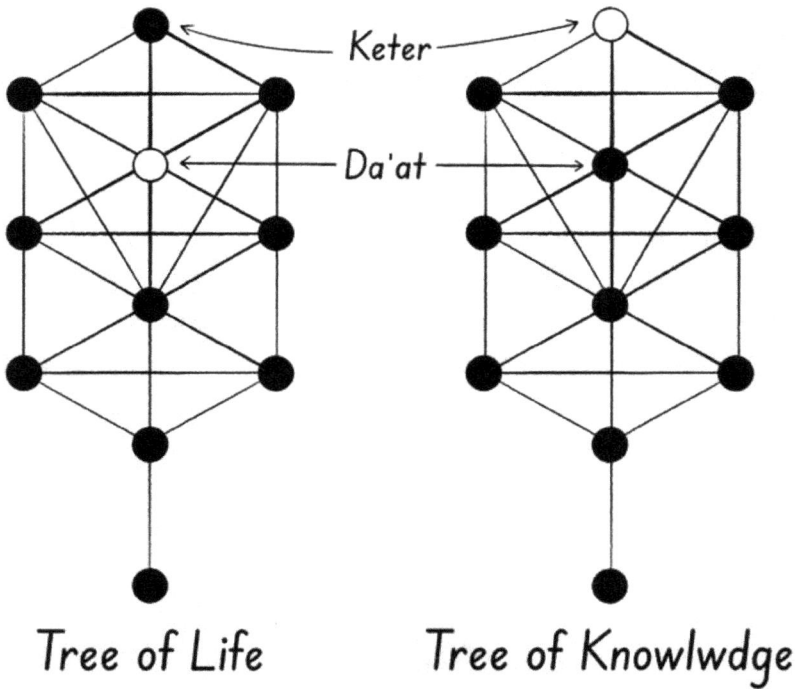

Tree of Life   Tree of Knowlwdge

Figure 32. There are ten sefirot in the Tree of Life, and Keter is one of them, while Daʿat is hidden. There are also ten Sefirot in the Tree of Knowledge, including Daʿat as one of the Sefirot, while Keter is hidden.

The story of Adam and Eve in the Garden of Eden is among the best-known narratives in the Bible. It is usually interpreted as a cautionary tale about disobedience to the divine command. However, a different

76 Sanford Drob, *Symbols of the Kabbalah*. 2000, page 218.

interpretation could be offered when viewed through the lens of Kabbalah. In it, the Tree of Life represents unity and the direct, unmediated light of divine illumination.

In the Garden of Eden context, this can be understood as a state of blissful ignorance with no perceived need for independent action or thought.

The Tree of Knowledge signifies awareness, knowledge, and discernment. It is the domain of duality, where good and evil, right and wrong, and truth and falsehood are known and must be navigated. By choosing to eat the fruit of the Tree of Knowledge, Adam and Eve chose the path of knowledge and understanding, accepting the burden of discernment and the challenges that came with it.

This interpretation reframes the act of eating the fruit of the Tree as a conscious choice for independence, exploration, and learning. Adam and Eve chose to step out of the limited, though blissful, existence in the Garden, where their needs were directly met without requiring any effort or understanding on their part. They chose instead the path of knowledge with all its challenges. By doing so, they opted to accept its opportunities for growth, exploration, and advancement.

Using the sefirotic metaphor, they chose the Tree of Knowledge over the Tree of Life. This choice signaled the transition from a passive existence to one of active participation, from being creatures of pure being to becoming creatures of knowledge. They chose to bear the costs and challenges of independence for the sake of the potential it offered them. Their ability to learn, grow, and understand, ultimately enabled them to become co-creators in the divine process of continuous creation and revelation. Indeed, this is a profoundly different perspective of the Genesis story, casting the choice of Adam and Eve in a different, more empowering light.

Sanford Drob, the director of Psychological Assessment, commented on this choice in *Symbols of the Kabbalah*, "The Kabbalists held that it is only a knowledge of particulars that fail to place things in a wider context of life ... The kabbalists used a variety of metaphors of separation ("chopping away the plantings," "separating the chutes," and "causing destruction") to symbolize an act of contemplation that separates knowledge from the ultimate values."[77]

Twenty-two Connecting Pathways

There are 22 paths, or channels, that connect ten sefirot. This number exactly corresponds to the number of Hebrew letters, which we will learn in Chapter 20. People who start learning Kabbalah are fascinated by the clarity of sefirot as the primary channels of energy and information. However, they often miss the importance of the letters as an integral part of the Tree of Life.

Bruce Friedman makes a symbolic depiction of the relationship between the sefirot and the letters by taking each Sefirah to be a "node." He then arranges these nodes in a particular tree-like pattern. Friedman likens the connectivity between these nodes to that of "trusses" a structure in engineering that consists of two-force members only, where the members are organized so that the assemblage as a whole behaves as a single object. These trusses are letters.[78]

The second verse of the book *Sefer Yetzirah* said,

"Ten Sefirot of Nothingness

and 22 Foundation Letters:

Three Mothers,

Seven Doubles

And twelve Elementals."

According to the system of *Olamot* (the words), the form, or visual structure of the written letter represents *Asiyah*, the word of actions. The letter's name, the full spelling, represents *Yetzirah*, the world of formation. The letter's number, as the pure abstract thought of the

[77] Sanford Drob, *Symbols of the Kabbalah*. 2000, page 355.
[78] Brice Friedman, *Mystery of Black Fire, White Fire*. 2016, pages 180-181.

letter, represents Beriah, the world of creation.

The ten sefirot and the 22 letters together form the Tree of Life's structural framework. Each element has a specific place and function, and all of them are arranged in a precise order, creating a beautiful, crystal-like structure of the Tree of Life ("marble"). Although we cannot perceive this intricate structure with our eyes, our knowledge of its "architectural" design allows us to visualize it, even if it remains invisible to our senses.

When delving into the realm of Kabbalah, we inevitably encounter a host of Hebrew terms and concepts that may initially seem complex or difficult to grasp. The objective of this chapter, however, is not to impose an additional burden of having to memorize these terms. Instead, the intention is to provide a conceptual toolkit that reveals a deeper understanding and exploration of biological life's complex and often unseen aspects.

Chapter 12

From the Crystal Structure to the Living Tree

The Tree of Life is an excellent example of a living entity with a hierarchical nature. This chapter explores the hidden architecture of the Tree of Life and its implications for studying biological life. This gives rise to the following questions, such as:

Can we comprehend a relationship between the simple structure of the Tree of Life and the complexity of biological life?

Can we link the interconnectedness of species within ecosystems to harmonious relationships between the Sefirot?

Can we relate the environmental adaptations of biological species to dynamic interrelations between the Sefirot?

The list of such challenging questions is potentially endless!

By exploring the hidden "architecture" of the Tree of Life, we can study biological life with an integrative mindset. Only by using this approach can we uncover the underlying patterns and principles that govern how complex systems function. By so doing, we can achieve a far more comprehensive understanding of life's intricacies and gain a deeper appreciation of the beauty and unity of all living things.

Thanks to Kabbalistic teaching, we may use the Sefirot and Hebrew letters to help us approach the unseen world of Nature! Aryeh Kaplan commented on the first words of the *Sefer Yetzirah,* "the letters and digits are the basis of the most basic ingredients of creation, quality, and quantity. The qualities of any given thing can be described by words formed out of letters, while numbers can express all of their associated quantities. Numbers, however, cannot be defined until there exists some element of plurality."[1]

The Sefirot as Dimensions

The first element of the plurality is the "ten sefirot." The sefirot are also called *Middot,* which in Hebrew means "measurements," or "dimensions." Despite their specific arrangement, these dynamic entities create Kabbalistic networks sensitive to every change in the outside environment. The idea of the Sefirot as measurements originated with the *Sefer Yetzirah,* the oldest Kabbalah text. Chapter 1:5 reads:

"Ten Sefirot of Nothingness:

Their measure is ten

Which have no end

A depth of beginning

A depth of end

A depth of good

A depth of evil

A depth of above

A depth of below

A depth of east

A depth of west

A depth of north

A depth of south."[2]

These descriptions are hardly concrete enough to measure biological phenomena directly. However, they do indicate the Sefirot's properties as "orientation points" to navigate the hidden worlds.

---

[1] Arieh Kaplan, *Sefer Yetzirah.* 1997, page 5.

[2] Ibid., page 44.

Rabbi Judah Loew ben Bezalel, or Maharal, a 16th-century Kabbalist in Prague, proposed a system of meanings of the ten digits. Each of them represents distinct concepts related to unity, duality, balance, and the interplay between the natural and supernatural realms. This numerical symbolism offers a framework for us to understand the principles that govern existence and the relationships between different aspects of reality. According to Maharal, each number carries a distinct significance, as follows:

1. UNITY: Number One represents the concept of oneness or singularity, the ultimate Source from which everything else is created.

2. DUALISM AND MULTIPLICITY: Number Two symbolizes duality, reflecting the existence of opposing forces or principles in the universe, such as good and evil, light and dark, male and female, etc.

3. UNITY BETWEEN TWO EXTREMES: Number Three signifies the harmonious balance or unification of two opposing forces, creating a stable, dynamic equilibrium.

4. MULTIPLICITY IN TWO DIRECTIONS: Number Four embodies the concept of multiplicity in two directions, representing the four cardinal directions (north, south, east, and west) that define spatial orientation.

5. THE CENTER POINT: Number Five signifies the central point that unifies the four extremes or cardinal directions, symbolizing harmony and balance within the world.

6. MULTIPLICITY IN THREE DIMENSIONS: Number Six represents the idea of multiplicity in three

7. Dimensions (length, width, and height), reflecting

the complexity of the physical world.

8. THE CENTER POINT OF NATURE: Number Seven symbolizes the unifying principle of Nature, as exemplified by the concept of Shabbat. This represents rest and spiritual rejuvenation, bringing balance to the natural world.

9. THE SUPERNATURAL REALM: Number Eight signifies the supernatural realm, which transcends and nourishes Nature, and the human desire to connect with this higher, transcendent reality.

10. COMPLETE MULTIPLICITY: Number Nine embodies the most comprehensive form of multiplicity, encompassing the division between the natural and supernatural realms, reflecting the full range of existence.

11. FINAL UNIFICATION: Number Ten signifies the ultimate unification of the natural and supernatural realms, representing the culmination of all creation and the harmonious integration of all aspects of existence.[3]

In *The Thirteen Petalled Rose*, Adin Steinsaltz, one of the leading scholars and rabbis of our time, wrote: "The sefirot are not just attributes or even spiritual or intellectual categories. They are dimensions of reality, and to study Kabbalah is to come into contact with those dimensions, to explore them, and to penetrate their secrets."[4] This citation emphasizes the idea that the Sefirot are not simply abstract concepts.

According to Sanford Drob, a clinical psychologist at Fielding Graduate University, the ten Sefirot can be recognized as the Ten

---

[3] https://en.wikipedia.org/wiki/Significance_of_numbers_in_Judaism
[4] Adin Steinsaltz, *The Thirteen Petalled Rose*. 2006, page 23.

Dimensions through which all things are created and comprised.[5] Drob repeatedly warned that this "dimensional" interpretation is not exclusive or absolute. It does simplify our understanding of the Sefirot as "molecular" components. To help us grasp the sefirot as "interactive elements," Drob reduced them to "points." He used them as analogies to the sefirot because of their complexity as "spheres." Accordingly, Drob has provided the following descriptions of dimensions derived from each sefirah:[6]

- Keter – Will/Consciousness
- Hokhmah – Conception/ Idea/ Essence
- Binah – Existence/ Non-existence
- Hesed – Spirituality
- Gevurah – Ethics, Morality, and Values
- Tiferet – Beauty/ Aesthetics
- Netzah – Length
- Hod – Breadth
- Yesod – Depth
- Malkhut – Time

Drob summarized the following reasons for connecting the dimensions to Kabbalism:

"Unlike the four-dimensional scheme, which limits the universe to only those objects that have a spatiotemporal form, these ten dimensions can account for the entire range of human experience. Any thought, object, or experience, whether referring to the material or the conceptual, to the existent or the non-existent, to the sentient or the lifeless, to value or event, etc., can be described within this scheme. In addition, any conceivable objects can be exhaustively described by appealing to predicates derived from one or more of the

---

[5] Sanford Drob, *Symbols of the Kabbalah*. 2000, page 229.
[6] Ibid., page 230.

ten dimensions, in their positive or negative forms."[7]

The idea of using sefirot as a measurement is exciting! When I began learning about the Tree of Life as the system of sefirot around 30 years ago, I was intrigued by the opportunity to explore the Sefirotic system as dimensions. This enabled me to learn more about biology. In the years since I have begun to see the sefirot as "points of orientation" rather than separate "dimensions." As a result, I have learned how to use relational connections between the sefirot to conduct such measurements.

For dimensions or measurements to be taken into account, they require at least two points of reference - the point from which they start and the point at which they end. For example, let's look at the Polar Star (North Star or Polaris). Since the Polar Star always stays in the same location, it is a reliable determinant of which way is North. This knowledge can be valuable if you have lost your way at night. However, it cannot be used to determine how *far* north you have to go in pursuit of this direction--whether one mile, a hundred miles, or a thousand miles. Interestingly, the Polar Star requires at least two reference points: where something begins and where it ends. In this case, there is the beginning of a relationship between points. Thus, each aspect is defined by reference to something else, rather than in terms of its intrinsic qualities.

On the other hand, using sefirot to measure positions in a particular direction might be more challenging. Each sefirah is infinite. Although you can follow its direction, you can never reach it! Mathematical tools and techniques such as limits, infinite series, and calculus can explain Infinity. We are told that using these techniques can help support our predictions and calculate values for specific infinite quantities, even if they cannot be directly measured. However, I am not a mathematician; I am a biologist. Furthermore, I am not concerned with abstract theoretical quantities. It is more interesting to me to measure biological systems and phenomena in

---

[7] Ibid., page 231.

terms of their uniqueness, diversity, dynamics, and meaningfulness.

I agree with Drob that the sefirot, as dimensions, are "continuously interacting with one another (uniting, competing, blending, breaking apart, reforming) ... and it is such a dynamic that lends significance and "life" to the Sefirotic scheme."[8] The 16th-century Kabbalist Moses Cordovero believed that each thing obtains its specific character through the *relative admixture* and dominance of the Sefirot of which it is comprised.[9] In the next chapters, I will demonstrate how sefirot and their relationships can help us identify new dimensions to measure the hidden sides of biological processes.

Alef-Beit Letters as Building Blocks

The late Lubavitcher Rebbe, Menachem Mendel Schneerson, wrote that the letters of the Alef-Bet are "the building blocks of creation."[10] His definition reflects the ancient Kabbalistic tradition, which emphasizes the role of Hebrew letters to name creatures, objects, concepts, and so on. From this perspective, the letters serve as channels that enable forces from higher planes to imbue objects with their existence and vitality.

Readers may wonder how letters can be used to measure natural processes. For most people, a letter is a character representing one or more sounds used in speech or a symbol of an alphabet. For example, in English, we have the words, "dog" and "cat." Historically, people agreed on which was a dog and which was a cat. The origin of the English word "cat" is thought to be the late Latin word *cattus*. However, the word "dog" presents a mystery, linguists have still not identified its roots. Regardless of the origin of these words, the question is: Why is one of these animals named a "cat," whereas another is a "dog?" If, thousands of years ago, they had been called the opposite, we would today call a "cat" a "dog." Why not? All languages around the world were developed as tools for human

---

[8] Ibid., page 2345.

[9] Sanford Drob, *Kabbalistic Visions: C.G. Jung and Jewish Mysticism.* 2010, page 94.

[10] Quoted by Berke and Schneider, *Centers of Power.* 2008, page 152.

communication. Letters within each word of these languages were placed arbitrarily to reproduce the sound of the letter.

However, this practice does not appear in the Hebrew language. In contrast to English and other conventional languages, a combination of Hebrew words is *not* arbitrary! Almost all Hebrew words are built upon root letters called *shoresh*, meaning "root," and are formed in ways where minor manipulations can create many different yet related meanings. According to Kabbalah, the fact that words possess the same root letters, even if in different combinations, indicates that they are inherently linked.[11]

Samuel Avital explains, "this is why the root of the Hebrew letters are of great importance to relate, unite, shape, create, which can make worlds or destroy them. This depends upon the degree of intelligence, awakened awareness, and consciousness that was invested within a person – the process of being and becoming... from a cell to an organ, the shape and form, to the whole organism... We use letters to form the world ... passing through time, space, and matter, manifesting reality from the invisible."[12]

Usually, a root consists of three consonant letters. Sometimes there are four, but rarely are there two. Each letter has a profound meaning. Hebrew letters are not conventional. Rabbi Joseph ben Shalom Ashkenazi, the Spanish Kabbalist living in the early 14th century, thought that understanding the essence of each letter revealed many secrets. However, the process of learning the meanings of the Alef-Beit letters takes long, attentive practice. In Chapter 20, I will introduce some meanings of each of the twenty-two letters selected for our search.

Knowledge of the Tree of Life and Structuralism

An interesting parallel between the Kabbalistic hierarchical system and the general trends in scientific methodology could be represented by the concept of "structuralism." Alan Barnard explains that "Structuralism in its widest sense is all about pattern: how things

---

[11] Berke and Schneider, *Centers of Power.* 2008, page 155.
[12] Samuel Avital, *The Invisible Stairway.* 2005, page 270.

which at first glance appear to be unrelated actually form part of a system of interrelating parts." [13] Applications of structuralism are more prevalent in human sciences and biology. The most evident interest in structuralism was sparked in anthropology by the outstanding French scientist Claude Lévi-Strauss.

Structuralism in biology could be seen as the development of an old trend oriented to the search for natural "laws of forms." In his influential book, *On Growth and Form*, D'Arcy Wentworth Thompson revisited the idea of "universal laws of form" to explain the observed forms of living organisms.[14] Structuralists have proposed different mechanisms that might have guided the formation of body structure. Günter Wagner, Professor of Ecology and Evolutionary Biology at Yale University, argued for structural constraints on embryonic development.[15] Stuart Kauffman, professor of biochemistry at the University of Pennsylvania, favored the idea that a complex structure emerges holistically and spontaneously from the dynamic interaction of all parts of an organism.

Jerome Levi, a professor of anthropology at Carleton College, argues that Kabbalah and structuralism share several theoretical foundations.[16] These include:

1. Surface diversity conceals underlying unity.

2. Truth is hidden within a layered model of reality,

3. Linguistic and mathematical relationships constitute elementary structures that enable diverse, seemingly interconnected orders to be correlated with each other systematically.

According to Levi-Strauss, there are two levels of reality: a "surface structure" and a "deep structure." Surface structures take on diverse

[13] Alan Barnard, *History and Theory in Anthropology*. 2000.

[14] D'Arcy Wentworth Thompson, *On Growth and Form*. 1992, 1116 pages.

[15] Günter Wagner, *The biological homology concept*. Annual Review of Ecology, Evolution, and Systematics, 1989, vol.20, pages 51-69.

[16] Jerome Levi, *Structuralism and Kabbalah: Science of Mysticism or Mystifications of Science?* Anthropological Quarterly, 2009, vol. 83, no.4, pages 929-984.

forms of empirical reality. Levi-Strauss says: "Behind this diversity there lies something deeper, something common to all its aspects. The effort to reduce a multiplicity of expressions to one language, this is structuralism."[17]

Structuralists teach that the apparent diversity of things hides a deeper unity. The idea that truth is hidden within a layered model of reality is also central to Kabbalah. Moses Cordovero wrote in Tzfat in the 17th century; "The essence of divinity is found in every single thing – nothing but it exists."[18]

A Holon Is Both a Whole and a Part

There are some demonstrable links between the concept of "holon" and Kabbalah. The idea behind "holon" was originally formulated by Arthur Koestler, an influential intellectual of the 20th century. His book, *The Ghost in the Machine*,[19] introduced an entirely new approach to confronting the dichotomy between parts and wholes.

The term "holon" describes a system or phenomenon that is both an independent whole and a part of a larger system. The word "holon" is a combination of the Greek word *holos*, which means "whole," and the suffix "on," which denotes a particle or part, as used in words such as "electron" and "neutron." Thus, the holon is simultaneously a whole and a part.

The systems of holons are nested like Russian dolls. Since a holon is embedded in larger wholes, it is influenced by and influences these large, whole systems. Each holon also contains subsystems, or parts, within it. Therefore, it is also influenced by and similarly influences these other parts. As a result, energy and information flow both in and out between smaller and larger systems.

Ken Wilber adopted and proceeded to further develop Koestler's holon construct. According to Wilber's *Integral Theory*, a holon is the unit by which explanations, analyses, references, and

---

[17] Quoted from Jerome Levi, 2009, page 936.
[18] Quoted from Jerome Levi, 2009, page 937.
[19] Arthur Koestler, *The Ghost in the Machine*. 1967, 384 pages.

measurements are made. Wilber stated the importance of this concept thusly: "Reality as a whole is not composed of things or processes but of holons (wholes that are parts of other wholes)."[20] In his Integral Theory, the holon is the basic point of analysis, explanation, reference, and measurement.

In its simplest form, the holon can be viewed as a reference point in a hierarchy of relationships between entities that are *self-complete* wholes, as well as those that are *dependent parts* of a whole. As an investigator's focus moves down or up, the perception of what is whole and what is part will also change, given the whole range of the hierarchical structure. This hierarchy of holons is called a "holarchy." Koestler wrote: "Whatever the nature of a hierarchic organization, its constituent holons are defined by fixed rules and flexible strategies."[21]

Koestler pointed out that holarchy is an inbuilt feature of biological life and, thus, a fundamental aspect of development. "... the different levels represent different stages of development, and the holons ... reflect intermediary structures at these stages."[22]

On the other hand, a holarchy presents a different perspective. Each entity or 'holon' is a whole in a holarchy, with its unique integrity and autonomy. Rather than a top-down relationship of command and control, there's a nested relationship of inclusion and cooperation. The focus is on both horizontal and vertical relationships, emphasizing the interdependence and interconnectedness of the system's parts.

Holarchy in Kabbalah
The holon concept is deeply kabbalistic. Indeed, the Tree of Life can be seen as a hierarchical system. Each sefirah is an independent unit with unique properties and characteristics. However, it also belongs to a larger whole, including the relationships and interactions between the sefirot that are crucial to the functioning of the entire

---

[20] Ken Wilber, *Sex, Ecology, Spirituality: The Spirit of Evolution*. 2001, 880 pages.
[21] Arthur Koestler, *The Ghost in the Machine*. 1982, page 55.
[22] Ibid., page 61.

Tree of Life. Each Sefirah is both an independent entity and a link in an interconnected chain. This symbolizes the dual nature of entities in complex systems, like the holons in a holarchy.

Each of the Sefirot contains the entire structure of the Tree within it, including ten Sefirot in each "subtree." (Figure 33). For example, it could be a "Hesed of Gevurah," or "Hod of Hesed," and so on.[23] Robert Haralick, a Distinguished Professor in Computer Science at the City University of New York, likened the hierarchical interaction between upper and lower Sefirot to the phenomenon of "resonance." To "resonate" means to vibrate sympathetically. Haralick wrote: "Here we resonate, in turn, all the "Cheseds" [the spelling used by him for Hesed in plural] on the subtrees, then all the Gevurahs on the subtrees, then all the Hods on the subtrees, then all the Yesods on the subtrees, and all the Malchuts on the subtrees."[24]

---

[23] Raphael Afilalo, *Kabbalah Concepts*. 2005, page 39.

[24] Robert Haralick, *God Consciousness, The Exercises: Working the Sefirot and Netivot*. 2014, page 240.

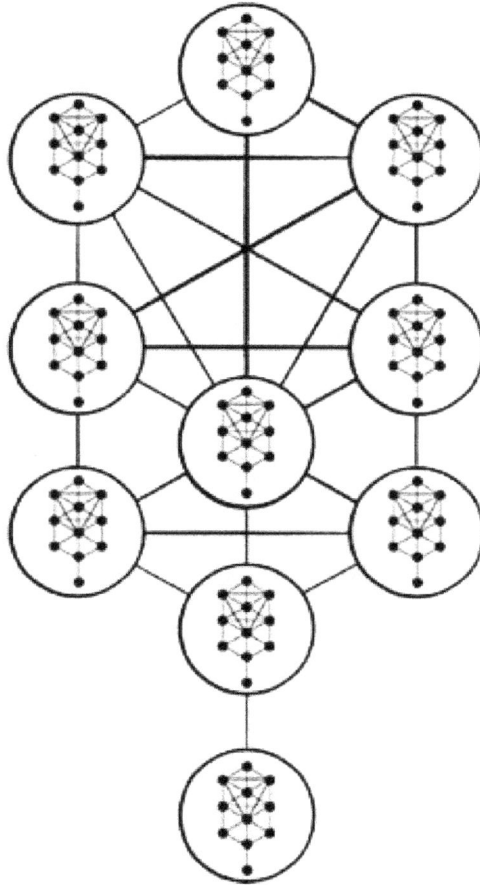

Figure 33. Each of the ten Sefirot reflects the entire structure of the Tree.

Another potent metaphor is *Ein Sof* ("no-end"), or the Infinite. In Lurianic Kabbalah (the teaching of Isaac ben Solomon Luria Ashkenazi), *Ein Sof* represents the original essence that permeates all existence and yet transcends it. This concept parallels the idea that the whole is *more* than the sum of its parts (the individual elements). Emergent properties arise from the complex interactions between the parts.

Kabbalah master Joel David Bakst proposed to use the holarchy as a powerful model to highlight the reality of four worlds (*olamot,* plural of *olam*) as they connected to dimensions. He proposed a model of three lower "cubes" representing the three lower olamot – *Briah, Yetzirah*, and *Asiyah*--nested one within the other. All of them are nested within the fourth cube representing the higher dimension, *Atzilut*.

Bakst commented on this metaphorical imagery: "From the moment of the higher-dimensional eating from the high-dimensional Tree of Knowledge, our entire reality has been turned "inside-out."[25] As a result of this "cosmic catastrophe," the three lower "cubes" (representing *Briah, Yetzirah*, and *Asiyah*) "have *prolapsed* and *inverted* one upon the other." As a student of Bakst, I know how highly he estimated the importance of holon and holarchy to grasp the essence of Kabbalah.

In Figure 34, I present the visual model of interactions between the four worlds (*olamot*) and Sefirot inside each world (*olam*). Imagine a nested telescope model, where each tube-like component slides inside another. In this model, each of the four tubes of the telescope represents a specific olam as a "whole" entity with ten sefirot. For instance, the smallest tube might represent *Asiyah*, the next *Yetzirah*, the third *Briah*, and the fourth, the largest, *Atzilut*.

Each world (*olam*) exists as a separate entity when the "telescope-like" model is extended (the condition of "the Fall"). However, when these components are nested within each other and aligned correctly, they form a larger "whole" – the Tree of Life (*Etz Haim* in Hebrew).

A similar visual model was used by Hyman Schipper of McGill University to depict a world where the lowest aspect is enclosed by its superior counterpart immediately underneath. For this reason, Schipper visualizes an extended telescope pointing downwards. He

---

[25] Joel David Bakst, *Beyond Kabbalah*. 2012, page 288.

refers to the Kabbalistic principle of *hitlabshut* ("enclothment").[26]

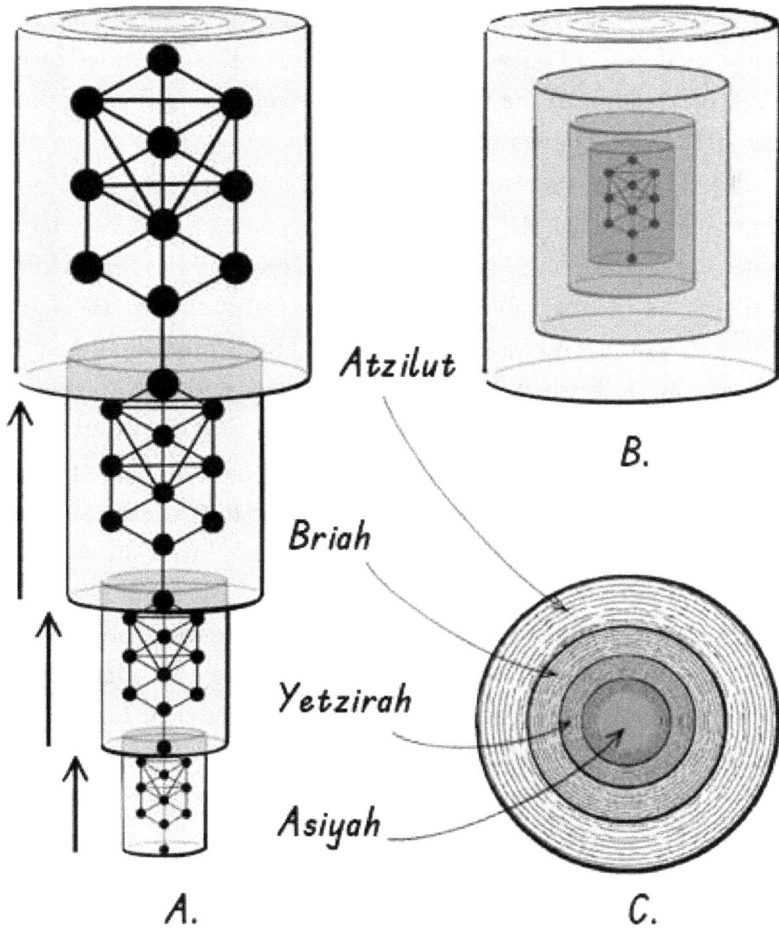

Figure 34. A tubular holarchical model of the Tree of Life: A. The Tree of Life expanded to show each sefirah within the Four Worlds (Atzilut, Briah, Yetzirah, and Asiyah), an arrangement nicknamed "Jacob's Ladder"; B. The nested Four Worlds model of the Tree of Life with the Four Worlds embedded in each other; C. Ten sefirot in each world arranged in concentric circles.

---

[26] Hyman Schipper, *Kabbalistic Panpsychism: The Enigma of Consciousness in Jewish Mystical Thought.* 2021.

A screen (divider) separates one World from another, and from this screen, the ten Sefirot of the Lower World emerge from the ten Sefirot of the Higher World.[27] However, there is no interruption of the energy flow from both directions. All the worlds are joined by the sefirah of Malkhut of the Higher World, overlapping the Sefirah of Keter of the world beneath it.

This is where the concept of holarchy comes into play: each component (the "whole" at one level) becomes a part of a larger whole at the next level (the complete Tree of Life). Each olam maintains its identity and function, yet contributes to the larger whole's function. The assembled Tree of Life is more than just the sum of the individual olamot. It's a new entity with unique functionality, like a telescope, that allows us to view distant objects.

Thus, the Tree of Life nicely illustrates the interconnectedness of different elements within a complex system. It is a diagrammatic representation of the sefirot, structured to show their interdependencies and relationships. This concept resembles biological networks, where nodes (representing genes, proteins, cells, etc.) are connected by edges representing interactions or relationships. We can use the metaphorical model of holarchy in relationships between olamot and sefirot to assist us in describing and interpreting connectivity and interdependency in complex biological systems.

Holarchy in Biology

The concept of holon has become a useful thought construct for biological investigations. By recognizing that each level of biological organization is both a whole and a part of a larger whole, we can use the holon concept to clarify the relationships between biological objects, from individual cells to entire ecosystems. For example, at the cellular level, organelles within a cell can be regarded as holons, with each organelle being both a whole in itself and a part of the larger whole of the cell. Similarly, an individual organism can also be

---

[27] Raphael Afilalo, *Kabbala Concepts*. 2005, page 83.

understood as a holon, with each organ system both a whole in itself and a part of the larger whole of the organism.

Thus, the concept of holarchy is gaining influence in biology, offering a nuanced understanding of complex biological systems. However, to fully appreciate its potential, one must differentiate it from the more traditional concept of hierarchy. In a traditional hierarchical system prevalent in many aspects of biology, each element in the system is subordinate to the element above it and superior to the one below it. It's a top-down model where higher levels control or command the lower levels. The focus is on vertical relationships and power dynamics between levels. An example of a biological hierarchy is the organization of the human body, which is divided into cells, tissues, organs, and organ systems. Each level has a specific function and is subordinate to the level above it.

Each biological system is characterized by "feedback loops." A feedback loop occurs where the output of a system either amplifies the system (positive feedback) or inhibits the system (negative feedback). Thus, holarchy, "a holonic hierarchy," presents a more flexible and adaptive system. In it, each level can respond to environmental changes and communicate with other levels. In contrast, a hierarchy is more rigid.

An example of a biological holarchy is the organization of an ecosystem. This would consist of individual organisms, populations, communities, and ecosystems. Each level would be both a whole in itself as well as being part of a larger whole (Figure 35).

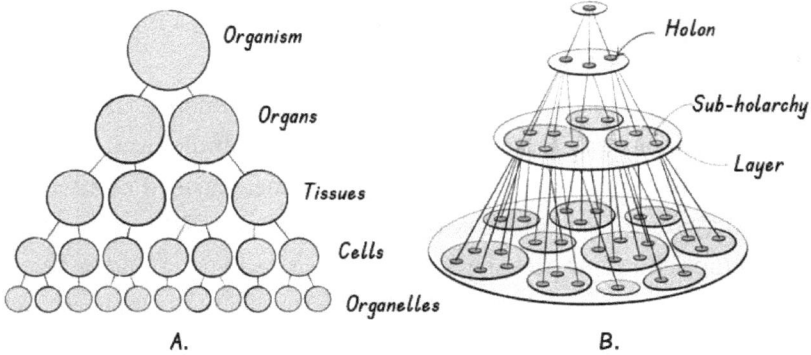

Figure 35. *There is a slight but significant difference between a traditional hierarchical system (A) and a holonic hierarchy (B). The holarchy presents a more flexible and adaptive system. Each level can respond to environmental changes and communicate with other levels, while a hierarchy is more rigid.*

The concept of holon can be linked to the concept of "semiotic agency" developed in the field of biosemiotics. According to Alexei Sharov, a research biologist at the National Institute on Aging, agents can be defined as "living systems capable of performing functions for the purpose of reaching certain goals."[28] These functions are encoded and controlled by a set of signs.

Agents often have a hierarchical structure and contain subagents in a nested structure termed "boundaries within boundaries.[29]

Given their unique symbiotic nature and complex organization, lichens are an excellent model to help us understand the concept of a holon. As a system, it is simultaneously a whole in itself and a part of a larger system. At the most basic level, a lichen is a partnership between a fungus (mycobiont) and a photosynthetic partner, typically an alga or cyanobacterium (photobiont). As individual entities, these two organisms (fungus and algae) are complete and functional. They are, in themselves, holons. However, when these organisms form a symbiotic relationship, they create a "lichen." This

---

[28] Alexei Sharov, *Functional Information: Towards Synthesis of Biosemiotics and Cybernetics.* Entropy, 2010, vol. 12, no. 5, pages 1050-1070.

[29] Jesper Hoffmeyer, *Biosemiotics: An Examination into the Signs of Life and the Life of Signs.* 2008, 300 pages.

is more than just the sum of its parts and exhibits new capabilities. The lichen, as a composite organism, is also a holon.

Within the lichen holon, the mycobiont provides structure and protection, absorbing water and minerals from the environment. Meanwhile, the photobiont captures light energy to produce nutrients via photosynthesis. This interaction results in a highly resilient organism that can survive in harsh environments that would otherwise be impossible for the fungus and the alga separately. Moreover, lichens can be a part of larger holons, such as a forest ecosystem. There are many trophic levels in the food chain, and lichens play an important role in nutrient cycling. They also serve as producers of other organisms that feed on lichens, such as reindeer, gastropods, nematodes, mites, and springtails.

In this way, they are integral components of larger ecosystems, contributing to the overall health and functionality of their environments. Once we can examine the structure and functions of lichens, it will be clear that individual organisms can function as both independent entities and interdependent components of larger, more complex systems.

Individual Development and Evolution are Hidden in an Organism

Once we examine the holarchical nature of the Tree of Life, we can more easily understand any biological system that is, simultaneously, "a whole" and "a part." It is both self-contained and contained within other systems. This concept can be instrumental in deepening our understanding of the hidden dimensions of biological life. Just as in the Tree of Life, where each Sefirah is a distinct entity yet also part of the overall structure, each cell and organ in a biological organism represents a unique entity that is also part of a larger whole. Seeing biology as a holarchy allows us to appreciate the intricate balance and interdependency that characterize life at all levels - from the cell to the organism to the ecosystem.

The intellectual resources provided by contemporary science are often insufficient to fully grasp the hidden past stages embedded within an existing form of life. In the previous chapters, we discussed different levels of "hiddenness" in the biological system. *The most challenging level is the hidden knowledge of what is hidden!* To illustrate the perspectives Kabbalistic Biology can offer to grasp hidden knowledge, I will take on this challenging task!

Consider a living organism, such as an animal. It comprises numerous cells, tissues, and organs, each of which is a distinct entity with specific functions and resource needs. Cells within the same organism may even compete for the same energetic resources. As each organ fulfills its specialized role, it requires resources to maintain its function. Yet, despite these inherent dynamics of competition and specialization, the various parts coexist in a harmonious balance, collectively contributing to the overall survival and well-being of the organism. This cooperative behavior of cells and organs allows them to act as part of a biological individual, or a "hidden obvious" entity. This phrase, coined by Samuel Avital, refers to a concept that is both incredibly fundamental and yet often overlooked or taken for granted.

It is easy to forget that, amidst the complexity of individual parts, these components of a single organism are working together to support the life of the whole. By adopting a holarchical perspective, we can expand our understanding of life. This approach can reveal the subtle complexities and unappreciated wonders that lie beneath the surface of the biological world. This allows us to not only appreciate the unity and harmony inherent in the complexity of life, it also underscores our need to respect and protect the delicate balances that sustain life in all its myriad forms.

Each organism, from the smallest bacterium to the most complex primate, carries within it a record of its personal development--from fertilization through successive embryonic stages, all the way to the adult form ("ontogeny"). This ontogenetic journey is essentially a repository of the organism's life history, a concept that might seem mind-boggling but is undeniably essential. Without it, we could not explain how traits are inherited from parent organisms or how the specific developmental pathways, as discussed in Chapter 8, are genetically imprinted.

Remarkably, every organism also encapsulates the evolutionary history of its species, tracing back to its most distant ancestors ("phylogeny"). This concept, though difficult to fully grasp, is a cornerstone of our ability to understand life. It explains how each individual organism can repeat or embody characteristics typical of its species or even higher taxonomic categories such as genus, family, order, or class. This notion of "phylogeny" -- the evolutionary history and relationships among species - is a bedrock principle in biology. It states that different species of plants and animals share common ancestors.

However, this ancestral lineage often remains partially obscured. This is because most species throughout history have become extinct, and the fossils preserving their existence are relatively rare. Yet, despite these gaps in our knowledge, the principle of phylogeny provides a powerful framework to understand the interconnected web of life that spans billions of years and countless generations. Each organism is not just a single life form, it's a living testament to the intertwined narratives of ontogeny and phylogeny, individual development and evolutionary history.

Ernst Haeckel, sometimes referred to as the "German Darwin," published his seminal scientific work, *General Morphology of Organisms*, in 1866. In this comprehensive study, he introduced his "biogenetic law," also widely known as the "theory of recapitulation." Haeckel's theory suggests that the embryonic stages of more advanced species reflect characteristics of their evolutionary ancestors.[30] For instance, Haeckel argued that human embryos might display traits akin to the gills of fish or tails of monkeys, signifying our shared evolutionary history. While this theory of recapitulation had a profound influence on biological thought, it was not universally accepted. In fact, critics pointed out that Haeckel's perspective

---

[30] Georgy Levit at al., *The Biogenetic Law and the Gastraea Theory: From Ernst Haeckel's Discoveries to Contemporary Views*. The Journal of Experimental Zoology B, 2021, vol. 338, page 13.

seemed simplistic and ignored the complexity and diversity of both ontogeny and phylogeny.

A more nuanced understanding suggests that embryos of advanced species might not closely resemble the adults of their ancestral species, but instead reflect the embryonic stages of these ancestors. This notion acknowledges the intricacy of both individual development and evolutionary heritage, providing a more comprehensive view of the interplay between ontogeny and phylogeny.

Building upon Haeckel's theory of recapitulation, it's essential to underscore that *all living organisms carry within them a resonance of their unique development and evolutionary lineage.* In other words, each organism, at various stages in its life, embodies traces of its embryonic and evolutionary past. Haeckel's assertion that humans contain a "fish" within might be simplistically construed, but the essence of his argument is valid. At certain points in our development, humans do exhibit characteristics of other life forms in our evolutionary tree, such as monkeys, reptiles, and, indeed, fish. A developmental stage in a human embryo that exhibits "fish-like" traits is just as significant as a later stage when the embryo resembles a baby or, ultimately, an adult human.

To better understand this challenge, let's examine it through the lens of the previously proposed "telescopic model." This clarifies the holarchical relationship between olamot and sefirot as per Kabbalistic thought. A human embryo, before its birth, passes through various stages. These include a "monkey-like" stage, a "fish-like" stage, and even a stage that resembles a colony of unicellular parasitic microorganisms, such as protozoa. These are not actually monkeys, fish, or parasitic protozoa: they simply present *images* as a monkey, fish, or parasitic protozoon. These "images" (*tselem* in Hebrew) resemble such organisms. The word *tselem* derives from the words "to carve" or "to cut."

When *tselem* is coupled with the word *demut,* these words together describe the same capacity. The Hebrew word *demut* means a resemblance or to be like something else in action or appearance. Derived from the parent root "*dalet-mem*," meaning blood, the child root *damah* means "to resemble" one descended from the "blood" of another often resembles the one descended from.

The early embryonic stages vary in their level of complexity. However, all of them represent biological systems whose forms can be seen with specific laboratory techniques. In contrast to these techniques, the holarchical model represents these images simultaneously (Figure 36-A). In a formed human embryo, we are unable to "see" these images; they are "hidden" in the later image of the human embryo (Figure 36-B).

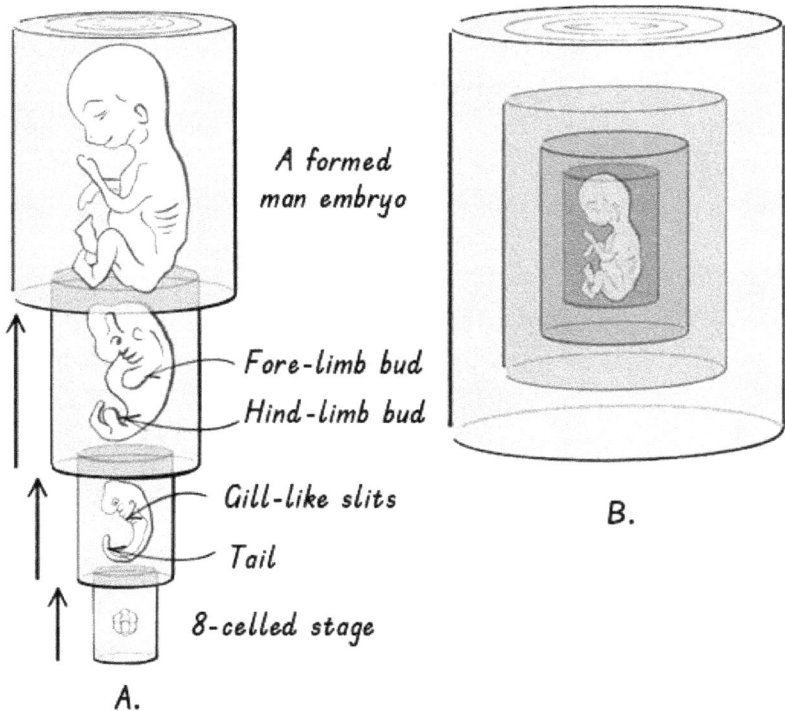

A formed man embryo

Fore-limb bud
Hind-limb bud

Gill-like slits

Tail

8-celled stage

A.

B.

Figure 36. In the holarchical model, the images of "primitive mammal-like," "fish-like," and "unicellular microorganism-like" stages of a human embryo's development are represented simultaneously (A). These images are "hidden" in the later image of a formed human embryo (B).

This concept echoes the insight offered by the American scholar R.G.H. Siu in his influential book, *The Tao of Science*. Siu posits that, "it is the reservoir of the evolutionary vestiges [a trace of something that is disappearing or no longer exists] shared by all, as they differentiated into their respective genera and forms. There is something of primordial man in the fish; there is something of primordial fish in man."[31] Later in the same book, Siu wrote, "…an embryonic leaf of the petunia is just as complete with its potentialities, apparatuses, and diverse structures as the oldest leaf of the giant sequoia."[32]

Each phase of development is not only a stepping stone to the next, but a complete expression of life's potential in its own right. Components at each level appear and disappear, but the continuity of main properties is preserved. Clifford Grobstein, the Dean of the School of Medicine at the University of California, San Diego, compared this capability of biological continuity, despite continual turnover, to that of a candle flame or a waterfall.[33]

Klipot as Agents of Separation

Kabbalah uses powerful metaphors to illustrate the connectivity and interdependency between the interconnected elements of complex dynamic systems. Each element contains a whole system. Kabbalah also provides a deep understanding of the origin and nature of the separateness of existing entities, including individual biological objects.

One of the crucial metaphors introduced by Lurianic Kabbalah depicts the "universe catastrophe" called *Shvirat HaKelim*, or "breaking of the vessels." The seven lower Sefirot recipients were unable to contain the original Light of Creation and were shattered into numerous sparks of the original Light.[34] Some of these sparks

---

[31] R.G.H. Siu, *The Tao of Science*. 1964, page 76.
[32] Ibid., page 88.
[33] Clifford Grobstein, *The Strategy of Life*. 1974, page 80.
[34] Raphael Afilalo, *Kabbala Concepts*. 2005, page 57.

reunited with their higher lights and helped the recipients rise and reunite with their own lights, while others fell into the lower worlds.

This vision is connected to the concept of *klipot* (*klipah*, singular; sometimes spelled *qlipah* or *kelipa*). In Hebrew, this word means "husks," "shells," or "peel," as in the peel of a fruit. The *klipah* conceals within it a spark. It is the fundamental metaphor of the "shell" that separates one living entity from another. From a traditional perspective, the word *klipah* describes barriers that obstruct the vitalizing force. This can create a negative connotation. However, Kabbalah delineates two distinct types of *klipot*: *Klipah Nogah,* which can be illuminated, and *Shalosh Klipot Hatmayot,* or "three totally impure *klipot.*" *Klipah Nogah* can be uplifted and refined. However, the only form of reformation or redemption for the three impure *klipot* is their destruction. Joel Bakst noted that "the klipah/shell always precedes the fruit."[35] Bakst explains that any form of life is always covered by a klipah that must be removed. The external husk precedes a nut, and the embryonic sac precedes a newborn baby.

The concept of *klipot* in Kabbalah is a powerful metaphor to help us understand the dual role of biological boundaries, such as outer layers. The function of *klipah nogah* in animals can be seen as a barrier necessary for the existence of independent organisms. The morphological features, such as skin in mammals, scales in reptiles, chitin exoskeleton in insects, and shells in crustaceans, serve a crucial role in the biology of animals. While they indeed restrict the direct exposure of an organism's internal environment to external conditions, this limiting factor is a necessary trade-off for the essential protective functions they provide.

In fact, protection is one of the primary biological roles these features perform. They shield the organism's internal tissues and organs from physical damage, invasive pathogens, and harmful environmental conditions such as extreme temperatures or radiation.

---

[35] Joel Bakst, *Beyond Kabbalah.* 2012, page 250.

However, their function extends beyond mere protection. These outer layers also regulate the exchange of matter, energy, and information between the organism and its environment. They selectively allow specific substances to pass through while blocking others, maintaining a delicate balance essential for the organism's survival. For instance, mammalian skin prevents excessive water loss while allowing for the excretion of waste products through sweat. Reptilian scales help retain moisture in dry environments. The chitin exoskeleton of insects provides protective armor and facilitates the insect's growth through a process known as "molting." Crustacean shells offer protection, and also play a role in locomotion and sensing the environment. In some species, the shells assist in processing food.

The skin is the largest and most diversified organ of the human body. If the skin of a grown person is stretched on the ground, it would cover up to two square meters. Jesper Hoffmeyer, a Professor of Molecular Biology at the University of Copenhagen, noticed, "The skin keeps the world away in a physical sense but present in a psychological sense." Skin gives us the experience of belonging, as it allows us to feel the world. Receptors (sensory cells) in the skin register touch pressure, pain, cold, warmth, pH, and various chemical influences.[36]

Thus, these barriers serve as interfaces that receive sensory information from the environment, acting as the organism's receptors for the world. They help the organism adapt to changes in the environment and respond appropriately, underlining their importance in the survival and evolution of the species. Therefore, while such biological *klipah nogah* impose some constraints, they are vital for the organism's survival, growth, and interaction with its environment.

When we apply this concept of *klipah nogah* as a protective barrier to the cellular level, the cell membrane comes to the forefront as an exquisite biological example. Analogous to the skin, scales,

---

[36] Jesper Hoffmeyer, *Biosemiotics: Signs of Life and Life of Signs.* 2009, page 19.

exoskeletons, and shells in larger organisms, the cell membrane delineates the boundary between the cell's internal environment and the external world. The cell membrane provides a protective layer that shields the cell's contents from the external environment. This barrier function is essential to keep the cell's complex internal machinery intact and prevent potentially harmful substances or organisms from entering the cell.

Much like *klipah nogah,* the cell membrane's role is not merely to isolate but also to facilitate interaction with the environment. It accomplishes this through "selective permeability," a feature allowing certain substances to pass across while blocking others. This process is vital to maintain the cell's internal balance or homeostasis. It ensures that the necessary nutrients and other molecules can enter the cell, while waste products and potentially harmful substances can be removed. Moreover, the cell membrane is not a passive barrier, but an active participant in the cell's interactions. It is studded with many proteins that serve as gatekeepers, communicators, and sensors. These membrane proteins can transport specific molecules across the membrane, receive signals from other cells or the environment, and even initiate a cell's response to those signals. In this way, the cell membrane plays a central role in cellular communication and signaling.

At their most fundamental level, *klipot* functions as "defining boundaries." These boundaries act as a demarcation line that distinguishes what belongs within a system and what lies outside of it. But these boundaries don't merely act as barriers; they also act as interfaces; the contact points where different systems converge. This is the paradox of *klipah* --a barrier that not only limits but also enables interaction. This concept is as crucial to the survival of cells as it is profound in the mystical tradition of Kabbalah.

These points of convergence, or "intersections," facilitate negotiation and mutual coordination. They allow for the exchange of energy, information, and resources between systems. In a biological context, the cell membrane is a boundary that delineates the cell from its environment. Still, it is also permeable, enabling the exchange of nutrients and waste. *In this context, boundaries can both separate and connect systems.* They separate by maintaining each system's integrity and identity, ensuring its unique structure and function remain intact. However, they also connect by allowing interaction and exchange, facilitating the coexistence and co-evolution of interconnected systems.

Thus, *klipot* don't merely restrict or confine; they also enable dynamic interaction and exchange, making them integral to the existence and functioning of complex systems. Understanding *klipot* isn't just about appreciating the limits they impose; equally important is the vital role they play in interacting and cooperating within and between systems.

System Biology as a Tikkun

The inception of systematic thinking was spurred by recognizing divergent components acting simultaneously in independent, yet cooperative structural configurations. Austrian biologist Ludwig van Bertalanffy conceived of visualizing biological phenomena as "interconnected systems" in the mid-20th century. He proposed General System Theory to explore the inherent wholeness of complex and dynamic systems. Fritjof Capra equated systemic with holistic, highlighting the fundamental tension between the "parts" and the "whole."[37]

Systems Biology, as defined by the National Institutes of Health, emphasizes the importance of understanding the broader picture, whether at the level of an organism, tissue, or cell by integrating its parts. Such an approach is in sharp contrast with the reductionist approach to biology. This attitude dominated prior decades, emphasizing the dissection and analysis of individual components.

---

[37] Fritjof Capra, 1996, *The Web of Life*. 1996, page 17.

Various approaches exist to formulate a concept of systems. The general theme is that a system is comprised of:

- Elements, which are all the components that make up the whole.

- Interconnections, which are the processes and relationships that bind the parts together in the context of the whole.

- A boundary, which delineates the limit determining what is within and outside a system.[38]

Systemic thinking acknowledges the intricate connections and influences between the components of a system, rather than viewing them as isolated entities. This perspective encourages the understanding of interconnectedness and interdependencies within a system, rather than viewing the components as detached units.

Thus, the crux of systemic thinking is not about dismantling boundaries between objects; instead, it focuses on regulating the permeability or transparency of these boundaries. In other words, it involves adjusting the level of interaction or exchange between the components. This regulates their mutual influence and visibility within the system. Systems Biology aims to comprehend the function and behavior of the entire system, rather than isolated parts of an organism or a biological process.

Systemic thinking, at its core, recognizes that objects or components in a system are not isolated. Indeed, they are intricately connected to and influenced by each other! This perspective helps us appreciate the interconnectedness and interdependencies within a system, rather than viewing the components as separate entities. It is less about breaking boundaries between objects and more about adjusting the permeability or transparency of these boundaries! That means we must adjust the level of interaction or exchange between the objects, regulating the degree to which they influence each other

---

[38] Bob Williams and Richard Hummel, "*Systems Concepts in Action.*" Stanford University Press, Stanford, 2011, page 16.

and, consequently, their visibility to us in the system. Systems Biology aims to understand the functioning and behavior of the entire system, rather than isolated parts of an organism or a biological process.

"Network theory," a subset of systems theory, is a method used to examine the relationships between distinct components within a system. It specifically focuses on mapping and studying the connections, or "edges," between entities, or "nodes," within a network. The "nodes" could represent anything from individual organisms to cells, and the "edges" represent the relationships or interactions between them.

Holarchy is similar to network theory. In Chapter 8, we learned about biological networks as an approach to exploring biology's invisible world. As we have discussed, a holarchy emphasizes the nested and interconnected nature of systems. It recognizes that each entity possesses its own autonomy while contributing to the functions of larger systems. Network theory underscores the relationships and interactions between different elements within a system, highlighting the importance of connectivity and interdependency.

Both holarchy and network theories detach from traditional reductionist approaches that view systems as simply the sum of their parts. Instead, they propose that the relationships between the parts are equally crucial, if not more so. This assessment would determine the system's properties and behavior. Although both theories present a valuable analogy, they are still very different. While holarchy emphasizes the vertical relationships and nested structures within a system, network theory focuses more on horizontal relationships and connectivity patterns. Combining these perspectives can offer an even deeper understanding of complex biological systems.

One of the words introduced by Lurianic Kabbalah became quite popular during the last few decades. This word is *tikkun*. It means "a reparation of the world to raise the sparks" created during the "breaking of the vessels," and unite them to their original status.

There is a certain resemblance between the idea of systems biology and that of *tikkun* from Kabbalah. *Tikkun* reflects the restoration of a whole system by reuniting fragmented parts, sometimes referred to as "broken sparks." As a matter of fact, these "broken sparks" can also be interpreted as individual biological components or processes that were studied individually in the context of biology.

The progression of systems biology can thus be seen as a form of *tikkun*. By striving to bring together these "broken sparks" as the diverse individual components of a biological system. Systems biology aims to restore our understanding of the whole organism or process. It offers a more holistic view that appreciates the interplay between various biological components and the emergent properties arising from their interaction.

Kabbalistic Thinking

The most significant impact Kabbalah can provide to help us grasp the invisible aspects of biological life is to change our way of thinking. Understanding the structure of the Tree of Life as the system of "olamot," "sefirot," "partzufim," and the "paths-letters," may appear complicated at first. However, the Tree of Life structure is as simple as a crystal. It simply requires some basic education about kabbalistic resources and how to practice using these as tools of discovery.

Thus, there is a "complexity-simplicity" ("simplexity") gradient on the journey to the Tree of Life. The critical point is to form mental constructs within the framework of these concepts. The opposite situation can be experienced while exploring the Tree of Knowledge. In biology, we study "wood," "forests," and the world characterized by diverse organisms, the variability of living forms, and the constant change in living processes. In that arena, scientists collect massive amounts of data and produce sophisticated information based on it. The challenge arises when we attempt to "know" biological phenomena. To know them, we must first develop mental concepts that reflect multi-dimensional and interconnected biological systems.

Metaphorically speaking, we can bring the Tree of Knowledge closer to the Tree of Life by simplifying biological concepts. First, we must admit that we can never overlap the Tree of Knowledge and the Tree of Life. However, we can bring them close enough to allow a resonance between them!

In Chapter 2, we argued that thought experiments could go beyond 3D vision. Alas, we cannot see the invisible sides of biological life by relying exclusively on data. In Chapter 3, we discussed the limitations of linear thinking to comprehend complex biological systems.

Linear thinking restricts our ability to understand the "cause" of a biological process behind any "effect." For example, an epidemic happens because of a specific virus or bacterium, or evolution happens because of natural selection. The "cause-effect" mentality is deeply embedded in Western thinking. We saw in Chapter 4 that formal Greek logic is limited if one is trying to address the extremely high diversity and unpredictable dynamics of biological life.

In summary, linear thinking and formal logic can be adequate for collecting biological data and processing information. However, they pose limitations when it comes to recognizing "knowledge" about biological life. Accepting Kabbalistic thinking is vital prior to exploring the invisible domains of life. Before referring to specific dimensions, however, I wish to emphasize a few major points about learning Kabbalistic biology.

First, whenever we explore a biological phenomenon, we should practice freely shifting our perspectives between olamot and sefirot. We must connect them in a pathway symbolized by Hebrew letters to continue our investigation.

Second, in order to accept the coexistence of multiple points of view, we need to understand the system of orientation within the invisible worlds.

Third, knowledge of all realms higher than a fully manifested one (*Assiah*) does not progress linearly, but rather by forming a zigzag-like connection between the right, left and middle pillars. The terms for these three different realities in Kabbalah are *akudim* ('bonded'), *nekudim* ('pointed'), and *berudim* ('connected').[39]

Joel Bakst stated that it is not only WHAT you are looking at, but also from WHERE you are looking and WHO is looking.[40] Quantum physics challenges one-point-of-view, determinism. Accepting multiple perspectives on the same natural phenomenon can reveal its previously hidden aspects. However, a potential side-effect of multiple perspectives is the possibility of losing one's direction. If we follow such a way of thinking, no direction is better or worse than another. Each will lead to another point with a different outcome. As in Jorge Luis Borges' *The Garden of Forking Paths*, we cannot know where each path leads.

In contrast, the Kabbalistic system of thinking requires constant orientation, which requires developing and practicing strict navigational tools. Physical dimensions do not guide reality's invisible worlds, and the linear relationships between visible natural phenomena do not cover the invisible landscape that dominates biological life. The task is to practice using one's navigational system to travel in multi-level and multidirectional worlds. It means we must know three points: (1) the point from which we come, (2) the point at which we are in the current situation, and (3) the point that we intend to achieve.

While engaging with biological phenomena, it's essential to acknowledge the multiplicity of perspectives that can co-exist and to recognize the underlying unity inherent in each biological entity. Consider, for example, the study of a rabbit. There are numerous ways to investigate it. One could focus on a cell, tissue, organ, or organism, or even across broader contexts such as population dynamics, community ecology, or ecosystem interactions. Yet, amidst this diversity of perspectives, it's crucial to remember that these different viewpoints are focused on investigating the same

---

[39] Joel Bakst, *Beyond Kabbalah*. 2012, page 165.
[40] Ibid., page 281.

entity – the rabbit. It would hardly be constructive to argue which approach is better!

The phrase "don't tell rabbit and rabbit, if this is the same rabbit" reflects Rabbi Akiva's quote, "don't tell water, water" as a reminder of this unity amid diversity. While we might study different facets of the rabbit or interpret it through different lenses, we must never lose sight of the integral rabbit entity that underlies all these explorations. The rabbit remains a coherent, unified organism, regardless of how we choose to dissect its complexities.

This encapsulates a fundamental tenet of biological study and systemic thinking: while we can investigate various aspects or levels of a biological entity, we must always remember the integral wholeness that underpins it. The seemingly fragmented perspectives we take are merely facets of the same, singular entity. Reconciling the experience of observable multiplicity in biological phenomena and acknowledging an underlying unity calls for a new, cognitive approach. The Kabbalistic mode of thinking encourages us to embrace the inherent tension between the diverse manifestations of life and the unifying essence within them.

Kabbalistic thinking recognizes the coexistence of seemingly contradictory elements and teaches us to navigate this paradox without being forced to solve or eradicate the conflict. Instead, we are encouraged to delve deeper into the issues, expanding our understanding and broadening our perspectives. As noted by the Kabbalah master, Joel Bakst, the goal is not to find a solution, but to explore the depths within, unravel the layers of complexity, and cultivate a deeper understanding of the interconnectedness of life.[41] Kabbalistic thinking offers us new transformational potential when exploring the biological sciences.

Bakst proposed a simple visual model to demonstrate such a paradox, the Möbius strip introduced by German mathematician Augustus Möbius in 1858. The Möbius strip is a one-sided surface

---

[41] Ibid., page 290.

that can be constructed by affixing the ends of a rectangular strip after giving one of the ends a half twist. This image exhibits exciting properties, such as having only one side and remaining in one piece when split down the middle. The Möbius strip exists in three dimensions but has only one surface. If a line is continued on the surface of the strip without letting the pen leave the surface, you will find that when you are halfway around, you are writing on the back of the paper, even though you are still on the same surface (Figure 37).

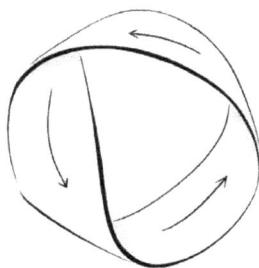

Figure 37. The Möbius strip exists in three dimensions, yet has only one surface.

In this regard, the Möbius strip offers a deceptively simple, yet very clear example of one of the most fundamental shapes that determines the world's hidden landscapes. From this perspective, the Sefirot are surfaces of one continuous side. In Bakst's words, "This paradoxical One Side is the more encompassing super unity of what is known in the tradition as the Ain Sof – the "NO END-ing" source of all sources that continually transcends conceptualization as being "this side" or "that side."[42] By using this model, we can remove the limits of being either a side of unity or a side of multiplicity. There now exists a third alternative that includes two aspects of complete unity, the "dual-unity" of the *Ain Sof* ("no end").

Elliot Wolfson, a professor at the University of California, Santa Barbara, argues that "kabbalistic esoterisms is grounded in the paradoxical interplay between concealment and unconcealment such that the mystery can be revealed only to the extend that the mystery

---

[42] Ibid., page 291.

is hidden … if the truth that is veiled is not unveiled as the truth that continues to be veiled, then the truth that has been unveiled is not the truth unveiled as the truth that was veiled."[43]

The remaining ten chapters of this book will introduce new dimensions that are focused on investigating biology's invisible worlds in depth. If you have reached this point by attentively following previous chapters, you will find yourself equipped to learn with the tools presented to you. Good luck on your journey through the invisible world! I will be taking it with you.

---

[43] Elliot Wolfson, *Phenomenology, Theosophic Topography and the Structures of Being: Unveiling the Seventh of Scholem's Ten Historical Aphorisms on the Kabbalah,* Journal for the Study of Jewish Mystical Texts, 2022, vol. 55, pages 7-71.

PART FOUR:

Dimensions of
Kabbalistic Biology

Chapter 13

First Dimension: The Unification-Fragmentation Axis

The process of exploring new trends in science to help us measure the *unseen* aspects of biological life can be both exciting and satisfying. In Part 1, we explored new trends in science that can open a multi-dimensional investigation of natural phenomena.

However, in Part 2, we began to recognize that the more we learn about the world of biology, the more uncertain we may be regarding how much *more* knowledge remains hidden. While researchers must accept the hierarchical structure of biological knowledge, they are often uncertain about how to find their footing in the more challenging landscapes of biological life.

In this chapter, we reveal the potentially game-changing results when an unexpected source helps us achieve this goal. Kabbalistic experts have, over the centuries, succeeded in developing practical tools to help investigators discover and enter the hierarchy of invisible worlds. We must take full advantage of this source of wisdom (both ancient and modern) when it is available to us.

In fact, the central theme of this chapter is to encourage readers to keep an open mind! This will help us avoid being tempted by contradictory, often opposing, perspectives of biological phenomena. Instead, science will be better served if we regard any paradoxes we encounter as "opportunities" to employ new options. Paradoxes are often advantageous to research; they may even represent a wide selection of unexpected variants aligned in a new dimensional direction!

When we choose a "holistic" approach to biological research, the emphasis is on seeing the living organism as an "integrated whole." Thus, its properties cannot simply be reduced to its parts: organs, tissues, cells, and so on. "Holism" identifies organisms by their integrated patterns. By following this concept, researchers studying the morphology of animals, plants, or microbes will consider these organs *necessary*--not just conditions of the body's existence and functionality. Thus, the properties of cells cannot be reduced to organelles, the subcellular structural elements; nor can the properties of ecosystems be reduced to organisms, and so on.

The holistic view is that *everything exists in a state of connection and meaning*. Therefore, any change or event can cause a realignment throughout the entire pattern. A "whole" is ultimately defined by the subject's pattern of relationships. Thus, we often say that "a whole is *more* than the sum of its parts."

"Reductionism" stands in sharp contrast to "holism." Here, the emphasis is on analysis, dissection, and strict definitions of the guidelines needed to understand biological systems. From a holistic perspective, reductionism can give us only a partial view of separate systems. Therefore, holism considers that results from a reductionist standpoint will be incomplete and may lead to serious miscalculations.

As much as I appreciate the holistic approach, I have found that strict adherence to this principle is not always workable. It all comes down to a difference in perspective. When using the holistic approach, a researcher tasked with solving specific biological problems must deal with the reality of endless details within each biological system. For all practical purposes, we cannot limit ourselves to strict holism. Instead, it would be more productive to make distinctions and differentiations. This option would allow us to build a system of orientations within the invisible realm of biological systems. Marc van Regenmortel, a virologist known for his work on virus classification, agrees that "extreme holism, according to which everything is connected, certainly does not provide a methodological alternative."[1]

There is, however, a fascinating middle ground and a possible reorientation. Let's imagine holism and reductionism as two extreme poles in the same direction, taking us from achieving fragmented information to accessing more unified knowledge. Now, let's represent this direction by a two-way line ("axis") going "up" and "down." By using two poles in the "unification/fragmentation continuum," we can select any point to begin our investigation along this "up and down" axis. Now that we have reserved a direction that can orient us, we can define an additional dimension to measure a biological system and discover how it expresses itself.

The act of selecting positions on the proposed "directional line" may be partly intuitive, but the biological system is real! While we can develop some formal approaches to estimate each position, the ideal solution would be to compare the relationship between two or more relatively close points. However, the researcher's decision would be based on his/her educational level, knowledge of literature, professional experience, and specific research tasks. Thus, the proposed dimension provides both orientation and an opportunity to select a position along the directed line.

COVID-19 and the Stairway to Heaven

Let's explore how this approach might be used to assess a contemporary situation that has changed the lives of nearly everyone on this planet. In the fall of 2022, humanity was still deep within the COVID-19 pandemic. These viruses had a unique structure that challenged human immunity and humanity's ability to survive the disease. Ordinary people suddenly became amateur virologists and epidemiologists, challenging specialists who were professionally

---

[1] Marc Van Regenmortel, *Reductionism and complexity in molecular biology. Scientists now have the tools to unravel biological and overcome the limitations of reductionism.* EMBO Reports, 2004, vol. 5, No. 11, page 1019.

trained in virology and epidemiology. The result? The array of contradictory opinions not only frustrated researchers, they also contributed to the public's mounting fears and insecurity.

We can consider various biologists' views on COVID-19 as points on a straight vertical line on the "fragmentation/unification's" directional line. According to such a model, at the bottom of the axis were views that considered viruses ultra-microscopic particles with short nucleic molecules coated by proteins. At the top were abstract concepts about viruses that can change the Earth's biosphere and affect humanity's history.

To illustrate this intriguing direction, consider the following "thought experiment." It invites us to enter the mind of an imaginary young biologist named "Dr. Jacob." If you sense a hint of a "ladder" in the background, your intuition is correct! Dr. Jacob, who had completed his graduate and post-graduate programs, was eager to delve into the mysteries of the SARS-2 coronavirus and the pandemic it had unleashed. An avid researcher, he immersed himself in a sea of literature, absorbing every available resource on the subject. He engaged in intense debates with his colleagues, hoping to gain clarity and guidance. However, the more he explored, the greater the number of contradictory opinions he encountered.

One fateful night, Dr. Jacob fell asleep and found himself on a vivid dream journey. At one point, he was standing atop a desert mountain, gazing at the remarkable sight before him. A tall ladder stretched from the ground and appeared to be reaching into the heavens above. What captivated him most were the myriad messengers gracefully traversing the ladder, each carrying a special message about the coronavirus that had caused this devastating epidemic. As Dr. Jacob observed the thought messages, a fascinating pattern emerged; a diverse spectrum of possibilities spanned the entire breadth of understanding viruses. These viewpoints, each distinct, were likened to points on the stairway leading up to Wisdom.

The steps of the stairway represented points along the "fragmentation - unification" axis. By using the steps, the messages became *visual representations* of the "fragmentation - unification" axis. Each message represented a point on the continuous stairway. The messages allocated to the lowest stairs viewed viruses as ultra-microscopic particles at the base of this metaphorical ladder. At this level, Dr. Jacob envisioned viruses as creatures of minuscule size whose genetic material was surrounded by proteins protecting them. The messages from the steps above unraveled the intricate mechanisms of viral replication, focusing on the molecular intricacies of these elusive entities. In the process of moving the messages upward, a shift occurred in Dr. Jacob's vision. His perspectives began to ascend, transcending the physical confines of viral particles. This shift forced Dr. Jacob to reflect on the broader implications of viruses as shapers of the Earth's biosphere and their influence on humanity's history. These floated to the top of the ladder (Figure 38).

Dr. Jacob soon realized that these messengers were none other than the reflections of his thoughts, which he was experiencing during his intensive studies. Each messenger embodied a unique perspective, representing a specific idea or insight. Dr. Jacob saw his thoughts in harmony, coexisting peacefully, each representing a fragment of the larger picture. In this dream-like realm, there was no discord or contradiction. Instead, it was a serene tapestry of diverse viewpoints woven together to form a more profound understanding.

Figure 38. Dr. Jacob's dream about the Stairway to Heaven reveals new potential knowledge of viruses and viral infections.

The Hebrew word, *malach*, means "messenger," commonly translated as "angel." Rabbi Samson Raphael Hirsch (1808-1888) noted that the word *malach* relates to the word *malachah*, a thinking man's ability to bring his ideas into reality through creative work.[2]

---

[2] https://en.wikipedia.org/wiki/Samson_Raphael_Hirsch

As the dream eventually faded, Dr. Jacob awoke with renewed vigor and clarity. He understood that the diversity of opinions he had encountered was not a hindrance, but an opportunity, a testament to the multifaceted nature of scientific inquiry. With a newfound sense of purpose, he continued on his research journey, eager to contribute his insights to the ongoing quest for understanding.

Of course, the story about "Jacob's Ladder" is symbolic, but the direction along which ideas were arranged is real. Such a directional line represents a powerful tool for estimating the progression of possible outcomes of concrete research. It can also help in a strictly practical way by reflecting on ever-changing situations and available information. Indeed, applying an imaginary scheme is still dubious ground from which to select practical actions. Thus, we need more tools and skills to investigate vast territories where every possible thought and action may have its place.

Up & Down

When we take such an "up and down" approach, we have a whole spectrum between two extremes, such as "part vs. whole," "structure vs. process," etc. The notable American evolutionary biologist Richard Lewontin weighed in on this challenge: "The problem for biology is that neither a radical holism nor an organic reductionism captures the actual structure of causation in the living world."[3]

Basarab Nicolescu, the honorary scientist at the French National Center of Scientific Research, commented: "While one can select and apply diverse approaches to exploring living things, we must remember that simply investigating one or more aspects of a biological body does not represent the whole organism. This is a paradox we have yet to overcome."

Let me repeat a major point of the holistic view: *Each biological system is NOT limited to the composition of separate parts.* We might admit that we are unable to solve the obvious paradox. Still, we can plan and continue our research. How? *By using the paradox as a*

---

[3] Richard Lewontin, *The Character Concept in Evolutionary Biology.* 2002, page 19.

tool. This paradox also reflects the fact that, although each biological object is represented at different levels in the hierarchical biological system, it remains the same. Thus, the uniqueness of each biological object unifies the images reflected from every level of the investigation.

A good example of higher or lower levels of hierarchical organization is "tissue," a specialized structure of a body's organ. Within such a hierarchical organization, each higher level exists only when all lower levels are intact. A single organ does not include the level of an organism, but it does present a part of the organismal system. Each higher level provides unique properties that constitute the emergent properties of hierarchical systems.

A biological hierarchy can be used as a specific dimension to estimate its level. For example, we can study how cells interact by concentrating on analyzing specific molecules. Still, to learn how these cellular interactions affect the function of cells or organisms, we must change our perspective from the molecular level to the cellular or organismal level. One might say that we are moving *up* from the molecular level or the cellular level to the organism level. If so, we must *acknowledge* making this move, as it affects key aspects of the study's design. These would include selecting the methods, calculating the sample size, interpreting the results, and avoiding any arguments about which level is preferable. After all, this is not an obvious task! The holonic model of the hierarchy (holarchy) that we discussed in in the previous chapter demonstrates the unity of each point along different hierarchical levels.

Seeing the Forest for the Trees

The most famous metaphor appropriate to this discussion is the "seeing the forest for the trees" dilemma. Usually, the understanding is that trees should not always conceal the forest. But the opposite is also true: a forest should not conceal a tree. This is actually a biological principle: if you study a specific tree, you will concentrate on investigating that tree. You can later take twenty more trees for statistical confirmation to pursue information about the surrounding forest.

The point is that *this tree* is the object of your attention. This single object ("a tree") is just as important as the system ("the forest"); only the level is different. At the same time, a biologist's concentration on one object can also include studying the object's relationship to the object class to which it belongs.

For example: if this object is an oak tree, then this particular tree represents a plant species of the genus, *Quercus.* When the same biologist approaches "the same" tree the next day, he will compare the relationship between "today's tree" and "yesterday's tree." This is not an easy task. Jacob Bronowski wrote that the progress of science is the discovery at each step of a new order that unifies what had long seemed unlike.[4] He also observed that "all science is the search for unity in hidden likenesses."

When we study relationships between biological objects, we need a unique system of coordinates to view them. Biologists cannot be limited to exclusively investigating biological entities as physical objects. This would not be biology, but the physics of biological objects. For biology researchers, multiplicity and diversity are the main objectives.

Some representatives of the so-called "exact sciences" claimed that theoretical biology was impossible because biological data was unstable and variable. Later, the tone dramatically changed, and exceptions are now being made to allow for laws and strict patterns in molecular biology. Ironically, the praised patterns more often belong in the realm of physics and chemistry, leaving biology with the dominance of data that present endless variations, peculiarities, and changes.

In fact, despite the fragmentary nature of most biological data, various patterns can still represent phenomena at more general levels. Patterns of biological diversity, for example, can be studied at various levels. The more intriguing question is, "Do underlying patterns of diverse and ever-changing manifestations represent a basis for

---

[4] Jacob Bronowski, *Origins of Knowledge and Imagination.* 1979, 160 pp.

unification?" Henrietta Bernstein observed: "The multiplicity of nature may be studied in order that the student can become aware of the underlying unity behind the diversity of nature ... use the visible only as a means of knowing the invisible."[5]

Biology is not about investigating a particular object from a particular perspective; it can be viewed by a scientist, an artist, or another investigator. *Biology is about the indivisible relationships between biological objects.* In his book, *Space, Time & Medicine*, Larry Dossey described parts as "illusions," which are only understandable with all other parts.[6]

However, parts are real and natural, although the importance of information obtained from investigating parts is greatly overestimated. When we talk about an organism, the first image is that of a combination of separate organs. We might know that an organism is a complex system that changes every moment. However, it's harder to *perceive* this principle. Importantly, it's also much harder to investigate and describe it as such.

Up to the Wholeness

"Wholeness" is not just a concept found in metaphysics, and there is no doubt that quantum physics has played a huge role in bringing such a concept into the scientific mindset. Physicist F. David Peat wrote: "The world acts more like a single indivisible unit, in which even the 'intrinsic' nature of each part (wave or particle) depends ... on its relationship to its surrounding."[7]

In biology, we can consider multiple criteria to define a system's wholeness. The three most significant of them are:

- A *holistic* unity that characterizes the level of relationships between parts of the biological system.

- An *ecological* unity that reflects the interdependence of biological objects with their environment.

---

[5] Henrietta Bernstein, *Cabalah Primer*. 1984, 192 pages.
[6] Larry Dossey, *Space, Time & Medicine*. 1982, 248 pages.
[7] F. David Peat, *Infinite Potential: The Life and Times of David Bohm*. 1997, page 353.

- A *research* unity that connects the observer with studied biological objects.

It is rare for modern biologists to investigate invisible elements in a biological system. This is not because biologists are unaware of them or reject hidden parts as "not applicable." Most of them are quite aware of the range of unknown information in any branch of biological science. The existence of missing information is too obvious for them *not* to have made this assumption.

Kabbalah teacher Samuel Avital saw this situation as an example of the "*hidden obvious.*" He explains, "A heart and a kidney are separate organs, but they work together because they belong to one body." Experimental biology can produce a lot of information on how organs can interact with each other: through the nervous system, metabolites, hormones, by sharing the same viruses and microscopic organisms, etc. But Avital had a distinct perspective. "There are many organs, but only one body," he said, "there are many branches but one tree, many pages but one book, and so on." His message is clear: what we call "many" is actually "one." From one side, organs constitute the body; from the other side, they inhabit the body. We can say that these organs are "entangled" in the body, as a physicist would describe particles that transcend space.

Figure 39 represents the manifestations of unity and wholeness: a) each biological system (e.g., organism) is *not* a composition of separate parts. It does *not* exist separately from the biological system of a higher level of consideration, b) each biological object does *not* exist separately from its space (environment), c) and each biological object is *not* separated from the researcher (biologist). Together, these three points comprise the uniqueness and incompleteness of each biological system.

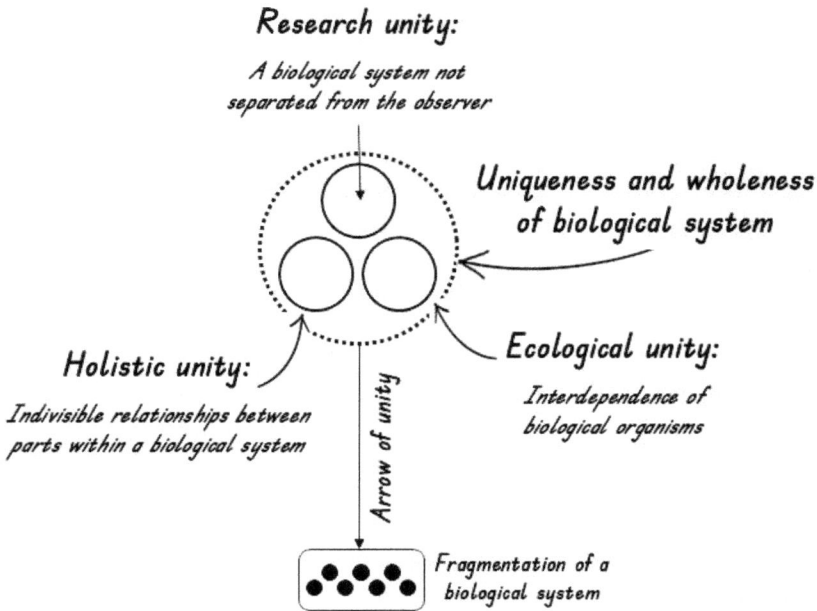

Figure 39. Multiple criteria define a biological system's wholeness. The three most significant are: 1) a holistic unity that characterizes the level of relationships between parts of the biological system; 2) an ecological unity that reflects the interdependence of biological objects with their environment, and 3) a research unity connecting the observer with studied biological objects.

At this point, these factors must be considered as belonging to this system; they are all specific to and undivided from it. Take the example of the microbiome discussed previously. Here is the third point reflecting the unity of any biological object as a whole system: the observer is a part of the biological system. It sounds like a reminder of the familiar fact in quantum physics that "the act of observation changes what is observed." However, this also happens at macro-observable levels of biological systems. A complementary interaction between "biological object" and "biologist" leads to the concepts of symbiosis, cooperation, and homology.

These concepts are applied to relations between biological objects or parts of the biological system. *In this sense, a biologist is part of a biological system!* As each pair of genes affects a particular phenotypic characteristic, the union of "biological object" and "biologist" creates a unique combination of qualities that affects the biological object. As Kabbalist Z'ev Halevi put it, "the beginning of the apprentice's training is to see his own true position."[8]

The two assumptions a biologist can make before starting an investigation are that 1) information already exists about the system to be explored, 2) and there is more to be learned. This information can be obtained at any time and could even be known within another scientific community. However, for this specific biologist, it is still unavailable.

Experimental science has traditionally tried to eliminate the subjective role of the observer, as we extensively discussed in Chapter 1. However, attempts to ignore subjective components of the research process can lead to misleading conclusions. Victor Mansfield emphasized that "the esthetic or spiritual experience is not publicly available for replication."[9] This was widely emphasized in the light of quantum physics but was ignored by modern biologists. As Erwin Schrodinger famously put it: "While all building stones for the world-picture are furnished by the senses qua organs of the mind, while the world picture itself is and remains for everyone a construct of his mind and apart from it has no demonstrable existence, the mind itself remains a stranger in this picture, it has no place in it, it can nowhere be found in it."[10] The concept of "observer" introduced by quantum physics must not be misunderstood to imply that some type of subjective feature is to be brought into the description of Nature.

David Bohm, speaking of "the living world," used the example of a seed. He pointed out that the environment produces almost all the matter and energy that emerge as the seed grows. "Who is to say that life was not imminent prior to the unfolding of the seed in its growing form, then the growing seed becomes more than the mere matter from which it began, as it takes on life itself. The growing seed has

---

[8] Z'ev Halevi, *The Way of Kabbalah*. 1991, 224 pages.

[9] Victor Mansfield.1998. *Synchronicity, Science, and Soul-Making*. 1998, 270 pages.

[10] Erwin Schrodinger, *What is Life? With Mind and Matter and Autobiographical Sketches*. 2012, 196 pages.

become more than the behavior of constituent molecules."[11] This "life-energy" (a term Bohm uses) belongs to that unseen totality that underlies the external world of things and events.

Ecological Unification

From an ecological perspective, a biologist focusing on studying animals, plants, or microbial communities observes entire networks of relationships. These networks connect separate species and describe populations' relationships and their environment. Because no biological object can exist separately from its space (environment), we call this "ecological unification."

Even so, professional ecologists must limit a few ecological factors that could potentially affect the biological system since the list of such factors is endless. Every living organism interacts with its environment, including other organisms within that environment. The same is true of biological systems belonging to other levels, e.g., populations and species. The study of biological systems is challenging because of the multitude of interactions between organisms, as well as between organisms and the environment.

In each ecosystem, every organism, population, and species is linked, directly or indirectly, with many other organisms, populations, and species. Plants provide food, shelter, and nesting sites for other organisms. Many plants depend upon animals to help them reproduce, such as insect-pollinating flowers, and provide specific nutrients, such as minerals in animal waste products. All animals are part of food webs that include plants and animals, either within their species or from members of other species. For example, parasites get nourishment from their host organisms, but these interactions can cause suffering to their animal hosts. On the other hand, scavengers and decomposers feed only on dead animals and plants.

---

[11] David Bohm, *Postmodern Science and a Postmodern World.* In *The Reenchantment of Science: Postmodern Proposals*, ed. David Griffin, Albany, 1988, 190 pages.

Different Unity Words

Discussions about the unification-fragmentation polarity may appear philosophical and unrelated to everyday biological research conducted in laboratories and fields. However, biologists must choose whether to select a general view or a specific one. A general statement can be seen as a manifestation of the "unification principle," while a specific statement is a trend toward the "fragmentation" principle.

How can this knowledge help? We cannot fully reproduce the unique relationship between a biological system and a biologist. Nor can we precisely measure how every part of a biological system is connected. We cannot even estimate the influence of all external factors on the development of a biological system. For this reason, our objective should not be that our viewpoint is *better* than somebody else's; instead, we are wise to select a well-defined thought that can help us explore a specific situation.

When I use the word "unification," I am not using it as a strict scientific term. Philip Kitcher of Columbia University sought to define this term and even proposed using it as a criterion to establish reliable scientific findings. In his work, Todd Jones of the University of Nevada explained "unification" as the attempt to "reduce the number of types of facts we have to accept as ultimate."[12] I appreciate how Hyman Schipper, a clinical neurologist and neuroscientist with a deep knowledge of Kabbalah, uses the word "unicity." In fact, I will often list more specific terms that are familiar to those seeking information in the biological arena. At the same time, I try to emphasize a deeper, more essential meaning, regardless of using multiple names.

To define "the unification-fragmentation axis," I can use different words: unification-fragmentation, generalization-specialization, oneness-multiplicity, wholeness-separation, system-details, etc.

---

[12] Todd Jones, *Reductionism and the unification theory of explanation*. Philosophy of Science, 1995, vol. 62, page 21.

People use these words both as scientific terms and in normal parlance. Nevertheless, the above words are intuitively familiar to describe the trend from a more detailed status to a more general one. Here, I wish to move from the fragmented manifestation of separate biological objects to more unified forms and functions. However, regardless of the terminology I use, my goal is to identify a practical, workable approach to getting myself oriented along the Unification-Fragmentation axis.

Kabbalah as the Foundation of the Unification-Fragmentation Axis

It is said that Kabbalah speaks an "essential language!" Whatever words we use to represent essential relationships, the true meaning is clear. From a Kabbalistic standpoint, every word describes the First Dimension, indicating the relationship between a pair of the Sefirot, *Keter,* and *Malkhut.* Their relationship provides us with the direction: "Up." We perceive Malkhut as the physical world, the Kingdom of Earth. Human beings live vertically on Earth, the head up toward Keter, the feet beneath on the Earth. Evolution has adapted our organs of perception to observe physical objects as natural phenomena. Our eyes can see visible objects, including biological objects. However, we cannot see "biological life" with our eyes. Biological life exists only at invisible levels, only the *physical* manifestation of biological life is visible to us.

The first dimension of Kabbalistic biology is the directional line we can use to orient us and enable us to observe the movements "up" and "down." This is quite a straightforward practice. The problem is with our early education, which encouraged us to stick with one of the extreme poles of "up and down." Unfortunately, this is extremely limiting because life exists within the whole spectrum *between* the points signified by Keter and Malkhut. A conscious decision to select a point on the line connecting the poles will define our point of view, still, we need to practice improving our orientation along this axis.

Thus far, we cannot identify any biological essence because we have been limited by the deeply ingrained perspectives of physics and chemistry. There are substantial differences between researchers in physics and other "exact sciences" when they apply general laws to any investigated object vs. the perspective of biologists. For the latter, multiplicity, diversity, and variations become the focus of their research.

Let's examine the search for underlying manifestation patterns that can offer us a basis for unification. Some general biological principles convey the idea that there is only one "Source" of life. The most notorious example of these principles is genetic coding. Every organism shares the DNA and RNA macromolecules' replication and translation mechanisms. Clearly, this is a universal language common to all biological creatures. Another excellent example of unification is the cellular organization common to all biological organisms. Other evidence illustrating this principle include evolution, environmental adaptation, self-reproduction, self-protection, etc.

The arguments provided by Kempes and Krakauer about the *multiple* origins of life on Earth do not contradict the idea of a profound "oneness." [13] I agree with these authors that Life is *not* a universal homology, and I agree with their theme of multiple paths and three levels of analysis. Kabbalah considers "oneness" at the level where analysis is utterly impossible. In Kabbalistic biology, a change of direction from oneness to multiplicity and from multiplicity back to oneness can give us a specific dimension to orient our thoughts, words, and actions when working with biological systems.

Shoni Labovitz, a rabbi in Fort Lauderdale, noted that "things that seemed separate were only steps in the total cycle of growth... think of the ways in which you separate one thing from another in order to understand them better, then notice how you need to bring them together in order to effect change." She was observing that

---

[13] Kempes and Krakauer, *The Multiple Paths to Multiple Life*. Journal of Molecular Evolution, 2021, vol.89, no.7, pages 415-426.

when somebody thinks, somebody else notices.[14]

From the point of authentic Kabbalah, one source of life with multiple natural expressions is clear and straight. The Hebrew name of this oneness is *EHAD* (alef-het-dalet). Torah often illustrates this point, emphasizing separation: light from darkness, sea from dry land, one species from another, water inside a container vs. the water of organic fluidness outside the container, etc. All these impose order where there would have been chaos and randomness.

One source of origin is the basis for all the underlying multiplicities of biological life on Earth. Note that this is *not* a belief system or a play on words. In Kabbalistic literature, a cell inside a membrane or an organism inside the skin manifests as *rakia* (firmament, separation).

Like the monotheistic belief that One Source is responsible for both the creation and governance of Nature, modern scientists instinctively believe that a fundamental unity underlies all of Nature. However, not all biologists or physicists believe that this unified theory can adequately describe *all* of Nature! Science has no common terminology to confirm that all natural phenomena reflect an underlying cosmic unity. Instead, it is sufficient to say that the natural world is composed of a combination of unrelated forces.

This belief is hardly new; in fact, it was fiercely debated within the Judaic tradition. Norman Lamm addressed this controversial approach by debating the positions of two medieval Jewish philosophers, Saadia (882-942) and Maimonides (1135-1204). Both sages agreed that understanding the natural universe flows from the affirmation of unity; however, they held profoundly different views on the nature/identity of that understanding.

For Saadia, unity was so exclusive that nothing else could lay claim to its attributes. He stated, "When the substance of the tree is examined, it is found to include ... branches and leaves and fruits, and

---

[14] Shoni Labowitz, *Miraculous Living:*
A Guided Journey in Kabbalah through the Ten Gates of the Tree of Life. 1996, 334 pages.

all that is connected therewith. When the human body... is examined, it is found to be composed, besides the elements listed above, of flesh and bones, and sinews and arteries and muscles and all that goes with them."

In contrast, Maimonides draws the opposite conclusion from an identical premise. According to him, unity gives rise to both the world's manifold quality and its unitary nature. In his book, *Guide for the Perplexed*, Maimonides stated, "The whole being is one individual and nothing else... is one individual and at the same time composed of various parts of the body, such as flesh and the bones and of various mixtures." Unlike Saadia, Maimonides sees the entire cosmos as one large organism. The world is not "many," but one. The unity of the world is that of an organism rather than that of a "simple" substance.

Norman Lamm also quotes Michael Berenbaum, the modern American scholar who pointed to the contradictory, fragmented quality of modern life, its multiplicity, and diversity. He believed that "the more sophisticated we become, the more aware ... of the enormous complexity in nature..., the more we identify with the trees and not the forest." He referred to Zohar, which calls such a state *alma de'peruda* ("dis-integration").[15] The powerful imagery later introduced by R. Isaac Luria as *shevirat ha-kelim* ("breaking of the vessels") left the unitary quality of existence shattered at the beginning of Time.

The Zohar describes the steps leading to the process of unification called *Yichud*. It speaks of two types of *yichud*: *yichud mah u ban* and *yichud ava*. The first one represents a person's perception and attitude toward the unification process. The second, *yichud ava*, represents unity in nature as an objective reality. Jacob Immanuel Schochet, the Swiss-born Canadian rabbi, noticed that the word *mitzvah*, which means in Hebrew "a good action," comes from the root word, *tzavta*, which means "connection" or "union." Thus, it

---

[15] Norman Lamm, *The Shema: Spirituality and Law in Judaism as Exemplified in the Shema, the Most Important Passage in the Torah*. 2000, 222 pages.

introduces and establishes the ultimate ideal of unity and oneness on all levels.[16]

Rules of the General and the Specific

With the "*EHAD* principle" indicating the direction toward the abstract point of unification and *yihud* as the process, we are now able to learn a practical methodology called "Rules of General and Specific" (*klal-u'pratim*, the plural form of *klal* and *prat*). In Hebrew, *klal* means a "generality" (the opposite of a particularity) and a "whole" (the opposite of a part). On the other hand, *prat* means "a specific detail" or a particular aspect of a whole system. This is a common example of the "forest for the trees" relationship. In this case, the forest is *klal*, and the tree is *Prat*.

This may sound easy, but determining the relationship between outcomes in specific situations can be challenging. The tradition of moving along the *klalim-u'pratim* axis not only survived but was intensively used for thousands of years. This methodology is continuously developing. Michael Abraham and his colleagues from Bar Ilan University recently developed a sophisticated logical analysis of the rules of *klal-u'prat*.[17]

Jack Abramowitz, the Torah Content Editor at the Orthodox Union, has given simple examples of the *klal-u-prat* methodology.[18] If animals and goats are in the same context, "animal" is the *klal*, and "goats" is the Prat. Similarly, since carrots are a subset of plants, "carrots" would be *prat*, and "plants" would be *klal*. The rule of *klal-u-prat* is applied when a specific point follows a general statement to limit the statement because it applies to a specific situation. For example, the statement that "all mammals have seven cervical vertebrae" may be correct for most mammalian species. Thus, we

---

[16] Jacob Schochet. *The Mystical Tradition:*
Insights into the Nature of the Mystical Tradition in Judaism (The Mystical Dimension. 1990, 165 pages.

[17] Michael Abraham et al., *Logical Analysis of the Talmudic Rules of General and Specific (Klalim-u-Pratim)*. History and Philosophy of Logic, 2011, vol.32, no.1, pages 47-62.

[18] Jack Abramowitz, *The Tzniyus Book.* 2009, 108 pages.

can assume that a particular tiger has seven cervical vertebrates without taking the actual risk of counting them!

The methodology of *prat-u'klal*, opposite *klal-u'prat*, is used when a specific situation is followed by a general category that is all-inclusive and not limited to a specific observation. However, there is a risk. If you were describing several mammalian species, you could be proved wrong if you assumed that seven cervical vertebrates could be found on a "sea cow," the West Indian manatee (*Trichechus manatus*) spotted along the Florida coast. This particular species has only *six* cervical vertebrates. Similarly, you could be proven wrong with a similar assumption made for the three-toed sloth (*Bradypus tridactylus*), which may have eight, nine, or even ten cervical vertebrae.

A methodology described by Abramowitz, called *klal-u'prat-u'klal,* is used when a general description follows a specific example that is followed by another general statement. For example, mammalian species have seven cervical vertebrates, but sloths and manatees have different numbers of vertebrates. Therefore, though most mammals have seven cervical vertebrates, there are a few exceptions. Abramovitz has additionally listed *klal shenu tzarich l'prat/ prat shehu tzarich l'klal.* This is applied when, for the sake of clarity, *klal* requires *prat* or prat requires *klal.* Another rule is *yatzah min ha-klal l'lameid.* This rule states that if a part of a general statement is singled out, the intent is to go beyond teaching only about itself and to include information about its entire category.

From Deer to Plague

Both concepts "unification" and "complete fragmentation" are symbolic. However, the trend from fragmentation to unification can provide a workable system to measure fragmentation (down) and unification (up). This chapter aims to describe a dimension that can help us evaluate any research in terms of "fragmentary--unity." Such an evaluation can be measured on a conceptual level but can also be represented more formally. Therefore, *we define the trend between unity and fragmentation as an independent dimension.*

Let's take another example. One zoologist decides to investigate the characteristics of mule deer (*Odocoileus hemionus*) in the northern part of Colorado. For one specific project, this zoologist is seeking specific characteristics that will allow him to compare these animals to representatives of the same species in California, New Mexico, and Kansas. He plans to use the exact measurements of the animal's body and its parts and patterns of color variation that his colleagues measured in other states. He even calculates that he needs to measure 30 individual animals from each site within this region to achieve his statistical analysis. For his second project, this zoologist decided to compare populations of the same deer species from different locations in Colorado, testing the hypothesis that the pressure to adapt in different landscapes can affect the size and colors of these animals. He then conducted a third project in which he compared variability in the same region during the period required for a long-term mark-recapture study.

Here is another example that relates closely to my area of expertise. In the literature on plague ecology, there is a widely accepted claim that fleas are the main vectors in the transmission of *Yersinia pestis*, a bacterium that transmits plague.[19] More than a hundred years ago, French physician Paul-Louis Simond demonstrated that rat fleas were a primary vector for the transmission of plague between animals.[20] We support this claim after a hundred years of investigations that demonstrate either the presence of the bacterium in a flea's body or a successful passage of this bacterium by fleas fed on infected rodents and transmitted to uninfected rodents. However, of more than two thousand flea species, only a few specific species were proved to be competent vectors; the most important of which is the Oriental rat flea. Actually, the picture is even more

---

[19] Kenneth Gage and Michael Kosoy, *Natural History of Plague:*
Perspectives from More than a Century of Research. Annual Review in Entomology, 2005, vol. 50, pages 505-528.

[20] Marc Simond et al., *Paul-Louis Simond and his discovery of plague transmission by rat fleas:*
a centenary. Journal of the Royal Society of Medicine, 1998, vol. 91, pages 101–104.

complicated: plague transmission by fleas can occur under many specific conditions: a dose of vital bacteria, the number of flea bites, the feeding activity of fleas, and so on. Therefore, we can confidently claim the role of fleas in each specific situation. The general statement about fleas as plague vectors is still important and contributes to plague control worldwide, both historically and currently.

Let's now move to an even higher level of generalization. One of the Biblical plagues in Egypt was called, "The Plague of Vermin" or *kinim* (Exodus 8:13–14). The Hebrew word *kinim* is often translated as "lice, but it could also mean fleas or other small insects (it is not easy for a non-specialist to distinguish between lice and fleas). The main point we can learn from the Biblical portion is that one of the essential aspects of a plague is the involvement of "vector transmitters."

We can use the flexibility of moving along the unification axis to answer crucial questions for the conservation, use, control, and even elimination of a specific plant, animal, or microbial species. Based on the unification principle, we might assume that reductionistic points based on precisely conducted measurements of separated and static elements of biological systems are more practical than speculative hypotheses. In many situations, this is true, but the main question is, "How much difference can those accurate measurements produce?"

Beware of getting stuck at any high position on the proposed "unification-fragmentation" line. A practicing biologist should insist on a fact check of any theoretical constructions and often scrutinize his or her own views. As C.H. Waddington, one of the most remarkable theoretical biologists of the 20th century, concluded, " ... one must check whether the result of the suggestion is, in practice, what it was expected to be, or reasonably near it."[21] In other words, the investigator must always be ready to take a few steps "down" until the experiential and reasonable ground justifies an "upper" mental construction. My point here is simple: none of these studies can be

---

[21] C. H. Waddington, *Tools for Thought.* 1977, page 197.

regarded as "good" or "bad." They each answer a different question. However, we can evaluate them as points on an imagined continuous line.

Not Right or Wrong, but a Trend

Here is an example of understanding "the parts regarding the whole." It is borrowed from the research of Athel Cornish-Bowden of the French National Centre for Scientific Research.[28] He and his colleagues considered metabolism to be a complete system, not just a collection of components. From one perspective, metabolism is commonly described as "a set of chemical reactions catalyzed by separate enzymes."

This is a solid, well-described process. However, one should be aware of various potential complications. The one I wish to describe is the physical association of different enzymes. Therefore, specific properties of the metabolic system make sense only if a broader view of pathways is considered. As the authors indicated, a serious investigation into possible metabolic processes implies an infinite regress in which each set of enzymes is needed.[22]

From this more general perspective, the consideration of metabolism emphasizes certain systemic aspects that may be unnoticed when attention is focused on details. Regarding the example of the situation focusing on separate enzymes, Cornish-Bowden, and his colleagues offered a warning: "…most current research in biology is not directed toward solving fundamental problems, but towards accumulating more detailed facts, and unfortunately, this may cloud understanding as much as it improves it."[23]

Conclusions coming from different levels can appear to be mutually exclusive. Here, the question is posed: "Who is right?" The placement along the 'unification-fragmentation' axis does not ask this

---

[22] Athel Cornish-Bowden et al. *Understanding the parts in terms of the whole.* Biology of the cell, 2004, vol.96, no.9, pages 713–717.

[23] Sarah Djebali et al., *Landscape of transcription in human cells.* Nature, 2012, vol.489, pages 101–108.

question because no point is "better" or" worse." It only indicates a level of consideration based on the closeness to a more general, unified point. We cannot judge any position of defining points on the "unification-fragmentation" axis as "true" or "not true," etc. This dimension is designed exclusively to indicate the *direction*. It is crucial because most intense disputes reflect a difference in investigators' perception, when the scientists pretend that their opinions are superior to those of their colleagues.

Let us now examine the question, "Which microorganisms are considered pathogenic?" Some molecular biologists may provide data about the presence/absence of pathogenicity islands. These gene clusters incorporated in the genome have been shown to play a role in pathogenesis. Some experimental biologists can rely on the results of the presentation of clinical signs in animals inoculated with bacteria. Clinicians may depend mostly on their experience dealing with actual human patients. Population biologists and epidemiologists may prefer an analysis of data obtained from many samples collected from animal or human populations.

They could all agree that the perfect situation would have coherent information from all these studies, but this is rarely the case. The information obtained from different levels may not be available or can be inconsistent in the interpretation. How can accepting the described "unification-fragmentation" dimension be of practical help?

The most striking vision belongs to Rabbi Joel Bakst of the City of Luz, who sees the coronavirus as a "crisis of dimensionality."[24] The line between those extremes -- the lowest point representing the simplest biological objects, and the highest point looking at the universal explanation is dotted with intermediate scenarios, each with a different degree of unification between them.

This lesson teaches that a trend (direction/ vector/ arrow) can be recognized as enabling the movement from fragmented

---

[24] Joel Bakst, *Kabbalah of the Adamic Messiah*. 2020, 346 pages.

information to unified knowledge. In other words, it enables the movement from differentiation to unification. As far as this direction is concerned, there is no right or wrong position; it simply represents a particular position along the dimensional line of fragmentation and unification. Rabbi Hillel Rivlin of Shklov (1757–1838), a close disciple of the Vilna Gaon, wrote in *Kol HaTor*: "It is impossible to climb a ladder whose top reaches toward the heavens without first stepping on the rungs of the ladder that are stationed near the earth."[25]

A key message is that this dimension should be applied as a practical tool, not simply as an abstraction. First, we must indicate a direction and propose a dimension: from" down" to "up", from "fragmented" to "unified." Second, we can define some criteria to estimate the uniqueness of quantitative measurements at the unity level for concrete biological research. As we move through the following chapters, we will gain a deeper understanding of additional tools. At this point, to learn more, it is imperative to acknowledge the "unification-fragmentation" direction.

Make no mistake, defining a level of uniqueness and wholeness for any research task can be challenging and requires practice. Depending on the task and the situation, a biologist can select a point along this directional line. Furthermore, depending on the circumstances and the context, that same biologist should be prepared to change her or his perspective.

---

[25] Cited from Joel Bakst, *The Secret Doctrine of the Gaon of Vilna.* Vol. II, page 149.

Chapter 14

Second Dimension: Complementary Polarity

The entire physical world is based on the concept of duality. In fact, no new forms can exist without clear evidence of their separateness from each other. Therefore, when researching biological phenomena, scientists have traditionally felt the need to choose between two exclusive, opposite aspects of biological systems. Science has, in the past, regarded interactions between elements of each system as a *struggle* between opposing forces.

Consider the following examples: male sex vs. female sex; inside an organism vs. outside an organism; single-celled organism vs. multicellular organism; phenotype vs. genotype; RNA vs. DNA; animals living on land vs. animals living in water; population growth vs. population decline; gradual evolution vs. sudden mutational change, etc. "Primary opposition" between life and death is evident in most situations and is usually quickly confirmed. As we discussed in Part One of this book, such dualistic thinking is not confined to biologists. The perspective that any biological object is either "this" or "that" is very common.

The Coincidence of Opposites

This dichotomy (a division between two sides of one natural

phenomenon, which are opposed or entirely different) is so clearly revealed in Nature that it is impossible to ignore! However, I would like to use it as a tool to measure relationships between opposing biological forces that not only divide but also *unite* each other! Let's begin by considering each biological phenomenon as an entity with two interconnected poles.

Sanford Drob, a clinical psychologist and author of many books on Kabbalah, appropriately named this principle "the coincidence of opposites" after the Latin phrase, "coincidentia oppositorum." This was attributed to the 15th-century German polymath, Nicholas of Cusa.[1] Jewish scholars Gershom Scholem and Abraham Joshua Heschel also used the term. Regardless of terminology, the notion that oppositions dissolve when they unite in an object or a phenomenon is familiar to many systems of thought. The question is, how can we make this perspective work for us in biological research?

Though "polarity" is often used interchangeably with "duality," we need to examine the difference between how these terms are commonly perceived. "Duality" usually refers to a physical separateness between related yet opposite phenomena or modes of being. A classic example is the existence of the male and female sexes.

In contrast, accepting polarity as a singular phenomenon can provide a very strong *unifying* principle when we are describing interactions between opposite phenomena/elements. Polarity is represented by a vector (arrow) that shows the point of a directed influence, a dynamic quality with a preferred direction, and a flow. However, the point must indicate that one direction along the vector *differs* from that of the other direction. This is in contrast with the term, "duality," which stresses oppositeness and exclusivity.

I wish to offer the perspective that "polarity" could be more appropriate to link opposite sides instead of separating them. This position could actually reveal a "coexistence of communications" between opposite sides in any biological system. Two poles, in this

---

[1] Sanford Drob, *Kabbalistic Vision.* 2009, 313 pages.

case, could be complementary rather than conflicting. For example, suppose we accept the existence of two opposite forms ("masculine-feminine trend" instead of "male-female"). In that case, we can define a continuing range between these extremes, using each situation as a point between two poles.

In my experience as a biologist, I have rarely seen my colleagues use opposing tendencies as a powerful tool for their investigations. Instead, they consider opposites an unavoidable necessity used to reveal different results and consider conflicting views. Of course, attitudes will vary depending on the research task, the investigator's background training, and the research institution's priorities.

Hesed and Gevurah Represent the Opposite Pole

It is interesting to consider the Kabbalistic perspective regarding opposites: *opposite tendencies are NOT separate entities; they are manifestations of the existence of one source.* Kabbalists consider each object or process as an interaction between two opposite sides, illustrated by the two "pillars" of the Tree of Life (*Etz Hayim*).

Traditionally, these poles, or pillars, are represented by two sefirot, Hesed and Gevurah, located on the right and left sides, respectively. The Hesed side symbolizes a tendency to give, operate, and diversify, while the Gevurah side represents receiving, preserving, and formatting. None of the biological objects, systems, or processes can be explained by the influence of only one of these forces, represented by Hesed and Gevurah. Instead, they are defined by the relationship *between* them.

The concept of two pillars of the Tree of Life that represent "giving" and "receiving" should not be perceived as a duality. Yes, the right and left pillars are dual extremes. But the constant movement (interference) between the extreme points represents the dynamic interaction between these sefirot. In fact, the range of the interactions occupies the entire spectrum between the poles, which are the essence of the underlying *unity* of this opposing activity (*"driving"* and "*stabilizing*"). This unity in itself confirms that the poles are *not* acting as mutually exclusive pairs.

Polarity Is Rooted in the Kabbalistic Tradition

Below, I will provide references from both classical and new sources to illustrate that unity in polarity is deeply rooted in the Kabbalistic tradition. In the oldest Kabbalistic source, the *Sefer Yetzirah*, all ten sefirot are seen as five sets of opposites.[2] Azriel ben Menahem (also known as Azriel of Gerona, a Kabbalist born around 1,160) used the Hebrew term '*ha-achdut ha-shawah*' (the "indistinguishable unity of opposites"). He pointed out that sefirot depend upon each other and are united like links to the first one. Each sefirah possesses both a positive and a passive quality, which emanates and receives.[3]

The Zohar expresses this concept in terms such as: "*the head that is not a head.*"[4] The point is that each biological object or system should be considered from its opposing side; otherwise, it has not been completed. Joseph Gikatilla, the Spanish Kabbalist of the 13th century, called this principle, "the power of opposites."[5]

Many images express the everlasting dynamic of Hesed-Gevurah interaction in the Judaic tradition: *ratzo va'shov* ("run and return"); *yesh* and *ayin* ("being and nothingness"); *hitpashtut* and *tzimtzum* ("expansion and contraction"). Another pair of definitions that reflects how Kabbalah can contribute to understanding the interplay of opposite directions is *ribbuym* (plural of *ribbuy*) and *miutim* (plural of *mi'ut*). The *ribbuy* stands for forces of extension and amplification, while *mi'ut* stands for limitation and restriction.[6]

The Tree of Life presents a model of the "Complementary

[2] Aryeh Kaplan, *Sefer Yetzirah - The Book of Creation: in Theory and Practice*. 1997, 398 pages.

[3] Josef Bláha, *Azriel of Gerona: Commentary on the Ten Sephiroth*. 2015, 84 pages.

[4] Isaiah Tishby, David Goldstein, and Fischel Lachower, *The Wisdom of the Zohar: An Anthology of Texts*. 1989.

[5] Joseph Berke and Stanley Schneider, *Centers of Power: The Convergence of Psychoanalysis and Kabbalah*. 2008, 254 pages.

[6] Michael Chernick, *The Use of Ribbūyīm and Mi'ūṭīm in the Halakic Midrash of R. Ishmael*. The Jewish Quarterly Review, 1979, vol. 70, no.2, pages. 96-116.

Polarity" principle. In Chapter 10, we talked about two trees in the story of the Garden of Eden: The Tree of Knowledge (*Etz-haDaʿat*) and the Tree of Life (*Etz Hayim*). "The eating of the forbidden fruit" reflects the severing of something from the original entity. Thus, we have an act of separating two elements of the one tree: Life ("oneness") and Knowledge ("individuation"). As long as both realms are united, there is no danger of a sense of superiority of one over the "other."[7]

Back in the early 19th century, Aaron HaLevi ben Moses, a student of Rabbi Shneur Zalman of Liadi, noted: "All created things in the world are hidden … in one potential, in coincidentia oppositorum … the essence of perfection is that even those opposites are opposed to one another be made one … opposites become united in a single subject, and their differences are, in effect, nullified." And here's another quote from Aaron HaLevi: "… in their state of *Hashawah* … opposites become united in a single subject, and their differences are, in effect, nullified."[8]

In their book, *Centers of Power*, Joseph Berke of the Arbours Crisis Center in London and Stanley Schneider of the Hebrew University in Jerusalem stated:

> "There are always two (opposite) ways of looking at something."[9] "By having opposites and polar extremes, we are forced to acknowledge the entire range of being and doing… by having the entire spectrum open and available, we can then make an intelligent choice that accounts for the extremes and enables us to find our position on the continuum."[10] "Kabbalah's stress on the importance of seeing how both giving and receiving are not only related but are the same. The idea that there can exist two opposite poles does not have to negate the other ... rather we need to derive

---

[7] Gershon Winkler, *Kabbalah 365: Daily Fruit from the Tree of Life.*

[8] Sanford Drob, *Kabbalistic Visions.* 2010, 313 pp.

[9] Joseph Berke and Stanley Schneider, Centers of Power:
The Convergence of Psychoanalysis and Kabbalah. 2008, page 188.

[10] Ibid., page 190.

meaning from each pole ... to arrive at a position relevant to our understanding."[11]

Allen Afterman, in his book *Kabbalah and Consciousness*, described the act of Hesed as the "creation of new potentiality" and used the poetic image of a seed planting itself into Life even before the ground for Life had been prepared. He continued: "This is the archetypal masculine act: purposive, visionary but lacking in being."[12]

Shimon Shokek, Professor of Jewish Philosophy and Mysticism at Baltimore Hebrew University, wrote:

"The awareness of change and stillness, which is, in essence, the duality of existence that constructs reality... The duality of change and stillness could be translated into multiple existential conditions ... When we perceive that our reality is a totality of opposites, we begin to experience the true meaning of being; first, we acknowledge our opposites, and then we strive to integrate them. "[13]

Jerome Levi, the Harvard University anthropologist, wrote in his fascinating article, *Structuralism and Kabbalah*: "Not just (the) opposition but also the logical relations between sets of oppositions are important to the analysis of symbolic systems. Their logical relations may be expressed in terms of reciprocity, analogy, homology, reflection, inversion, isomorphism, and so on."[14] And here is a line from Arturo Schwartz: "For the Kabbalist ... the two poles of a polarity are in a complementary, rather than conflictual relationship."[15]

Rachel Elior, a professor of Jewish philosophy at the Hebrew University of Jerusalem, pointed out that opposites are not accidental—they are parts of a living system, and that system is a

---

[11] Ibid., page 191.

[12] Allen Afterman, *Kabbalah and Consciousness*. 2005, page 87.

[13] Shimon Shokek, *From Kabbalah and the Art of Being*. 2001, page 80.

[14] Jerome Levi, *Structuralism and Kabbalah: Sciences of Mysticism or Mystifications of Science?* Anthropological Quarterly, 2009, vol. 82, page 956.

[15] Arturo Schwartz, *Kabbalah and Alchemy*. 2001, page 4.

necessary part of life.[16]

Shirley Chambers, the founder of the Karin Kabala Center in Atlanta and my first guide into the miraculous world of the Tree of Life, in *Kabalistic Healing*, placed some biological processes, specifically "anabolism and "catabolism," on two pillars.[17] Anabolism is a set of enzyme-catalyzed reactions that synthesize complex molecules in living systems from simple structures. In contrast, catabolism is a set of metabolic reactions that break down large, complex molecules leading to the release of energy required for survival.

And here is one more quote that sounds poetic, as it should, since it was written by a poet. Peter Cole wrote in his book *The Poetry of Kabbalah*: "Wind between opposing aspects: darkness and light, concealment and revelation, absence and presence, upper world and the world of decay. Riding the tension between these extremes, the poems, through their composition and then by means of their recitation, participate in the process of cosmic restoration."[18] These lines do not seem to represent biological problems, but they are indeed relevant as well as beautiful.

Samuel Avital calls this relationship many things: "Beyond Opposites," "The Cosmic Accordion," "the Law of Polarity," "Bouncing Principle," or "the journey between simplicity and complexity."[19] He supports the principle that for everything that exists, there is an existing opposite. As an acclaimed mime actor, Avital expressed the "Law of Polarity" during his performances on the world stage. One piece of his art, called "Black and White," represents the tragic divisions as "the endless battle between ourselves." He first created this dichotomy in his make-up, having his left side (Gevurah) entirely black and his right side (Hesed) white. He continued this line

---

[16] Rachel Elior, *The Paradoxical Ascent to God: The Kabbalistic Theosophy of Habad Hasidism.* 1992,
302 pages.

[17] Shirley Chambers, *Kabalistic Healing.* 2000, page 75.

[18] Peter Cole, *The Poetry of Kabbalah: Mystical Verse from the Jewish Tradition.* 2014, 544 pages.

[19] Samuel Avital, *The Invisible Stairway.* 2005, 287 pages.

of division over his head. The climax of the Black and White performance came when two hands, the black and the white, approached and met each other.[20]

Everything can be expressed as the interplay between these two tendencies that Avital illustrated in the beautiful image of "the Cosmic Accordion" (Figure 14). Expansion and contraction, exploring and stabilizing, increasing variants and limiting them, and so on... None of these are bad or good, right or wrong...they are simply acts of giving and receiving. One of the mottos expressed by Samuel Avital states, "We divide in order to unite ... not to sustain the division."[21]

The Dance of Hasadim and Gevurot

Although Hesed and Gevurah represent two opposite pillars, they do not represent the entire range of forces that create and affect the systems being observed. The main Kabbalistic reference, the book of *Zohar*, uses two concepts: *Hasadim* (plural of Hesed) and *Gevurot* (plural of Gevurah) to illustrate the interplay between forces from opposite pillars.

The description of these forces may sound very dualistic. However, we can only explore the principles of Kabbalistic biology when we realize that such a view is deficient. To succeed in dealing with this issue, I will share important lessons I learned from another teacher, Rabbi Joel David Bakst. I studied Kabbalah with him until his sudden passing in 2021. I was very aware that during our weekly classes, he continued pointing out the importance of the relationship between Hasadim and Gevurot. He called this process "The Dance of Hasadim and Gevurot." Because I consider his teaching vital to our understanding of Kabbalah, I am including below many detailed quotes of Bakst from his books, lectures, and seminars.

The teachings of Joel Bakst are rooted in the wisdom of the most

---

[20] Samuel Avital, *Mime and Beyond: The Silent Outcry*. 1985, 174 pages. The performance can be watched on YouTube https://www.youtube.com/watch?v=ZZgTBJ_1IMg&t=285s
[21] Samuel Avital, *Mime Work Book*. 1977, 158 pp.

prominent sage of the 18th century: Vilna Gaon (Elijah ben Solomon Zalman, also known as GRA). This wisdom is reinforced in the writings of GRA's close disciple, Rabbi Shlomo Eliyashiv (known also as Leshem). Bakst abbreviated the system known as "Hasadim and Gevurot" into three letters: "*HuG*" (pronounced, "HOOG"). According to Bakst, HuG was neither a duality nor a simple singularity. Instead, HuG contains both a duality and a simple singularity, yet is more than both. Bakst considered HuG the inner mechanism at the very center of the Kabbalah: *We cannot fully know something without contrasting it to its opposite.* This principle is crucial when considering any biological phenomena from two opposite views.

Hasadim can be described as "an expansive quality loosely associated with masculine energy." In contrast, the Gevurot can be described as "a contracting quality loosely associated with feminine energy." Bakst formulated five underlying statements of HuG that can be observed:

1. All things have two aspects: a Hesed aspect and a Gevurah aspect.
2. Any aspect of HuG can be further divided into Hasadim and Gevurot.
3. The Hasadim and the Gevurot mutually create each other.
4. HuG defines each other.
5. HuG converts into each other.

The main characteristic is this: All these qualities are opposites, yet they describe relative aspects of the same phenomena. The qualities of Hesed and Gevurah exist only in relation to each other.[22] Interestingly, Bakst noted that although he used the term, "opposite," this word can be misleading because Hasadim and Gevurot cannot be characterized as being opposite each other. A clearer definition would be a "polarized singularity." HuG can be defined as "paired

---

[22] Joel Bakst, *Beyond Kabbalah: The Teaching that Cannot Be Taught.* 2012, 418 pp.

units," or just "pairs" (zugot in Hebrew, plural of zug, which means, "a pair"). Bakst also used the word "appositional," which means juxtaposing one thing to another--but not in opposition. The interplay of the HuG is not only a set of correspondences, it also represents a way of thinking, a mode of perception.[23]

Bakst powerfully expressed his vision in graphics with a picture indicating the position of the imploding Gevurot on the "*inside*" and the expanding Hasadim on the "*outside*," not only in opposition to each other (Figure 40). Based on his 30 years of teaching, Bakst noted that "regardless of one's IQ or breadth of one's knowledge, not everyone… is suited to master the paradoxical movements of this great cosmic dance."[24]

When we become grounded in the Kabbalistic tradition, we will discover how the manifestation of masculinity and femininity reveals the difference between the forces of giving and receiving. In fact, the role of the genders in biology is the clearest example of this phenomenon. Arturo Schwartz, an Italian artist, and scholar who studied Kabbalah for many years, noted that for a Kabbalist, two poles of polarity are in a complementary, rather than in a conflicted, relationship. He stated that "the male-female polarity is the fundamental model for all other polarities." [25]

Aryeh Kaplan, a remarkable Kabbalist of the 20th century, wrote: "We can see that the paradigm of a man is more diversified, whereas the paradigm of a woman is more unified. If you think of male and female biologically, the male is the giver; the female receives and nurtures and then gives much more than the man initiated. The man gives over a million sperm cells from which the woman selects only one. From her one single fertilized egg, however, she provides a complete infant. Not only did she receive, she also ended up creating and building something complete. Hence, the essence of femininity turns out to be much more complex. If masculinity is *giving*, femininity

---

[23] Ibid., page 72.

[24] Ibid., page 70.

[25] Arturo Schwartz, *Kabbalah and Alchemy*. 2001, 198 pages.

is *receiving* and *completing*."[26]

Giving and Receiving in the Biological World

In the biological world, the actions of Hasadim are reflected in the unlimited reproduction of organisms, the creation of an unlimited variety of biological types, and the free distribution of biological creatures around the world. The organic world is astonishing in its ability for organisms to produce new forms, generate new species, and occupy new ecological niches.

Yet, even as we acknowledge this magnificent potential, we must accept that not every imaginable biological form and function can--or did--exist on Earth. Mark Olson of the Universidad Nacional Autonoma de Mexico has found that a key question of biology is, "why, with the array of imaginable morphologies, are only some observed? And of those…an even smaller subset is commonly observed!"[27]

Some biologists admit that constraints (limiting factors) in the evolutionary process are a critical challenge for modern biology. Douglas Futuyma, the evolutionary biologist at Stony Brook University, sees the existence of substantial limits for the adaptation of biological organisms and species, as "paradoxical."[32] Futuyma outlined the following examples of challenges to the traditional view on the evolution of biological species:[28]

1. The theory that extinction affected the vast majority of biological species….

2. The fact that there are limits to the geographical and habitat range of species….

3. The emphasis on historical accidents or contingencies by the

---

[26] Aryeh Kaplan, *Inner Space: Introduction to Kabbalah, Meditation and Prophecy.* 1991, page 75 (italics is mine).

[27] Mark Olson, *Overcoming the constraint-adaptation dichotomy:*
Long live the constraint-adaptation dichotomy. In:
Perspectives on Evolutionary and Developmental Biology, ed. Giuseppe Fusco, Padova University Press, Padova, 2019, page 123.

[28] Douglas Futuyama, *Evolutionary constraint and ecological consequences.* Evolution, 2010, vol. 64, page 1865.

famous evolutionary biologists, Stephen Jay Gould and Niles Eldridge....[29]

4. The many "empty niches" (the definition of Futuyma) that exist in the distribution of adaptive forms....

5. The evidence of "phylogenetic niche conservatism" such as plant-eating insects--associated with particular plant families since the Mesozoic Era.

6. The "stasis" that dominated the history of most fossil species instead of gradualism....[30]

7. The rapid, short-term evolution that often occurs in intervals shorter than one million years; these changes are constrained and do not accumulate over time.

Because the term "constraint," often has a negative connotation, Alexei Sharov of the National Institute of Aging prefers the term, "regulated variation." This emphasizes the adaptive nature of phenotypic varieties that help populations and species survive.[31] I use the term, "constraint" because it and others embrace the innumerable manifestations of the force that, in the Kabbalistic tradition, is symbolized by Gevurot.

HuG Dance in Evolution

Following the principle of Kabbalistic biology, we must define each direction's essential core. By doing so, we can examine the various manifestations that normally contribute to developments in this direction. When we identify multiple evolutionary and ecological constraints as manifestations of Gevurot, we can see additional biological functions participating in the HuG dance.

Today, the theory of evolution is the main foundation of modern biology. Thus, it is not surprising to see how important it is to describe factors contributing to the evolutionary capability of

---

[29] Niles Eldredge and Steven Gould, *Punctuated equilibria: an alternative to phyletic gradualism.* In *Models in Paleobiology.* 1972, pages 82-115.

[30] Josef Uyeda et al., *The million-year wait for macroevolutionary bursts.* Proceedings of the National Academy of Sciences of the United States of America, 2011, vol. 108, pages 15908–15913.

[31] Alexei Sharov, *Evolutionary constrains or opportunities?* BioSystems, 2014, vol.123, pages 9-18.

organisms. Though "evolution" has not been well-defined, it refers to the tendency or ability to produce adaptive variations. A much less common but equally critical acknowledgment of constraints supports long-term "survivability." Michael Palmer and Marcus Feldman of Stanford University argued that the ability to survive is even more fundamental than the ability to evolve.[32]

From the position of Kabbalistic biology, any argument about which ability is more significant seems pointless. This is the interplay of Hasadim and Gevurot, and it takes two to tango (or "HuG dance" in the words of Joel Bakst). The action of Gevurot can be seen in the "stabilizing selection" in contrast to "directional selection." This is a manifestation of Hasadim and is more fully represented in biology textbooks.

The notion of a "stabilizing selection" was pioneered by Ivan Schmalhausen, a zoologist and evolutionary biologist of German descent born in Kiev, Ukraine. Schmalhausen described this phenomenon as "constructive," leading to stabilizing particular non-extreme traits.[33] The stabilizing selection causes the narrowing of the phenotypes seen in a population. Later, the development of a specific genotype under different conditions was named "canalization" by C.H. Waddington."[34]

Another expression of the "expansion and contraction" polarity is the pairing of "originality" and "directionality." These two distinct behaviors of biological systems were formulated by Victoria Alexander, a scholar noted for her unique vision of commonalities in humanitarian and natural worlds. Under "directionality," she refers to mechanisms to maintain order: stability, habituation, and automation. She also describes directionality as resulting in types, species, genres, and habits.

The second type, which she calls "originality," is the discovery of

---

[32] Michael Palmer and Marcus Feldman, *Survivability is more fundamental than evolvability*. PloS One, 2012, vol. 7, e38025.

[33] Ivan Schmalhausen, *Factors of evolution: The theory of stabilizing selection*. 1949, 327 pages.

[34] C. H. Waddington, *Genetic assimilation. Advances in Genetics*. 1961, vol.10, pages 257-293.

brand-new functions: change, creativity, and freedom. Expanding on these types, Alexander says: "With directional selection, slight differences are inconsequential, and sameness prevails. With (the) original selection, slight differences result in significant functional improvement and/or systematic reorganization."[35]

Niles Eldredge, the scientist who greatly influenced evolutionary biology in the 20th century, continuously repeated in his book, *Why We Do It,* that "all organisms... have an economic and a reproductive side to their lives. Every living creature is part of at least two larger-scale systems that are hierarchically structured."[36] Eldridge considers organisms as either "ecological interactors" or "reproducers" as two phases of any biological phenomenon: "There is a yin and yang here, not a one-way street."[37]

Another paired relationship that requires the complementary approach is that between genetic components when developing a biological system and the ability of the same system to interact with the environment. These lead to adapting to specific conditions. I phrase this as the interplay between "ecological potentiality," represented by Hasadim, and "genetic canalization," represented by Gevurot.

Kabbalah of Biological Sex

It is impossible to find a better illustration of the "HuG dance" in biology than the interplay between biological sexes. Traditional Kabbalah abounds with sexual imagery, characterizing the right pillar (Hasadim) as masculine and the left pillar (Gevurot) as feminine. Indeed, when discussing masculine-feminine polarity, Kabbalah goes deep into primary forms. Every physical characteristic and function mentioned serves as a symbol to underscore this essential polarity.

Rabbi Eliahu Klein pointed out that the root of the word *zachar*

[35] Victoria Alexander, *The Biologist's Mistress: Rethinking Self-Organization in Art, Literature, and Nature.* 2011, page 97.
[36] Niles Eldredge, *Why We Do It: Rethinking Sex and the Selfish Gene.* 2005, page 26.
[37] Ibid., page 94.

("male" in Hebrew) originally meant "to pierce." From this we have the concept of memory, "to pierce" as "to fix in one's mind." *Zachar* as "male" is anatomically associated with piercing. The word, *nekeva,* ("female" in Hebrew) relates to "puncture." [38] The feminine always represents the receptive aspect of creation. *Zachar* means "expansion of light," and *Nekeva* means "reception of light."

In the Lurianic Kabbalah, everything in Nature emerges in the union of its parental patterns ("masculine" and "feminine"). This metaphoric view relates to the Talmudic midrash: "Adam and Eve were created back-to-back." Next, these primordial twins were separated and "re-emerged" as the model for a real partnership or a "face-to-face" relationship. Klein interprets this metaphor as the stage when "two primary vehicles" -- "direct light" and the instrument that can receive this light must turn toward each other. Metaphoric language is plentiful in original kabbalistic texts!

The evolution of the sexes is still acknowledged as one of the most controversial issues in modern biology. There are several reasons why the origin of sex presents such a problem, with the most common argument being that it leads to a waste of resources in producing males. Since a sexually reproducing female gives birth to an equal number of male and female offspring, only half of these would be able to have more offspring. Although sex is almost universal in higher animals and plants, both the evolution of and support for both sexes and the mechanisms of sexual reproduction are difficult to explain. Sexual reproduction in most biological species has always been the most challenging part of biology since the era of Darwin and throughout modern biology.

Matthew Hartfield and Peter Keightley, evolutionary biologists at the University of Edinburgh, proposed some theories to solve this dilemma:

1. The ability of sex to fix multiple novel characteristics of organisms that enables it to survive and reproduce better than

---

[38] Eliahu Klein, *Kabbalah of Creation: The Mysticism of Isaac Luria, Founder of Modern Kabbalah.* 2005, 302 pages.

other organisms in a given environment.

2. Sex as a mechanism to stop mutations that increase one's predisposition to a specific disease or disorder.

3. Recombination leading to the creation of novel genotypes that can resist infection by pathogens and parasites.

4. The ability of sex to reduce the frequency of deleterious alleles prompted by inbreeding.[39]

As the literature on the biology of sexes is abundant, I will concentrate on the teaching developed by one Soviet evolutionary biologist whose teaching greatly influenced me. His name is Vigen Geodakyan, and during several decades of the last century, he concentrated on what the phenomenon of female-male differentiation means from an evolutionary perspective. Although he was not the first biologist to ask such a question, his conclusions fit nicely with ancient Kabbalistic perceptions, though he would be surprised to hear this.

Geodakyan regarded sex differentiation as reflecting two integral but contrary aspects of evolution: "conservative" and "operative."[40] He described a series of mechanisms that could justify the role of women in supporting the conservative (*"generative"*) stream of information. In his words, it would be the transfer of genetic information to generations *"from past to future."* In contrast, males are more important in sustaining the *"operating"* streams of environmental information *"from present to future."* Males, when compared to females, experience more mutations, inherit fewer properties from their parents, are more aggressive, and demonstrate riskier behavior. According to Geodakyan, all these properties move the male sex to the front line of evolution. The second group of properties includes the incredible plethora of male gametes, their small size and high mobility, the excellent activity of males, their

---

[39] Matthew Hartfield and Peter Keightley, *Current hypotheses for the evolution of sex and recombination.* Integrative zoology, 2012, vol. 7, pages 192-209.

[40] Vigen Geodakian, *The Role of Sexes in the Transfer and the Transformation of Genetic Information.* Problems of Information Transfer, 1965, vol. l. no. l, pages 105-113 (Russ).

inclination towards polygamy, and other behavior patterns.[41]

Having dealt with the question of the evolutionary roles of the sexes, Geodakyan extended his proposition to other questions. For example, what do connections between right and left-brain hemispheres, DNA /RNA protein pairing, the existence of X- and Y-chromosomes, autosomes, sex chromosomes, nucleus and cytoplasm, genotype, and phenotype have to the sexes?[42]

Ultimately, Geodakyan concluded that all biological systems (populations, brains, and cells) consist of two subsystems ("*operative*" and "*conservative*") created in the process of evolution. The "conservative" subsystem is responsible for preserving the whole system, while the "operative" subsystem is responsible for changing it. Notably, he considered these two subsystems ("operative" and "conservative") as two sides of one biological system rather than as two separate static forms. The "operative" subsystem corresponds to the masculine form, while the "conservative" subsystem corresponds to the feminine form.

Here are some examples to illustrate such a view. There are two distinct scales of time relative to the development of a species: First is "ontogenesis" (developing an individual from egg to adult); second is "evolutionary history" as the genealogy of a biological species. During ontogenesis, the egg, fetus, infant, child, and adult male are in different phases of developing a male subsystem. Most evolutionary changes occur during extended periods (the phylogenetic time scale), and the subsystems can be viewed as "phases," rather than as two different static forms. Following this reasoning, any system, by adapting to variable environments, will divide into two connected subsystems that are specialized according to *conservative* and *operative* trends in evolution and personal development.

Overall, the so-called "asynchronous evolution" increases the system's stability. Beyond the differences between male and female

---

[41] Vigen Geodakyan, *The Evolutionary Theory of Sex.* Priroda, 1991, issue 8, pages 60-69 (Russ).

[42] Vigen Geodakian, *About the Structure of Evolving Systems.* Problems of Cybernetics, 1972, vol.25, pages 81-91 (Russ).

sexes, Geodakyan lists other examples of binary conjugated differentiations: DNA-proteins, autosomes-sex chromosomes, right- and left-brain hemispheres, lateral asymmetry (unequal size or structure of right and left sides of organisms), and much more.[43] (Geodakyan, 2015).

The graphic representation of the relationship between the "conservative" and "operative" subsystems closely resembles the visual representation of Gevurot-Hasadim relations by Joel Bakst (Figure 40).

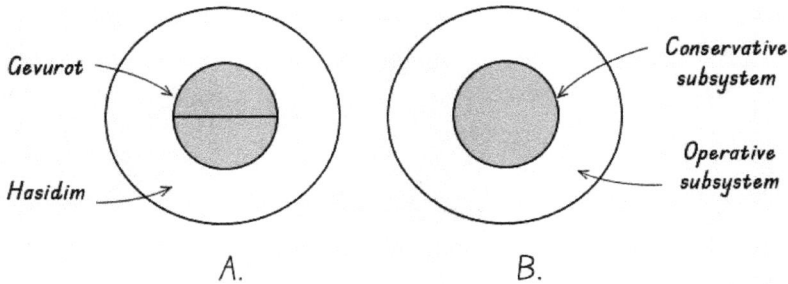

Figure 40. A comparison of views on the complementary polarity expressed by Kabbalist, Joel Bakst and evolutionary biologist, Vigen Geodakyan: – Bakst: With the power of the imploding Gevurot on the "inside" and the expanding Hasadim on the "outside" (modified from a slide from Bakst's lecture in 2021)– Geodakyan: Each biological system is divided into "stable nucleus" and "labile shell" (modified from Geodakyan 2015)

Two Subsystems in any Biological System

The good news is we can use the HuG code to handle any biological system! Here is a clear example of the nervous system, specifically, the autonomic nervous system, which automatically regulates multiple bodies and enables organisms to respond immediately to various environmental changes.

The autonomic nervous system is described as being composed of two subsystems: the sympathetic and parasympathetic.[44] This pair

---

[43] Sergey Geodakyan, *The evolutionary theory of asymmetry by V. Geodakyan*. International Journal of General Systems, 2015, vol. 44, no.6, pages 686-704.

[44] Eric Kandel et al., *Principles of Neural Science*. 2012, 1760 pages.

of subsystems innervates body organs and reflects evident "push-pull" interactions. Specifically, one side *increases* the heart rate while the other side *decreases* the heart rate; one side *dilates* the eyes while the other side *contracts* the eyes, and so on. Overall, the sympathetic system energizes the so-called "fight-or-flight" response that generates the ability to quickly adapt to environmental stimuli and behave as an arousal system.

In contrast, the parasympathetic system is considered a calming or inactive system. It maintains the balance of bodily functions and conserves the energy of the organism. Understandably, the sympathetic and parasympathetic systems are described as being "antagonistic" or "inhibitory" to each other. Thus, increased activity by one system tends to produce decreased activity from the other. Therefore, each system behaves to inhibit the functional expression of the other. Both of these systems are arms of the nervous system, which can, under normal conditions, innervate and regulate almost every organ and function of the organism being analyzed.

The main point here is the interaction between these two systems of primary importance and their inhibitory opposition. The sympathetic and parasympathetic systems "dance" together to use Bakst's expression. New studies have demonstrated that the interaction between the sympathetic and parasympathetic systems cannot be reduced to simple antagonistic oppositions.

The development of organisms is a complex process involving an apparent competition between organs for limited metabolic and informational resources. The growth of individual characters appears independent or "modular," as patterns of expression and transcription are often highly localized. Mutations have trait-specific effects, and gene complexes can be co-opted as a unit to produce unique traits.

The role of each organ is precise and irreplaceable. However, body parts are known to interact during ontogeny, and these reciprocal exchanges can be an essential determinant of developmental outcomes. Genetic and epigenetic mechanisms direct cell and tissue behaviors that prompt organs to communicate and cooperate. Together, the organs create adaptive, mutually beneficial dynamics expressed by physiological and biophysical processes.[45] The present-day view of organ interaction as "competitive" is limited, and closely related coordinative interactions across scales are seen as complementary and necessary. As Samuel Avital likes to say: "There are many organs, but only one body."

We can see a similar situation in the structures of biological communities and ecosystems, where multiple species of animals, plants, and microbes compete for resources. But above all, they should be seen as partners in the constant "HuG dance." Here, they act as organizing networks of relations; these complex systems demonstrate sustainable functions and adaptations.

Each biological system represents a unique system with two subsystems. Note: *there are never two separate systems.* This principle is, indeed, a declaration, but it is also a style of thinking. Here is an example: in May 2015, around 12,000 saiga antelopes (*Saiga tatarica*) were found dead in Kazakhstan. It was a typical example of an epidemic caused by an emergent disease. The preliminary results suggested that the antelopes had died of a *Pasteurella* infection that affected the lungs. However, bacteria of the *Pasteurella* species are already widely distributed in animal populations and usually do not cause massive deaths among naturally infected, healthy individuals. As a first step, we can assume that these bacteria cause acute illness and rapid death when an animal's immune system is compromised by another infection, poisoning, stress, or malnutrition.

We can also concentrate on determining which strains were isolated from the dead animals, thus helping us to identify the bacterium's unusual characteristics and then determine virulence factors. In any case, all the attention was on the emergence of bacterial infection, animal deaths, and the epidemic itself. In this

---

[45] R. Gawne et al., *Competitive and coordinative interactions between body parts produce adaptive developmental outcomes.* BioEssays, 2020, vol.42, no.8, e1900245.

situation, it is easy to overlook the miraculous phenomenon that saigas can live in the desert despite very harsh conditions. For example, saigas can eat some plants that are poisonous to other animals, including other antelope species.

Do you see how our perception can turn in one specific direction? Of course, in the midst of an epidemic, one must act quickly and efficiently to save the animals. However, this is not an excuse to limit your thinking regarding the animal population from both an emergent and stabilizing perspective.

It is very important to accept "opposite processes" in such investigations. Let's take a topic that has troubled society for years: the overuse of antibiotics. Recently, science has recognized the importance of a healthy microbiome, which further demonstrates the potential harm done by antibiotics. These substances kill pathogenic bacteria and severely damage a whole microbial community that had coexisted with host cells. "To use or not to use antibiotics" is a modern modification of the Hamlet question raised by health providers and consumers. A cautious use of antibiotics accompanied by actively considering all possible risks could lead to the productive interaction of *both* approaches.

The HuG dance has endless ways of manifesting in the biological world. One of the most visible is the "complexity-simplicity" dialogue. The familiar sources of biological complexity could be variability in organisms or the diversity of biological species. A mutation is one of the ultimate sources of genetic variation in biological populations. It is essential to recognize that new genotypes are constantly being formed, even without mutations. The source of increased variability or diversity is an expression of the giving tendency (more variants, higher complexity). Correspondingly, decreased variability and diversity indicate a receiving tendency (fewer variants).

The critical point from a Kabbalistic perspective is that opposite tendencies are *not* separate entities; they are manifestations of one Source. Once, during our class, Samuel Avital described Kabbalah as

335

a "way from complexity to simplicity, and from simplicity to complexity." By accepting the existence of two dualistic forms (male-female), we can define a continuing range between these extremes, seeing each situation as a point between two pillars. Each biological object and phenomenon can create the interplay between two forces, a) formative and b) diversifying.

Biological Force and Form as Hesed and Gevurah

The relationship between the pillars (Giving-Receiving/Hesed-Gevurah) can also be represented by an interplay between force-expressed and form-manifested tendencies in biology. Energy must manifest, and every biological system must come into play when manifested. A biological system cannot be analyzed based only on structure, e.g., morphological characteristics or nucleotide sequences. Remember, *biological objects are far MORE than patterns of form.*

Shirley Chambers pointed out that mankind's great problem is the identification with form. Scientists once identified objects by classifying the formation of these objects. When describing biological systems, it is far more appealing to measure their forms than to describe the activities and functions of their complexity and ephemerality.

When any biological object constantly changes its movement, condition, status, or form, a force is responsible for this change. At the same time, it is evident that biological processes are not simply reduced to mechanical processes. There are also specific biological phenomena such as breeding, migration, digestion, communication, and others. The force that created such processes can be defined as a "biological force."

There is nothing mystical about the above. When I use the terms "power'" or "force,"' I refer to very simple, testable, and irreplaceable categories. The terms are close to those used in elementary physics and are based on common sense. The "force" can be measured and investigated by studying potential mechanisms. However, the first step is to define and measure the effects caused by force.

Can we continue to study the forms and behavior of biological objects without asking ourselves, "What was the source?" Of course, we can list multiple descriptions of life's countless forms and manifestations, and these descriptions will allow us to accumulate complete information. However, it will be increasingly difficult to comprehend this information. Instead, let's use "force" in biology as a convenient concept to measure observed effects in changing biological organization.

Every effect has a source, and any biological process is a *type* of effect. Therefore, each biological process, movement, or development is caused by something. For example, climatic factors such as temperature cause a specific effect, such as the change in hormone status leading to the beginning of the breeding season. We can experimentally test this hypothesis, but this factor will remain a factor, not a source. We can learn about many other factors promoting similar biological effects. The act of describing the factors is a necessary part of biological science. It provides information that identifies internal and external factors contributing to the emergence of biological processes.

The question here concerns *forces*, those fundamental causes that follow traditions established in physics. There is no need to discuss other designations for the word, "force." Whatever word is used, this concept should help us understand biological systems and the measurement of "biological forces." The purpose, direction, and motivation of the action cannot replace the essence of biological forces. When an animal moves, it does not think, "This is time to eat or mate." Biologists can identify stimuli that cause behavioral patterns or physiological changes, but this is never the complete picture. In many ways, we can measure the value of biological processes by their current success in survival, breeding, extension of the area of distribution, etc.

Biological individuals, populations, and species express some change in activities from a force pushing in this direction. When energy passes through a system, it tends to organize that system. The point is that energy is not just passing through; *it is an inseparable aspect of every biological system.* Nevertheless, while it is a physically experienced phenomenon, polarity in biology remains completely non-objective, like the concept of force in classical physics. The "quantum leap" beyond physics into metaphysics recognizes force as the essence of any biological phenomenon. As Shirley Chambers wrote in *Kabalistic Healing*:

> "The right pillar relates to levels of force. The left pillar relates to levels of form... Male gender expresses both masculine (force) and feminine (form) polarities through its force... Every animate... form has two streams of life – the life of the form and the life of the essence ...While force cannot be seen, we can see its effects... Force drives, energizes and provides impetus... [Force's] nature is to enliven and energize but must be contained to be effective... Involution consists of the development of form; while evolution is the expression of force. ...The power has no value unless it is stabilized in a form or activity. Sometimes that period of stability is just a moment and other times it may be eons, but eventually the continuous inflow of power breaks through the containment of form and dissolves it... Force seeks freedom to express, yet, uncontained force dissipates...Since the force never stops flowing, it must be directed into containment which is suitable both in strength and motive. Like nature moving through the seasons, force moving in cycles, although its application remains constant".[46]

British theoretical biologist C.H. Waddington wrote in his book, *Behind Appearance*: "What could be a better symbol of sexual love than the double helix of DNA? And the most far-fetching of modern molecular biology, that of an 'allosteric' compound – action at one sensitive site on the same body to change its properties – might have been directly derived from erotic experience."[47]

---

[46] Shirley Chambers, *Kabbalistic Healing.* 2000, 160 pp.
[47] C.H. Waddington, *Behind Appearance.* 1970, page 73.

HuG Relationship as a Dimension

We can pick any example from the biological world to see the force/form interplay as a manifestation of Hasadim and Gevurot. Let's return to the nervous system and examine its elementary unit, the "neuron." Neurons are the basic functional unit of the nervous system, and they generate electrical signals called "action potentials," allowing them to quickly transmit information over whole bodies. Neurons are responsible for receiving sensory input from the external world, then responding to environmental changes by causing "muscular contraction or glandular secretions, and regulating all body activities. While this process is about the functionality of neurons, structurally, neurons are cells. While variable in size and shape, all neurons have three parts: a) the nucleus; b) a long axon that extends from the nucleus toward other neurons; and c) small extensions called "dendrites."

Neurons, as functional units, can be seen as manifestations of Hasadim in their mission to unify all manifestations as one system. On the other hand, the effect of Gevurot can be seen in specific structures of neurons, which allow them to conduct their work. All interactions between functional and structural properties should follow a well-orchestrated performance in receiving signals from other nerve cells and relaying those signals to other neurons. We need not try to remember all the details, it is enough to imagine them as elements of the "HuG dance."

The HuG Dance is accomplished by many cellular processes, including glucose metabolism to supply energy for the cells. This energy maintains an electrolyte balance throughout the length of the axons. The electrolyte gradient changes rapidly when the nerve cell conducts a signal down its axon. The electrical charge then spreads down the length of the axon until the signal reaches the axon terminal. Then, the neuron produces various neurotransmitters released across the synapse (the gap between the axon and the next nerve cell). The neurotransmitters migrate across this gap and enter specific receptors on the next nerve cell, which then signals to that cell to begin propagating its signal.[48] And the dance continues.

The point is this: any biological process can be seen as a manifestation of the "HuG dance," where partners must interact very closely and precisely. Lyall Watson wrote in the book, *Lifetide*: "Life is not in proteins, but in the music written on it; in its ability to recognize other molecules and to hold ordinary atoms in an extraordinarily precise way... Form is the shaping force of life. All form and shape contain information... in the very configuration of our molecules and cells is the essence of our identity."[49]

On one side of the structure is the masculine principle; on the other, the feminine. Then, there is a balance between formative and driving forces. The general conclusion is that the two theories, which were previously regarded as mutually exclusive, actually require a single theoretical construction. Niels Bohr formulated a methodological requirement followed by the complementary principle: "The integrity of living organisms... demands typically complementary means of description." [50]

In order to comprehend the investigation of biological systems, there should be no fear of accepting two opposite views. We must appreciate the opportunity to consider opposite forces in manifesting any biological phenomenon. As Fulvio Mazzocchi of the Italian National Research Council phrased the principle of complementarity in biology, "two mutually exclusive modes of description are equally necessary." [51]

Regardless of the proposed terminology, we can see the "giving" and "receiving" tendencies in every biological system. In population ecology, an appropriate example is the concept of "source and sink," which addresses the situation when a local demographic surplus arises in good-quality habitats (source), whereas a local demographic deficit occurs in habitats of poor quality (sink). A related situation on

---

[48] Andrew Newberg and David Halpern, *The Rabbi's Brain*. 2018, 435 pages.

[49] Lyall Watson, *Lifetide*. 1979, pages 40 and 103.

[50] Niels Bohr, *The unity of knowledge*. In *Atomic Physics and Human Knowledge*. 2010, pages 67-82.

[51] Fulvio Mazzocchi, *Complementarity in biology*. EMBO reports, 2010, vol.11, no.5, pages 339-344.

the ecological community level emerges with the problem of invasive species and their interaction with native species. In other biological applications, biologists prefer to use the terms "donor" and "recipient." Still, generally speaking, the meaning is the same: something comes from one biological entity (DNA molecule, cell, organism, population, etc.) to another biological entity that receives the material or information.

The view of opposite sides as duality and mutual exclusivity is rooted deeply in the European tradition of attaining knowledge. In contrast, the Kabbalistic approach presents a more cooperative view of any biological system existing side by side with opposite, yet complementary forms of expression. Examples are life-death, health-diseases, movement-stillness, development-regression, multiplication-reduction, diversification–selection, invasion-extinction, etc. The mastery of Kabbalistic Biology is based on switching back and forth between views of a biological object as a separate object and as a biological process in the object's integrity.

Those directions appear in the "HuG dance," and the biological world offers endless manifestations. In the Introduction to the main treasure of Kabbalistic knowledge, *The Book of Zohar*, Daniel Matt, who translated it, concludes, "to penetrate a realm beyond distinction …to navigate between conflicting meanings and determine the appropriate one – or sometimes to discover how different meanings pertain simultaneously." [52]

The last citation in this chapter is from the great Kabbalist, Rabbi Adin Even-Israel Steinsaltz. In *The Mystery of You*, written by Steinsaltz together with Ron Goldschlager, they wrote, "life is not a black-or-white matter; on the contrary, every event … is something on the gray scale…. In the scientific realm, too, there is a great advantage in discerning the shades… Understanding small differences and how they are created is the most basic tool for

---

[52] *The Zohar*. Pritzker Edition. Volume One. Translation and comments by Daniel C. Matt, 2004.

developing in-depth knowledge."[53]

Each biological phenomenon can be viewed from its components as *giving (impacting or driving)* and *receiving (adapting and stabilizing)*. The forces of the Hasadim side represent the "Giving, Operative, and Diversifying" tendencies, while the forces of Gevurot represent "Receiving, Conservative, and Forming" tendencies. When we recognize opposites and polar extremes, we must acknowledge the range of possible conditions of any biological phenomenon. This process allows us to arrive at the specific point of observation relevant to us at a particular moment.

By keeping the entire spectrum of possibilities open and available, we can make far more intelligent choices that account for extremes and enable us to find our position on the continuum. No biological objects, systems, or processes can be explained by a single force, *but we can define them by their relationships.*

In the previous chapter, we defined "fragmentation-unification" orientation. Here, we are adding additional coordinates. When we travel, we must have a system of coordinates (map, compass); otherwise, we are lost. Any investigation into life and living systems is also a kind of journey--a very complex and challenging journey. We will need other coordinates, but we are off to a good beginning with the support of these basic dimensions.

---

[53] Ron Goldschlager and Adin Steinsaltz, *The Mystery of You: A Journey through the Paradoxes of Life.* 2010, page 345.

Chapter 15

Third Dimension: The Included Middle

In the previous chapter, we discussed two opposite poles in biology and the two-way movement between them. These poles represent two bases, or "pillars," as they are called in Kabbalistic literature (*matzevot* in Hebrew, meaning "to stand"). Kabbalistic Biology reveals how we can use the axis *between* the opposite poles to measure how these bases relate to various biological systems. However, it would not truly be a Kabbalistic approach if we limited ourselves to simply measuring the effects of the relationship between two opposing forces.

Introducing a "Third Pole" into the Bipolar World

I propose taking on a more challenging task: introducing a third pole into the bipolar world! By doing so, we can use Kabbalistic biology to show us how to gain more knowledge and take better-informed action. According to formal logic, the Law of an Excluded Third (or Middle) states that "every proposition is either true or not true." However, with Kabbalistic logic, we can use the term, "an included third." We cannot claim that we "know" the essence of biological life until we accept the presence of a third point. Then, we can use it to gain additional information and achieve greater depth in our research.

The value of exploring the "included third" becomes particularly clear when doing biological research. It can lead biologists to the middle ("central") pillar (axis). The right axis can be defined as "a pillar of driving force toward development, evolution, and a variety of changes." In contrast, the left axis is "a pillar of consolidation of stable and long-lasting biological forms that have adapted to existing conditions." Thus, the role of the middle pillar is that of "The Pillar of Balance."

Let's define the "Third Dimension" and the cardinal difference between it and the Second Dimension. After all, *both* dimensions relate to two extreme poles. It is no accident that Basarab Nicolescu, the president and founder of the International Center for Transdisciplinary Research and Studies, called such a point "the hidden third."[54]

Tiferet is the Balancing Point

Here are some warnings about using the word "balance" in biology. First, this is not the same "balance" as that experienced in non-living systems where a "balancing point" can be fixed. In biological systems, finding and keeping a balance is challenging, due to its highly dynamic, unstable nature.

The second caution is even more critical. Unlike non-living systems, living systems possess both "functional" and "meaningful" balances. Biological organisms are complex systems that require a delicate balance of various internal parameters, such as temperature, pH, and nutrient levels, to maintain health and functioning. However, the presence of balance in an organism is often dynamic and subject to change. Most importantly, balance is crucial to the organism's survival and well-being.

In Kabbalistic thinking, the Third Dimension is easily defined by the concept of Tiferet, a sephirah commonly depicted in the center of the Tree of Life. Many books explain Tiferet as the balancing

---

[54] Basarab Nicolescu, *The Hidden Third*. 2016, 178 pages.

position between Sefirot Hesed and Gevurah. However, one must be careful to avoid the perception of the word, "balance." It is *not* a "compromise" or a "little bit of this and a little bit of that." The word, "middle," as in the middle pillar of the Tree of Life, represents an additional dimension (Figure 41).

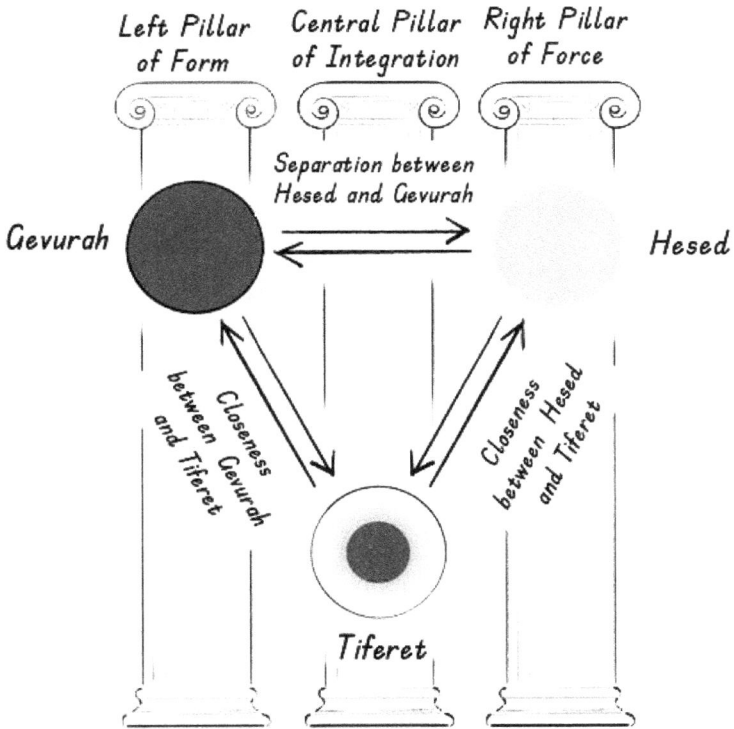

Figure 41. The middle pillar, "The Pillar of Balance," is represented by Tiferet. The sefirah of Tiferet closes the triangle that joins together the two polarized Sefirot, Hesed and Gevurah.

Kabbalah often refers to the importance of the number "Three." Whereas Number One represents "the unity of one dot," Number Two represents a line between two polarized extremes, and Number Three closes the triangle, introducing "wholeness." The third axis designates a position vastly different from either the right or left pillar--and simultaneously involves them both! It creates an entirely different position that supports activity in a specific developmental and evolutionary stage. Crucially important is that *each point between the extreme manifestations of Life indicates a certain proportion between the opposite forces.*

We can trace the recognition of the power of the third point to the oldest Kabbalistic source: *Sefer Yetzirah*:

"Three:

Each one stands alone

one acts as an advocate

one acts as accuser

and one decides between them."[55]

In the Zohar, the most fundamental source of Kabbalistic teaching, we are told: "The right entered that perfect pillar in the center, embracing the mystery of the left..., grasping the power of three points... All were united in the central pillar, generating the foundation of the world... The central pillar ...entered between them [right and left], mediating the conflict, reconciling the two sides." [56]

Rabbi Moshe ben Maimon (Rambam, also known as Maimonides) speaks of avoiding extremes and choosing the sensible way between them. The middle ("golden") way between extremes, the medium between contradictions, is the basis of his ethical system. Representing Maimonides' view, Adin Steinsaltz states that any tendency towards the extreme, even if that extreme is considered good, is "equivalent to casting off the yoke to create a disappointing deviation. In contrast, the medium way is good and genuine."[57]

Berke and Schneider stated in *Center of Power: The*

---

[55] *Sefer Yetzirah* 6:5 (Aryeh Kaplan, *Sefer Yetzirah. The Book of Creation.* 1997, page 250 (italics is mine).

[56] *The Zohar*, Be-Reshit 1:17a. Pritzker Edition, Vol. One, 2004, pages 126 and 128.

[57] Maimonides wrote indispensable works of Jewish philosophy and Halacha, but he was not a kabbalist.
The reference to Maimonides's view on the middle way is from the book of the great kabbalist of the 20th century, Adin Steinsaltz, *"Dear Son to Me,"* 2011.

Convergence of Psychoanalysis and Kabbalah that the function of the central trunk of the Tree of Life is to maintain order and balance, since "any branching (orientation) too far to the right (softness) or too far to the left (severity) needs to be counterbalanced."[58]

In Kabbalah, a common way to designate points indicating the pillars is to use the names of the three sefirot associated with them. As we learned in previous chapters, the right pillar is Hasadim (plural of Hesed), and the left is Gevurot (plural of Gevurah). The central pillar is symbolically marked as Tiferet. I have determined some points (states) that will help define how we can measure biological systems by using a kabbalistic orientation system. The system of "sefirot" is not equal to the system of "pillars." Instead, we will use the names of these three sefirot as markers, with Tiferet representing the balancing point.

Berke and Schneider opined that in Kabbalah, emanation and flow do not operate as a one-way system. Instead, there is a back-and-forth movement that is often oppositional, bringing us back to the center so we can move forward. That is Zohar's concept of using the left and right poles to arrive at a center, a more balanced position.[59] The central pillar, Tiferet, mediates between the opposites of Hesed and Gevurah.

The Aesthetic Balance as Beauty

While the third point may not be conventionally associated with biology, nevertheless, it is essential. Harmonious relations between an element of a whole biological system imply that the third point is balance. Tiferet, the archetypal model of the third point, literally means "beautiful."

In his book, *The Future of Art in a Digital Age*, Mel Alexenberg, an American-Israeli artist and writer recognized for his pioneering

---

[58] Joseph Berke and Stanley Schneider, *Center of Power: The Convergence of Psychoanalysis and Kabbalah*. 2008, page 107. Joseph Berke is the director of the Arbous Crisis Center in London. Stanley Schneider is professor and chairman of the Program for Advanced Studies in Integrative Psychotherapy at Hebrew University, Jerusalem.

[59] Ibid., page 189.

work in exploring the intersections of art and science, discusses the concept of "aesthetic balance." He cites Paul Weiss, Head of the Laboratory for Developmental Biology at Rockefeller University, who described beauty as a dynamic balancing between the polarities of freedom and order: "The aesthetic experience is to me the attempt of man to depolarize extremes and reconcile them; to learn the dynamics of moving from the whole to the part and from the part back to the whole."[60]

Alexenberg explains that aesthetic experience fits between "seeing freedom in the small" and "finding order in the gross." As his excellent example points out, "We can see that a maple leaf is clearly not an oak leaf. No two maple leaves, however, are congruent and share the same venation pattern. Each leaf displays the order in the gross and freedom of excursion in the small. It is a unique, one-of-a-kind expression of a general pattern of a genetic type called a maple leaf."

Alexenberg continued, "According to an anthropologic metaphor in ancient kabbalistic texts, Beauty is the pulsating heart energizing all the other parts of the body symbolized by the sefirot. The Hebrew letter pathways are the arteries and veins bringing the vital flow from the Beauty's heart to the body's extremities."[61]

The Central Column Combines the Right and the Left

Tiferet harmonizes Hesed and Gevurah in a dialectic relationship without flattening their characteristics. Tiferet is not a reductionist approach to reject any of the opposite forces. In fact, Y. David Shulman referred to the function of Tiferet as the word root, "pier," which means a "tiara" or "the crowning branches of a tree."[62]

In another of his books, *Photograph God*, Alexenberg wrote:

"Aesthetical balance between Hesed and Gevurah gives rise to the

---

[60] Mel Alexenberg, *The Future of Art in a Digital Age*. Chapter: Aesthetic Experience in Creative Process, pages 197-202.

[61] Ibid., page 140.

[62] Y. David Shulman, *The Sefirot*. 1996, page 139.

sephirah of Tiferet ...emerging from dynamic interplay, creative dialogue, and elegant integration between the sephirot...Hesed opens the floodgates to encourage overflow ... Gevurah is the strength to control the flood by routing the boundless flow through a narrow channel, and Tiferet controls the flood by offering it freedom to flow through alternative waterways."[63]

The "Central Column" combines the positive right and negative left columns into a single unified whole, creating a circuit. The practice leading to the unification process between male and female aspects is denoted in the Kabbalistic tradition as "Yichud," denoting male-female "seclusion." Moshe Haim Luzzatto wrote in the "Secrets of the Future Temple":

"The movement of creation toward this rectified state is characterized by various kinds of "joining" and "pairing" of the various *partzufim*, some of which are involved in giving, while others are more associated with receiving. The supreme example of creative interaction, giving and receiving, is when Man and Wife align and join together in holy union – Yichud – to parent a child who will unite in one being essential qualities of both."[64]

Christopher Benton wrote *In Search of Kohelet*:

"The pattern of opposites and an intermediary can also be modeled mathematically by positive and negative numbers on a number line with zero standing in between. The concept of zero was a late addition to the current number system. This is understandable since in all cultures numbers were originally created for counting, and as it says in the *Sefer Yetzirah*, "Before one, what do you count?"[65]

In modern mathematics, Benton argued, all of the positive and negative numbers can be created from logical manipulation of the

---

[63] Mel Alexenberg, *Photograph God*, page 74.

[64] Moshe Chaim Luzzatto, *Secrets of the Future Temple: Mishkney Elyon*. Translation and overview by Avraham Greenbaum, 1999, page 40.

[65] Christopher Benton, *In Search of Kohelet*. 2003.

"null set," the set-theoretical version in mathematics of zero and the void. Eduard Shyfrin noted in *From Infinity to Man* that the Central Pillar ("Median Balance") does not represent a choice between two alternative poles (a "yes" or "no"), but is based on the acceptance of both poles.[66]

As Himon Shokek, Professor at Baltimore Hebrew University, wrote in *Kabbalah and The Art of Being*: "In Kabbalah, creation is always the culmination of the struggle between the thing and its opposite, but at the same time it is the integration between "the thing and its opposite."[67] The unification of the opposites' movements stands at the center of any process in Nature. According to Kabbalistic tradition, this movement reveals the inner wholeness of any system into a divisible structure of ten differentiated Sefirot aligned in two opposite pillars. The second movement signifies integration, allowing opposites to interact in harmony.

The co-existence of these two simultaneous movements is vital to the development and growth of individual biological systems: cells, organisms, populations, etc. Shokek continues: "The internal principle of the organism is a developmental principle that strives toward varying levels of potentiality and actuality. It is a principle that has within itself a force for future growth and development ... in the realm of change and stillness."[68]

It's important to understand that the Third (central) point is an actual position where the direction of any biological process can change to the opposite side. I like comparing the Third point to a checkpoint at the state border. Although you have left one state, you have not yet entered another state, so you do not belong to either one. (A border checkpoint cannot be manned for long enough to prevent all other movements from being blocked!) Still, you can measure the distance from the border while moving forward and back.

---

[66] Eduard Shyfrin, *From Infinity to Man*. 2019, page 32.
[67] Shimon Shokek, *From Kabbalah and the Art of Being, The Smithsonian Lectures*. 2001, page 49.
[68] Ibid., page 79.

An Attribute of Nothingness

Dov Ber ben Avraham, better known as "the Maggid of Mezritch," taught: "Nothing can simply change from one reality into another without first attaining the level of Nothingness. An egg must first cease to exist as an egg before the chick can come forth from it. So, it is with everything in the world."[69]

The example provided by this Hasidic sage who lived in the 18th century is very specifically biological! There is a point when an animal, during its delivery, is no longer an animal embryo connected to its mother but is still not an independent individual organism. It is both "a thing and its contradiction." Two "things" do not exist independently, but they can co-exist at the same time!

The situation discussed by the Maggid of Mezritch was used by the great Kabbalist of the 20th century, Aryeh Kaplan, who called it "the attribute of Nothingness." He referred to this attribute of the Hebrew word "Hyle" as "the state between potential and realization." Kaplan wrote, "...an instant when it is neither chicken nor egg. No person can determine that instant, for in that instant, it is a state of Nothingness... Nothing can change from one thing to another [without first losing its original identity]. Thus, for example, before an egg can grow into a chicken, it must first cease totally to be an egg. Each thing must lose its original identity before it can be something else."[70]

As Joel Bakst noticed in *The Jerusalem Stone of Consciousness*: "...both views are superimposed one upon the other generating a third reality... is not a "thing" but rather it is a "method."[71] Thus, the middle point is a superposition of two pillars. From a Kabbalistic perspective, the dilemma of "the chicken or the egg" does not make sense. Instead, the solution is to accept "the chicken AND the egg" as one entity. The Sefirah of Tiferet serves the role of the "AND," the middle point that connects both; the chicken and the egg. We cannot

---

[69] Martin Buber, *Tales of the Hasidim: The Early Masters*. 1947, page 104.

[70] Aryeh Kaplan, *Meditation and Kabbalah*. 1982, page 300.

[71] Joel David Bakst, *The Jerusalem Stone of Consciousness*. 2013, page 117.

see such a point with our physical eyes. This is the reason why the third point is called "hidden." However, it is a necessary component of each biological object or process. We cannot see it, but we can learn and use it!

Homeorhesis vs. Homeostasis

"Balance" is a good word for describing the central pillar, but it can be a bit misleading. When we talk about balance, we might initially associate the word with "equilibrium" or "homeostasis." Homeostasis is the tendency of biological systems to maintain relatively consistent conditions in their internal environments. At the same time, each biological system constantly interacts with and adjusts to changes originating within or outside the system.

Examples of homeostatic mechanisms include controlling body temperature, blood osmotic pressure, and hydrogen ion concentration to keep them within the normal range. Such mechanisms allow nutrients to be supplied to cells as needed and waste products to be removed before they reach a critical level of toxicity.

When considering homeostasis and homeorhesis, we must understand the difference between these two processes. Both represent the balance between two opposite forces, but the latter is more important in biological life and closer to the definition of the third dimension in Kabbalistic biology.

The term "homeorhesis" is derived from the Greek word, *rheusis*, meaning "similar flow." Originally, this term was proposed by British biologist C.H. Waddington as an alternative to Cannon's homeostasis. It conveys the energy of flow, rather than stasis. This concept encompasses dynamic systems that return to a trajectory, unlike systems that return to a particular state known as "homeostasis."

Rene Dubos, the famed microbiologist and experimental pathologist, pointed out that homeostasis says, "Whatever is, it is right." However, Dubos prefers the concept of "homeokinesis" --

basically, a term identical to Waddington's "homeorhesis." Dubos' definition of this process is "stabilized flow, rather than a stabilized state."[72]

The "homeorhesis phenomenon" was proposed to demonstrate the dynamic nature of embryonic development, but it is not limited to developmental biology. In physiology, homeostasis is the ability of a system or living organism to adjust its internal environment to maintain a stable equilibrium. One such example is the ability of warm-blooded animals to maintain a constant body temperature. The term "ecological homeorhesis" was applied in several scientific publications as a stage of the evolutionary process and as a mechanism of pathogen virulence.[73]

Mel Alexenberg made an interesting connection between Tiferet and the work of C.H. Waddington. In *The Future of Art*, he wrote: "The cell, the basic element in living systems, provides a meaningful analogy for sephirot in relation to the pathways leading in and out of them. The Beauty Sephirah exhibits what biologist Waddington calls "homeorhesis" in his paper, "The Character of Biological Form."[74] In an unexpected (for an artist) twist, Alexenberg wrote:

> "He explains that organic form, whether the structure of a population, individual organism, or cell, "is produced by the interaction of numerous forces which are balanced against each other in a near-equilibrium that has the character not of a precisely definable pattern but rather a slightly fluid one… Organic form like the Beauty sephirah simultaneously exhibits balance and the lack of it, homeostasis and homeorhesis. Homeostasis is balanced wholeness, integrity, and self-regulation. Homeorhesis is a fluid structure of growth, transformation, and evolution. A cell's integrity is

---

[72] Rene Dubos, *Environment*. In *Dictionary of the History of Ideas*, 1973, vol.2, page 126, cited from Morris Berman's The Reenchantment of the World.

[73] Alexander Dubov, *Ecological Homeorhesis as the Stage of Microevolution*. European Journal of Natural History, 2007, no.2, pages 142-145.

[74] Alexenberg refers to C.H. Waddington. *The Character of Biological Form*. In *Aspects of Form.*, 1968.

maintained by a semi-permeable membrane. The same membrane that contains the protoplasm is also an organic system that regulates what flows into and out of the cell. Although the cell is an integral unit, it can grow and reproduce itself. Encoded in its DNA, an individual cell contains both information for its own process of homeostasis and homeorhesis, as well as the entire set of information for the balance and growth of the entire organism."[75]

I found a fascinating point related to our discussion of homeostasis in Viktor Frankl's seminal book *Man's Search for Meaning*. Harold Kushner called this "one of the great books of our time." In his book, Frankl wrote:

> "It is a dangerous misconception that ... in the first place is equilibrium or, as it is called in biology, "homeostasis", i.e. a tensionless state… but rather the striving and struggling for a worthwhile goal, a freely chosen task… is not homeostasis but ... "Noö -dynamics," i.e., the existential dynamics in a polar field of tension where one pole is represented by a meaning that is to be fulfilled and the other pole by the man who has to fulfill it."[76]

Noö-dynamics is Frankl's term for the tension between what one has already achieved and what one ought to achieve. The "noological dimension" can be seen as the space between stimulus and response, where humans are free to choose their reaction to their biological, psychological, and environmental conditions.[77] Frankl says that healthy people must live in a state of tension between the past and the present. Of course, this could be said about any biological system!

The third point holds the "hidden" position in the simplicity-complexity continuum, which indicates the change in the function of

---

[75] Mel Alexenberg, *Future of Art*, page 139.

[76] Viktor Frankl, *Man's Search for Meaning*. 2006, page 105.

[77] McDonald and Perry, *Franklian Existentialism and Transformative Learning: Unlikely Co-Captains in Uncertain Times*. Journal of Transformative Education, 2023, vol.21, no.4, pages 556-573.

a whole biological system. Sky Nelson-Issacs wrote in *Living in Flow*: "Sometimes flow involves letting go of attachment to a central path; sometimes it involves pushing through resistance to stay on a central path. *Flow is a middle path between rigidity and spontaneity.*"[78]

The Third Point and the Optimality Principle

Below, I will provide several biological processes in which the balance is more dynamic and fluid, rather than simply a fixed homeostasis. One of the main properties of the third point is "optimality." This is the condition in which any biological process works at maximum efficiency. The idea that Nature pursues optimality (economy) in all Her workings is one of the oldest principles of biological science.[79]

Optimality in biology can manifest in different ways. Here are some fundamental expressions of the principle of optimality that I propose as parameters of the Third Dimension of Kabbalistic biology:

1. The coherent coordination of all elements within a biological system that facilitates the most harmonious expression of the system's functionality.

2. The coexistence of a biological system with its environment that improves adaptative mechanisms and leads to optimally effective ways of using external resources.

3. A choice of variants that are the most adequate for the evolutionary trend, which is meaningful for a particular biological system for a specific time and space.

Let us illustrate the first of these principles by using the process of "metabolism." As mentioned above, metabolism includes a range of biochemical processes occurring within any living organism. It consists of "anabolism" (the buildup of substances) and "catabolism" (the breakdown of substances). Innumerable biomolecules of

---

[78] Sky Nelson-Isaacs, *Living in Flow:*
The Science of Synchronicity and How Your Choices Shape Your World. 2019, page 68; italics are mine.
[79] Robert Rosen, *Optimality Principles in Biology.* 1967, chapter General Considerations, page 1.

different kinds regulate metabolic processes that represent one of the most complex, dynamic systems in Nature. Who can guide a performance of such a complex orchestra? System biology proposes an answer, and surprisingly, *without* a conductor!

When calculating the design principles of a metabolic network, Jan Berkhout of the Systems Bioinformatics in Amsterdam and his colleagues compared this complex biological system to an engineering strategy.[80] To design a particular device, an engineer needs a system specification explaining what the system should do and enumerating the many conditions that should be considered, e.g., the production costs and other constraints. When one studies a biological design, the reverse approach is required. Both *the function and the relevant constraints must be balanced.*

We can identify a designing principle when similar functions occur across organisms or networks, and similar networks achieve them. In other words, *a design is a mechanism for system functions that have succeeded in the evolution of organisms.* In biology, it makes sense to identify design principles as the result of an "optimization" process of fitness. If the process is too simple, it might not work correctly. However, if it is too complex, resources can be wasted, leading to its elimination by selective pressure (an evolutionary force) from a certain environment.

Another example is the trade-off between growth and defense in plants. Plants have evolved into carrying an array of defenses against pathogens. However, mounting a defense frequently comes at the cost of reducing growth and reproduction. These carry critical implications for natural and agricultural populations. As plants evolve, eventually, they will reach the point of developing *an optimal point of balance* for both immunological protection and growth. Indeed, such a trade-off depends on both its underlying mechanism and its environment.

---

[80] Jan Berkhout et al., *Optimality principles in the regulation of metabolic networks.* Metabolites. 2012,
vol. 29, no.3, pages 529-552.

Talia Karasov of the Max Planck Institute for Developmental Biology in Tübingen, with her colleagues, clearly illustrated the "optimal" relationship between growth and defense in plants:

"Imagine a young plant that is energy limited, suffering attack, and finds itself in a highly competitive environment where early growth is essential. The best strategy for this plant may well be to forego defense and grow, at least provided the pathogen is not too virulent. By contrast, that same plant in a sparse field or at an older age may do better to defend itself. In other words, the "optimal" allocation to defense is not static but should be responsive to ecological and phenological conditions."[81]

The Challenge to Define "Health"

Now, let's consider a complex concept such as health. As I mentioned earlier, defining "health" is a crucial but challenging task. Any complex biological system, such as an organism, is not just a collection of parts. Rather, it's a system of complex relationships that can bring every component into one coherent union. Analyzing any one parameter apart from the whole system can lead to significantly misleading conclusions, thus limiting any forecasting. Therefore, complex systems are characterized by unavoidable uncertainty. Problems arise when any decision about health as a complex system is based on linear thinking and considering elements of these systems as "separate parts."

There is often a tendency to consider health as the "normal" status of an organism. Often, one assumes that there is a kind of "norm," a range of biological parameters that can be defined as "normal." The most common examples are considering the "normal" body temperature of adult humans as ranging from 97°F to 99°F, "normal" sugars from 70–99 mg/dl, "normal" hemoglobin levels between 13.5 and 17.5 g/dl, and so on. There is no argument against such statements. Such an approach to describing biological systems

---

[81] Talia Karasov et al., *Mechanisms to Mitigate the Trade-Off between Growth and Defense*. Plant Cell. 2017, vol.29, no.4, pages 666-680.

is quite legitimate on some levels. However, these parameters are not sufficient to confirm one's health status.

The critical message here is that the third point is *not* limited to any average rate or "normal" range. It is more indicative of functions balanced under specific conditions. Each of the listed parameters will be accurate in different situations, depending on whether the organism is resting or actively involved, before food or after, and so on.

As we're talking about health as a property of complex biological systems, let me introduce an approach to define health proposed by David Krakauer of the Santa Fe Institute.[82] In his paper, *Complexity and the Ultimate Meaning of Health & Disease*, he suggested four general principles that can characterize the status of health:

1. *The equilibrium property* represents a tendency to define "health" as a normal state, a kind of baseline.

2. *Stability* is the ability to return to a stable state over a reasonable period. The environment inside and outside human beings is constantly changing. The health of people and the environment is not gauged by keeping a standard rate. It is measured by the ability of organisms and the environment to restore balance when upset.

3. *Multiplicity of memory* means the ability of a healthy system to perform multiple functional states.

4. *Adaptation through learning* refers to long-term adaptation as the reconfiguration of the condition under the new physiological regime.

No biological system exists apart from its surrounding environment. We must consider the environment as a part of any biological system. When ecologists talk about "the environment," they often concentrate on specific variables, such as temperature, humidity, pH of the substrate, etc. However, *any* natural environment is a complex system, not just a combination of separate ecological factors.

---

[82] David Krakauer, *Complexity and the Ultimate Meaning of Health & Disease*. 2014, no.3, pages 143-157.

The "Third Dimension" reflects a degree of the dynamic balance between the functionality of each biological species and its ever-changing environment. Remember, the elements of each biological system are constantly changing. Here, I refer readers to physician Larry Dossey, who wrote in *Space, Time & Medicine*: "A single DNA molecule... exists only for a few months... Over a period of months, our entire genetic structure is renewed... The entire body participates in this astonishing dynamism." Dossey coined the term "biodance" to describe the endless exchange of the elements of living things with the Earth.[83]

The "Biodance" between Growth and Decline

To illustrate the principle of "biodance," I will use the dynamics of rodent populations as an example. Many hypotheses were proposed to explain the mechanisms responsible for dramatic fluctuations in rodent populations over the years. In a stable population, ecologists talk about a kind of "homeostasis" when the sum of birth and immigration are relatively equal to the sum of death and emigration.

In reality, the number of individual rodents in the same population can increase tenfold or decrease. The natural selection of self-regulatory behavior in animal populations states that cyclical fluctuations are caused by the prevalence of certain genetic types in rodent populations. These lead to an increase or decrease in numbers, a critical part of self-regulation.[84] However, during the same period, many other investigators questioned this explanation, claiming that other factors were responsible for the fluctuation in population densities. These included weather conditions, food supply, diseases and parasites, predators, habitat transformation, and random events.

Here is the bottom line, *there is no long-lasting balance between the growth and decline of rodent populations*. However, there is a central point in population dynamics when internal and external

---

[83] Larry Dossey, *Space, Time & Medicine*. 1982, page 73.
[84] Charles Krebs, *Population Cycles Revisited*. Journal of Mammalogy, 1996, vol. 77, no. 1, pages 8-24.

factors jointly lead a biological population to *the status between growth and decline*. The most important factor here is the unique *position* from which the population can go up or down.

Let's take a situation where a biological organism is infected with bacteria. Immediately, the immune system races to protect the organism from outside invaders. However, the organism has a rich symbiotic microbiome (the community of microorganisms living in the body) that plays a crucial role in many biological functions: digestion, production of vitamins, behavior, and many others. Therefore, being able to discriminate between "own" and "alien" microorganisms can be challenging.

Imbalances in microbiota-immunity interactions under defined environmental conditions may well contribute to today's prevalent immune-mediated disorders. The immune system should not be oversensitive to symbiotic microbiota. There is a specific point, even if it is rarely described at our present level of experimental capacity, where the immune system reacts equally toward eliminating internal microorganisms and allowing them to remain. The concept of immunity as a self/non-self-discriminatory process is not simple. The balance between eliminating harmful microbes and keeping symbiotic microorganisms depends on multiple factors that can create a unique context at each moment.

The Measurement of Meaningful Points

The third important characteristic of the Third Dimension is its ability to reveal how meaningful each point is in the biological process. This is particularly important in terms of ecological plasticity and evolutionary trends. The Third (central) pillar represented by Tiferet shows how each biological system or its element reacts with other biological systems, including those at higher levels. It also shows the investigator the essence of the biological element in question.

We can define the central pillar as a "sphere of integrity" regarding the meaning of biological organization. "There's certainly a place for meaning: the included middle," wrote Basarab

Nicolescu.[85] Warning! Here, we are on shaky ground. Biologists are often uncomfortable with the idea of "purpose" or "meaning" in Nature and are uneasy about its possible association with some form of "intelligent design." In philosophy, the doctrine known as "teleology" suggests that natural phenomena are guided not only by mechanical forces; they can also move toward specific goals of self-realization. Most biologists, however, do not like to be associated with this concept.

Such a conflict between the attitude of the biological community towards the idea of "purpose" is very controversial. Biologists consider the "purpose" of photosynthesis as the ability to transform water, sunlight, and carbon dioxide into oxygen and simple sugars. In organisms, the "purpose" of the kidney is to cleanse the blood of toxins and transform the waste into urine. Even if we replaced the word "purpose" with "function," it would not change this perception.

Nobody describes this dilemma better than J. B. S. Haldane, one of the most influential evolutionary biologists of the twentieth century: "Teleology is like a mistress to a biologist: he cannot live without her, but he's unwilling to be seen with her in public."[86] This phrase has given the name to a beautiful book entitled, *The Biologist's Mistress*, written by Victoria Alexander, the co-founder of the Dactyl Foundation in New York City. Alexander is not shy "to be seen with teleology in public," but tries "to rescue teleology from theology" using her own words. Alexander formulated teleology as "the emergent ordering tendencies of chance."[87]

Alexander has shown that "purpose" entails two distinct kinds of behavior. The first is associated with what she calls "directionality," the mechanisms that maintain order, such as stability and persistence. The second is associated with what she calls "originality" the discovery of new functions (change).[88]

---

[85] Basarab Nicolescu, *The Hidden Third.* 2016, page 58.

[86] From Ernst Mayr, *Teleological and Teleonomic, a New Analysis.* In *A Portrait of Twenty-five Years.* Boston Studies in the Philosophy of Science,1974.

[87] Victoria Alexander, *The Biologist's Mistress: Rethinking Self-Organization in Art, Literature, and Nature.* 2019, page 1.

[88] Ibid., page 97.

Tiferet as a Balance Between Persistence and Plasticity

In 1979, Ernest Nagel pointed out that goal-directed systems have two opposing properties, "persistence" and "plasticity."[89] Daniel McShea, Professor of Biology at Duke University, provided excellent explanations of these features. In McShea's terms, "persistence is the tendency to return to a certain trajectory after an assault or accident. If the cells in the developing embryo are diverted from a path toward the equator, whether by an experimenter or on account of some developmental accident, they will return to that path.[90] "Plasticity," in McShea's terms, is the ability to adopt a certain trajectory from alternative starting points.[91] For example, a bacterium will find a trajectory toward food, whether at the edge of the Petri dish or near its center.

From the perspective of Kabbalistic Biology, "persistence" (or "directionality," according to Alexander) represents a relationship between Gevurah and Tiferet. In contrast, "plasticity" (or "originality," according to Alexander) represents a relationship between Hesed and Tiferet.

The central pillar's critical role is to define the meaningfulness of information. Each point belonging to the central pillar could be designated, using Alexander's words, as "something truly new and yet meaningful." Imagine that the right pillar is a driving force ("energy" in the physical sense), and the left pillar is a formative expression ("matter" in the physical, not philosophical sense). In that case, the middle pillar is considered a *"measure of progress"* regarding information received (Figure 42).

---

[89] Ernest Nagel, *Teleology Revisited and Other Essays in the Philosophy and History of Science*. 1979, 352 pages.

[90] Daniel McShea, *Freedom and purpose in biology*. Studies in History and Philosophy of Biological and Biomedical Sciences, 2016, vol. 58, page 68.

[91] Ibid., page 68.

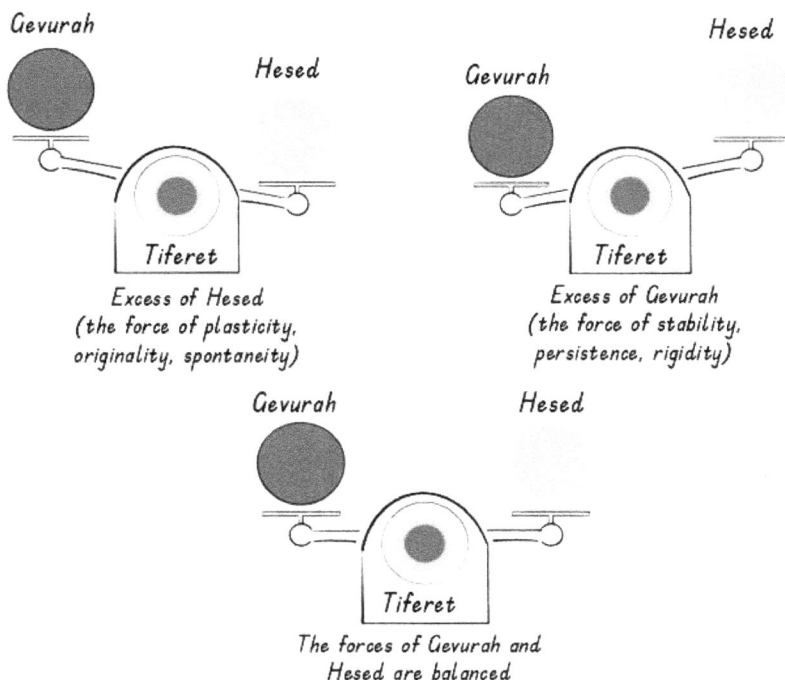

Figure 42. Tiferet is a balancing point between persistence and plasticity in biological systems.

A Journey from Fall to Growth

Part Two of this book describes many examples of "invisible" biological systems and functions. Figuratively speaking, the "Hidden Third" is the point where the invisible is made visible. The Third Dimension allows us to measure a degree of "invisibility."

Lawrence Kushner, a rabbi and storyteller, wrote in his book, *The River of Light*: "...life is a function of its balance, the relationship to one another, its inner arrangement of motion... Since the pattern is never still, it is more accurately called an arrangement of motions."[92]

Piet Mondrian, a Dutch painter and art theoretician, declared the great struggle for the annihilation of static equilibrium through continuous oppositions among the means of expression. He called it "dynamic movement in equilibrium." Mondrian wrote in *Plastic Art*

---

[92] Lawrence Kushner, *The River of Light.* 1983, page 89.

and Pure Plastic Art: "When dynamic movements established through *contrasts* or oppositions of the expressive means, the relationship becomes the chief preoccupation ... The right angle is the only constant relationship, and that, through the proportion of dimension, its constant expression can be given moment, that is, made living."[93]

The middle pillar can also be a metaphor for the phased transition between chaos and order when a biological system is in development. In Chapter 3, I provided a neat definition for this place in life, coined by Norman Packard as at the "edge of chaos." Here, I use an additional scientific metaphor to clarify the hidden meaning of the Third Point that provides the direction from the dynamic balance. According to this metaphor, the "Hidden Third" represents the contradictory relationship between "presence" (visible parts of biological systems) and "absence" (invisible parts of the same biological systems). The invisible parts are not absent—*we simply do not know about them!*

Let's go back to the example of rodent population dynamics. We defined the balance in this phenomenon as the point between the "growth" and "fall" of the population. Routinely, field ecologists measure the status of rodent populations by estimating the density of rodents (the number of animals in the area). However, the accuracy of such estimates depends on many factors. First, we don't know the actual number of animals, but only the number of animals captured or seen during a specific period. The results will substantially depend on such factors as the amount of effort to maintain them, seasonal effects, qualification of the field workers, etc. Nevertheless, we are aware of such limitations ("known unknowns") and can stay more or less close to the identification of the "balancing" point.

While many factors are critical for population dynamics, even an experienced biologist can only make a guess. Here are a few examples: the number of animals may be going up, but their

---

[93] Piet Mondrian, *Plastic Art and Pure Plastic Art*. 1945, page 10 (cited from C.H. Waddington *Behind Appearance*).

resistance (immunity) to pathogens is going down; the behavioral activity of animals can strongly affect the methods of detecting them; forage reserves may have become more favorable for animals, but the plant chemistry has negatively affected animal physiology. The knowledge of potential pitfalls is essential for researchers, and they should not be discouraged if the information is absent.

The Dilemma of Seeking Balance in Research

Kabbalistic Biology is based on the premise that you can choose how you perceive and measure the functionality of all biological systems. With the Third Dimension, you can choose between the strict requirements of modern science and a range of more flexible options free of such strict rules. This Kabbalistic attitude stems from the Judaic tradition of declaring one's inherent right to choose. A corresponding obligation accompanies each right. Making an unpredictable decision is contrary to relying on established norms and protocols.

As a scientist investigating epidemics biology, I often experience a similar dilemma. For example, I understand how the complexity of a biological phenomenon can result in epidemics; however, amid the epidemic of bubonic plague, for example, it is neither an appropriate time nor situation to offer public health experts my personal opinion on the complexity of the plague. More likely, these specialists will be looking for concrete recommendations. On the other hand, when I was invited to a conference at Cambridge dedicated to the complexity of epidemics, I enjoyed the opportunity to move on the side of Hesed (being open to every available point of view). The Third Dimension is a tool that can measure a degree of openness to potential variants at a meaningful moment of an investigation.

It is most important that *you* decide from which perspective you will undertake your research, considering the situation and your priorities. If you must urgently provide straightforward recommendations, move left (toward Gevurah). If you wish to more intensively explore the complex interactions of multiple factors affecting population dynamics, move right (toward Hesed). That is why I call this a "direction"--it allows one to measure the relationship with the central point and orientation in moving to the right or left.

Here are my suggestions to fellow researchers for being able to recognize the central point. Use your intellect, intuition, or personal preference. Once you seek the central point, you will be freed of uncertainty about making decisions. If you consistently practice the Third-dimensional approach and become aware that you are crossing the central line ("checkpoint"), prepare to explore the variants on both sides of the balance.Always remain aware of how far you may be from the balance point (central line).

Suppose a colleague criticizes you for seeking comprehensive solutions to a research problem instead of focusing on concrete goals. Smile and carry on. In another instance, should someone suggest you are "simplifying the situation" by making a direct decision, acknowledge the fact and keep going. You will know when the time is right to change course.

"You are right, too!"

As I indicated in the beginning, accepting the Third Point (direction) requires revising the logical practice of excluding the middle. Once you have included the middle, you will meet and be able to accept paradoxes, not as obstacles but as tools; in other words, you will be led to "paradoxical logic."

To introduce paradoxical logic, I will use an old Jewish story traced back to antiquity about a rabbi settling a dispute between two of his followers. The first man presented his complaints to the rabbi, and when he finished, the rabbi said, "You're right." Then, the second man presented his point. When he finished, the rabbi said, "You're also right." The rabbi's wife, who had been listening to the conversation, incredulously asked her husband, "These people say completely different things. How can they be both right?" The rabbi thought for a few moments and then replied, "You know, you're also right!"

I have told this story at several conferences, each time changing the context of the discussion. Here, I prefer to retell this story by referring to a debate between ecologists regarding the dynamics of rodent populations. One professor invited two visiting zoologists to his seminar. During the discussion, the first scientist argued that seasonal and annual changes in rodent physiology caused cyclic fluctuations in the vole population. The second presenter provided evidence that a population of the same rodent species strictly followed climatic changes. To both, the professor said, "You are right."

After the seminar, the professor's postdoc questioned the paradox and was told, "You know, you are right, too!" Joseph Witztum of La Jolla titled his article in the Journal of Clinical Investigation, "You Are Right, Too!" as he discussed the role of macrophage scavenger receptors in atherogenesis.[94]

This is not about the pros and cons of being a conformist. *This is about having a third point that acknowledges a paradox and acknowledges a contradiction.* The included middle leads to the acceptance of two possibly opposite trends. This is the point of bifurcation on Waddington's epigenetic landscape provided in chapter 8.

Onward to Tiferet!

Remember this: there is a dimension that does not measure the distance between polar positions. It measures between each extreme position and the central (third) axis. There is actually a point where both polar situations will potentially co-exist.

Let's go back to the example of fluctuating rodent populations. Each population is rising or declining. This can be measured by not only the number of individuals involved but also by parameters such as migratory behavior, aggression level, age structure, immune status, and more. It can also be defined as a specific trend in population dynamics between rising and falling, but there is an "ideal" point

---

[94] Joseph Witztum, *You are right too!* Journal of Clinical Investigation, 2005, vol. 115, no.8, pages 2072-2075.

when both tendencies are still possible. This is the third position.

Field and experimental biologists can select parameters to define the status of any biological process. The distance between outcomes can be measured by differences in forms, sizes, genetic similarities, and others. However, measuring distance requires some specifics regarding exactly what we are measuring: different species, different populations, different individuals, and conditions (stages, phases) of these objects, such as growing, stable for a while, and declining. "Proximity" is a better word than "distance" to represent relationships and connections between two objects or conditions.

I have used many words that might appear unrelated to define the third point: central pillar, "Hidden Third," "included middle," balance, homeostasis, homeorhesis, beauty, directedness, functional information, and meaning. This list is still inadequate to represent all aspects of this location. However, in Kabbalah, it is designated by only one word – Tiferet!

The Third Dimension is proposed to measure the proximity between Tiferet-Hesed and Tiferet-Gevurah, not between Hesed and Gevurah. Is it closer to a balance or farther from a balance? These points can indicate some "ideal" positions, but we now have a field with a spectrum of potential variants between those two extreme points. Positions of these variants are flexible and very sensitive to many internal and external conditions. Therefore, we have an available axis that provides a direction toward the central point (Tiferet). This axis is proposed as "the Third Dimension of Kabbalistic Biology."

Be aware that constant practice is required to master the field of biological research. We must sometimes admit that Nature is *not* perfect. On a positive note, however, moving *toward* an optimally effective status is natural. Although we may not yet see such a movement, we can trace its signs, and use these traces to achieve a clearer orientation in the invisible world of biology.

Chapter 16

Fourth Dimension: The Hierarchical Worlds of Life

Unlike physics and chemistry, biology faces unique challenges due to the inherent variability and constant change in biological systems. Variability has proven to be an essential aspect of biology that deserves closer study. Biologists have now developed conceptual models to explain their observations as biological organizational hierarchies.

The biological hierarchy spans a range of degrees of complexity, from the microscopic cellular level to the broader scale of ecosystems. Taxonomic classifications may vary and create questions about the reality of specific levels. However, the emergence of properties specific to each level remains a core challenge. Biologists understand that biological objects cannot exist independently from higher hierarchical system levels.

Although the proposed fourth dimension relates to the accepted biological hierarchy, it cannot be reduced to the "cell-tissue-organism-species" hierarchy. Understandably, reductionistic ("flat") perception is more prevalent among researchers working with biological objects at the molecular and cellular levels. However, the "flat" view of biological populations, communities, and ecosystems as independent entities is still very common among ecologists.

In accordance with the Kabbalistic framework of *Olamot* (the Four Worlds), the Fourth Dimension is available to help us comprehend natural phenomena from a multiple-world perspective. This chapter introduces each of the Four Worlds using a variety of biological research examples. Before beginning these exercises, I will introduce this idea by using a more metaphorical language adapted from the Kabbalistic tradition.

Expedition to the Biological Worlds of Pardes

Do you remember the Talmudic story of the four sages entering the Pardes Garden presented in Chapter 9? Below is a reproduction of the imaginary scientific expedition carried out by these individuals. With the assistance of thought experimentation, we can add academic degrees to each of the sages' names to avoid confusing them with the original legendary sages. Let's call the sages Dr. Simon ben Azzai, Dr. Simon ben Zoma, Dr. Elisha ben Abuyah, and Professor Akiva ben Joseph.

Imagine that once upon a time, these four distinguished scientists were commissioned by a scientific society to explore the Pardes Garden. Four doors led to the Garden, and each scientist chose a different one.

Opening a heavy door, Dr. Azzai entered the realm called "The Apparent Realm of what is Apparent." He looked around and recognized his favorite objects of investigation: animals and plants. He immediately began counting them, measuring them, and leaving notes for the Society that had commissioned him. Soon, however, he noticed that some animals, which he had already counted, had disappeared. Either they had become unrecognizable, moved away, or died. At the same time, newborn animals appeared! Even worse, when he repeated his measurements of the same animals, they were different from those previously recorded, the animals had grown! He was increasingly bewildered: his notes had now become inaccurate.

Dr. Azzai began correcting his calculations, but the number of animals and his measurements continued to change. Whether he pursued his work as an investigator for one day or for forty years more, he would not be able to complete his research. He had no more incentive to continue his professional life, and his disappointments as a scientist led to his early demise. His notes were eventually sent to the Society, but no one knew the value of those records because they were outdated.

To continue our story, another scientist, Dr. Zoma, arrived at another notion called "The Hidden Realm of What is Apparent." As had occurred with Dr. Azzai, this area seemed, initially, quite familiar to Dr. Zoma. He began his research by selecting some animals to investigate. The problem was, he could not decide whether these animals belonged to the same species! Some of them looked similar, some did not. Dr. Zoma could not figure out how to apply the results from a study of one animal to studies of others. He decided to not only study their morphology (appearance) but also to measure their internal organs and the shape of their bones, conducting biochemical analyses to obtain more information.

Each test reflected some real characteristics, but when he tried to put them together, he could not get a clear picture. Previously familiar animals disappeared. The data were real and objective, but when Dr. Zoma put them together, they made no sense! Instead of understanding the subject of his research, Dr. Zoma could find only numerous pieces of information. The more data he obtained, the less confident he felt about making a final conclusion. Soon his reports were so disjointed that Society members decided he had lost his mind. They called him a "mad scientist."

Our third doctor, Dr. Abuyah, entered a door named "The Apparent Realm of What is Hidden." To his surprise, he could not find any animals or plants around. He found it advantageous to avoid meticulously collecting information about these creatures afterward. Aware of Dr. Zoma's madness, he decided to concentrate on developing concepts before he began collecting information about real creatures. For example, it would seem to be more practical to define the concept of "biological species." Once done, he could establish criteria for how one species differed from another. Only then, he speculated, could he request some individual animals from Dr. Zoma or other colleagues to check out his concepts. He formulated many definitions and proposed hypotheses.

To his disappointment, when he received the requested specimens for identification, few fit his criteria. He decided to avoid working with concrete biological objects, claiming they were not good enough for his theories! He claimed that his concepts were created to maintain knowledge and were flawless.

His colleagues found his argument unscientific and arrogant. Ultimately, they excluded Abuyah from the Society and deprived him of a doctoral degree. Abuyah laughed in response: "I don't care about their limited scientific framework. After all, there is no objective truth anyway. My science is better because this is my creation, and that's it." Dr. Abuyah had made a grievous mistake: *he forgot that he is not the creator of biological life!* Indeed, all he could do was explore and formulate concepts about *existing* life!

Finally, we come to Professor Akiva. After entering the Garden, he walked straight over to the open space. Aside from his curiosity about animals and other organisms, he was also curious about the surrounding environment. There were many possible scenarios, but some of them were improbable, while others were very consistent with observed conditions. Prof. Akiva concluded that some organisms were likely to exist even if they remained invisible after one observed the space from different angles. In the end, he was able to determine what kinds of animals and plants could live anywhere in this world. He called this world, "The Hidden Realm of What is Hidden." After completing this work, Dr. Akiva walked through the other three worlds, returned home, and submitted a short report to the Scientific Society. This report stated that there were four worlds behind the doors, all interconnected with each other.

This story is, of course, purely fictitious, but the theory behind it is critical to our ability to work with Kabbalistic Biology. Akiva's lesson offers us two critical aspects. First, there is the Hidden Realm, where biological life is both invisible and essential. We must accept this reality to be able to work within it. By accepting the wisdom of Kabbalah, we can begin to work on this level. The second reality is this: all worlds are real, significant, and complementary!

The Four Levels of Existence of Biological Systems

For us to be able to work with Kabbalistic Biology, each biological object or phenomenon must be accepted in the Four Worlds. The four-tiered order is crucial, even if, at first, we may find it challenging to accept. At the very least, it allows us to avoid useless arguments about which research level is better! More importantly, it helps us orient ourselves when choosing how to describe a specific level of visibility of manifestation. When we realize that biological objects exist on different planes of existence, we can advance in our research. As a final benefit, this flexibility enables biologists to build relationships with the biological objects they investigate.

Many metaphysical schools view two-tier orders of the world: the higher one and the lower one. However, the Kabbalistic vision of Olamot ("worlds") takes it further. It proposes a four-tier system that provides researchers with more flexibility. This is because of its alignment with the four levels of the DIKW pyramid described in Chapter 5. The four planes of biological systems correspond to the four hierarchical worlds of "Olamot": Asiyah, Yetzirah, Beriah, and Atzilut, as described in Chapter 9. The planes refer to four levels of visibility in living objects, which we discussed in detail in Chapters 6 to 8.

According to Shneur Zalman in Nilton Bonder's interpretation (Chapter 9), these hierarchical worlds include:

- The apparent realm of what is apparent.
- The hidden realm of what is apparent.
- The apparent realm of what is hidden.
- And the hidden realm of what is hidden.

With this background, we can follow each world/level to explore biological life mysteries more thoroughly.

The Biology of Asiyah

The Asiyatic level is commonly called, the "physical" or "materialistic." If a biologist holds an animal, plant, or microbe firmly in his view via microscope, binocular, or with the naked eye, it is easy to observe particular visible characteristics and make some physical measurements. The obtained data are supposedly reproducible as long as the same tools are used. More advanced tools can deliver more precise measurements that provide additional data.

A collection of butterflies or bird songs recorded by amateur naturalists offers a clear example of the biological world at the Asiyah level. The most natural way to become a professional biologist is to move through various stages as an enthusiastic naturalist. Personally, I was inspired to become a naturalist by Charles Darwin's book, *The Voyage of the Beagle*. When Darwin was hired to be a naturalist on a ship named after a breed of dog, he was only 22 years old. He spent considerable time ashore collecting plants and animals from a particular spot. He placed his findings in a tray that hardened in the proper position. It was processed for long-term preservation, labeled, and delivered to the museum. As a result, we have a collection of disassociated biological objects. Darwin's insects are still available at the University of Cambridge's Museum of Zoology.

This is not an unusual practice: in their routine work, biologists deal with separate objects all the time. For example, when we trap rodents during a plague investigation, we take an individual captured animal and measure it. We keep a record of body weight, tail length, and other parameters, which could be useful for species identification or age indication. As bacteria grow on agar, we observe their shapes, colors, and other morphological characteristics. We run gel electrophoresis to separate DNA fragments extracted from isolated bacteria and stain the gel to make it visible. If we are studying an animal, a plant, a bacterium, or a DNA fragment, it is extremely imperative to record precise data. The first thing I taught students coming to my laboratory was to appreciate the importance of the accuracy of their data.

Asiah represents the data world as a whole! Here, objects are clearly visible. If the object cannot be seen with the naked eye, a microscope, gel electrophoresis, or DNA sequence could be used to help us "visualize" it. There is no value in data until it is linked to a specific object. Data could be incorrect if applied directly to another object. This does not mean, however, that data obtained from one object cannot be used to learn about another. Achieving such a goal requires transforming it into information. We must remember, however, that obtaining information takes place in a different world – in Yetzirah!

The Biology of Yetzirah

Yetzirah is the world of relationships. This plane immerses itself in the nuanced world of defining and classifying biological entities. Researchers must pinpoint the essence of measurement while grappling with the challenge of establishing precise categories for organisms. At this point, the interplay between objects takes center stage. The process of describing relationships between them involves naming, defining, and analyzing them.

To obtain information about any biological object, we must connect data belonging to one object to other data, such as measuring another object to help us determine the two objects' closeness. More important, we can evaluate how these objects are distinctive. We can compare data obtained from measuring the same biological object at different times or under different conditions. Such information can help us define the degrees of differentiation. As I mentioned in Chapter 5, "the elementary unit of information is a difference that makes a difference.[1]"

The Yetziratic plane of biology can reveal variations within both animal populations and microbial populations. This plane allows us

[1] Gregory Bateson, *Steps to an Ecology of Mind; Collected Essays in Anthropology, Psychiatry, Evolution, and Epistemology.* 1972.

to explore the subtle differentiation among individuals within a species, weaving together the threads that form their cohesive identity. Similarly, it directs our attention to the distinguishing characteristics that set closely related species apart, thus helping us recognize the subtleties that define each one.

In other situations, we can connect data from a concrete biological organism to data about where the object was investigated. Let's revisit Darwin's collection. Each specimen was accompanied by certain data, either attached directly or recorded in a notebook. This data corresponded to a numbered tag attached to the specimen. Ed Turner, Curator of Insects at the University of Cambridge's Museum of Zoology, claimed that Darwin sometimes lacked details when he was labeling his specimens. While an excellent collector, he sometimes collected for collecting's sake, rather than cataloging exactly where he had found a specimen.[2]

Darwin's labels were often vague; rather than saying where he found something, one label might say, "Brazil." It meant that specimens such as beetles collected from Brazil had the same information about their location. This could limit the analysis of specimens. Initially, young Darwin did not realize that different Galapagos Islands had different tortoises living on them. It was the Galapagos governor who pointed that out to him, enabling Darwin to re-label the specimens he collected. It was the same with Darwin's finches--he did not initially label on which island he had found them.

Here is an example of how the Yetziratic plane presents a field guide. Field guides offer us key characteristics that allow us to reliably identify species of animals, plants, or other organisms without using too sophisticated methods. One such example is that of identifying chipmunks in North American forests. This could be a challenging task! First, we must consider the region where a squirrel-like animal ran on the ground. We can record its habitat and behavior. When we hold an immobilized chipmunk, we can make the most accurate identification. Then, we can measure its body and tail length, record

---

[2] https://museumcrush.org/barnacles-beetles-finches-and-fishes-charles-darwins-specimen-collections/

color in different parts of the body, and describe striped patterns on the back. The Peterson Field Guide provides useful identification keys, such as which chipmunk species is bigger or smaller. It also describes whether stripes are rich on the face or the base of the tail, the color of hair on the neck and shoulder, and many other details.

When working with the Yetziratic plane, the question may emerge: *What are we actually observing?* This question applies to two areas:

1) the definition of the object or 2) the measured parameter, such as the movement of an organism or a specific characteristic, such as a chipmunk's stripes. This is the plane where we "name" the object, assigning it to a particular class. Let's say we measure the weight of a bird of a particular species. If we are still determining what species it belongs to, we prefer to know, at the very least, what *kind* of bird it is. Is it a bird of prey, a gull, or a hummingbird? To compare this bird with others, we need to select our criteria.

Let's imagine that the criteria for bird species identification have changed. In addition to morphological signs, scientists are now counting chromosomes. The birds, which were previously considered one species, are now considered as two species since discovering this taxonomic revision. Although we still have objective data, it is now used differently to produce information about various birds' taxonomic status. In fact, the whole observation plane is different. The results obtained at the Asiatic plane can be used on the Yetziratic plane, but the information is obtained solely from the Yetziratic plane.

However, these identification keys do not tell us what species we are examining. Instead, we can only estimate that the animal in question is closer to one of the described chipmunk species. This is how Yetzirah works! Since we must compare different animals, we are no longer in Asiyah, and we do not have a separate individual organism. Alas, we have not yet reached the concept of biological species, we must wait for the next level.

How we select our data sources will affect how we acquire our information. For us to analyze the relationship between elements of a biological system, we need terms and definitions. What are we comparing? Beetles? Are we looking at beetles in general, or those belonging to one species, or found on one island? This is not where we seek to identify what species it belongs to, or how many observations we need to make. Such endeavors are driven by formulated hypotheses subject to rigorous statistical scrutiny. For example, how many beetles must we compare? What confidence do we have when claiming a similarity between two groups of beetles?

The Yetziratic world is composed of connections between separate data. The more diverse the data, the more information there is. *If Asiyah is the world of data, Yetzirah is the world of information.* Those are indeed different worlds. One of them consists of separate biological objects, while the other consists of connections between these objects. Each world has its own inhabitants: the Data are dots in Asiyah, and the Information are Lines in Yetzirah (Figure 17).

The complexity of all biological systems is why there is always increased uncertainty during scientific investigations. While some pockets of information can be more complex or straightforward, biological life research is always very complex. To remind you of why this is so, one of the inevitable characteristics of such a system is "non-linearity."

A Lost Cat and Dog in Yetzirah!

Yetzirah is the world in which biologists work the most. Data collection is the first step before one can analyze and acquire information. While it would be highly desirable to prepare "field guides" for all biological objects, including viruses, this concept has become outdated. The following scenario may appear exaggerated, but it is created this way to illuminate the dangers of exploring Yetzirah's biology!

Let's take our favorite animals like dogs and cats. Everybody can distinguish between dogs and cats, right? Dogs and cats share many similar characteristics. As carnivores, animals belonging to the order

Carnivora share various features typical of such animals. They all have blade-like carnassial teeth, their fourth upper premolar and first lower molar that bite together to tear through food. Cats and dogs also have four legs, tails, and a horizontal body posture. They also have snouts with noses, whiskers, and toe pads on their feet. They both have fur to keep them warm in the winter months. Hunting is another major similarity between cats and dogs, as they rely on hunting and scavenging for food.

On the other hand, we can create a long list of parameters to distinguish cats from dogs. Cats and dogs interact differently with their owners. Dogs have always worked in groups, so they will most likely find a way to involve themselves in a person's daily routine. On the other hand, cats are very independent and may primarily rely on their owners for basic needs. Cats like being alone for hours, while dogs always want attention. Dogs and cats have different sleep schedules. Most of the time, dogs are active and playful during the day. On the other hand, cats rest most of the daylight hours. They are one of the laziest animals and sleep about 85% of the day.

There are many ways to compare cats and dogs. Dogs are usually larger than most cats. Claws are also a huge difference between the two animals. A dog's claws are duller than a cat's simply because they are always out, and the ground they walk on wears down their claws (Figure 46).

We can keep listing on and on… By the way, have you noticed that we have lost our dog and cat? We discuss some of their characteristics, but where are these animals? This is the danger that comes with working at the level of Yetzirah. No wonder Dr. Zoma lost his mind!

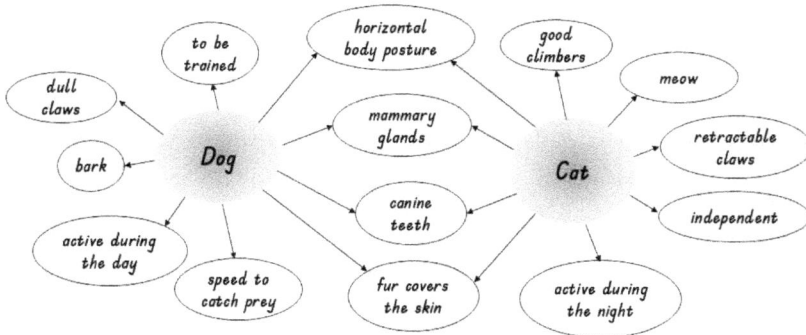

Figure 43. Comparing dog and cat characteristics can obscure our knowledge about these animals.

Trevor Goward, the curator of the lichen collection at the University of British Columbia, made an interesting point: "I often say that the only people who can't see a lichen are lichenologists. It's because they look at the parts, as scientists are trained to do. The trouble is that if you look at the parts of the lichen, you don't see the lichen itself." Commenting on the Goward insight, Merlin Sheldrake noticed: " ... it is a point so simple that it is hard to grasp."[3]

The Biology of Briah

"Briah" finds resonance in biological concepts. It is here that the focus shifts from the tangible and observable to the abstract and intellectual dimensions of living systems. By developing these conceptual frameworks, researchers can change complex biological systems into comprehensible models. These models offer a highly accurate view of the essence of biological phenomena without getting lost in intricate details. Briah is where these frameworks are created. This enables us to transcend the complexity of individual cases and gain clarity about the underlying principles that guide living systems' behavior and evolution.

What, incidentally, is a "biological species?" Asiyah depicts dogs as concrete animals, while Yetzirah describes "dog-ness" as a combination of some characteristics. In Briah, however, we can

---

[3] Merlin Sheldrake, *Entangled Life*. 2020, page 83.

formulate a concept about dogs as a species or subspecies. Chapter 7 provides a detailed discussion of this question. Similarly, any biological concept can be questioned. What is biological evolution? What is an organism's adaptation to the environment? What is immunological protection from alien organisms?

Ultimately, the interplay between Briah's world and biological concepts shows us how important it is for us to use abstract thinking and theoretical explorations. It reveals how our ability to conceptualize empowers us to penetrate biological complexity. We can recognize underlying patterns, principles, and relationships that might remain hidden when we simply focus on the physical and observable aspects of life.

The level of Briah reached will reflect a researcher's knowledge about the biological object or phenomenon being investigated. There is more to knowledge than handling data and obtaining information. It is woven from pieces of knowledge received from studying at school, communicating with colleagues, reading books, and, most importantly, from personal experience. In his influential *Steps to an Ecology of Man*, Gregory Bateson wrote, "It means that knowledge is all sort of knitted together or woven, like cloth, and each piece of knowledge is only meaningful or useful because of the other pieces."[4]

Bateson makes a fascinating point that highlights the difference between Briah, as the realm of ideas, and Yetzirah, using arithmetic terms. He wrote: "The first thing about being clear is not to mix up ideas which are different from each other. The idea of two oranges is very different from the idea of two miles. Because if you add them together, you only get fog in your head ... Combine the. But don't add them ... the thing to do is to multiply them by each other."[5]

In Yetzirah, we *add* information – the more information, the better. This is not so in the world of Briah, where all emerging knowledge is multiplied by previously accepted knowledge!

---

[4] Gregory Bateson, *Steps to an Ecology of Man:*
The New Information Science Can Lead to a New Understanding of Man. 1972, pages 21-22.
[5] Ibid., page 25.

The Biology of Atzilut

In Kabbalistic Biology, the concept of Atzilut unveils a world where biological entities, from individual organisms to their intricate connections, harmoniously coexist within a unified and interconnected system. This profound interconnectedness echoes the essence of the Atzilutic plane.

The Gaia hypothesis serves as a clear illustration of the Atzilutic principle. This hypothesis suggests that Earth's living organisms collaborate with the inorganic elements of their environment to form a sophisticated, self-regulating system, like a single organism. Just as Atzilut emphasizes unity and harmony, the Gaia hypothesis underscores the interdependence and interplay between all components of Earth's biosphere. It reveals how life forms weave together to sustain our planet's delicate balance.

The pioneering ideas of Lynn Margulis on symbiosis and cooperation between species are in line with the core principles of the Atzilutic plane. Her insights illuminate how life's unity emerges not just from competition but also from intricate collaborations among different organisms. The Kabbalistic wisdom of Atzilut reflects this perspective in which relationships and connections intertwine to form a living tapestry.

While the Atzilutic plane might seem abstract, it mirrors life's hidden but *real* existence. There is a deep connection between biological entities and an inherent unity. This unity is the underpinning behind the diversity of life that manifests in this abstract realm. Biology acknowledges the intangible but undeniable unity that emerges from interactions and relationships, and how they shape the living world. Similarly, Kabbalistic thought acknowledges the ineffable essence of Atzilut.

Using the language of Bateson, Atzilut represents "knowledge about knowledge."[6] Instead of trying to define the intangible world, Atzilut, it would be more productive to walk through some examples

---

[6] Ibid., page 24.

of it to show how we can distinguish the four planes of research in biology. The essence of Atzilut is nicely expressed by David Bohm, who "envisioned the existence of each part in hinge upon its intimate relationship to the whole, implying that individuality is only feasible if it unfolds from wholeness."[7]

A Bacterial Species in the Four Worlds

Practically, we can select any biological object or phenomenon to demonstrate its existence in all four worlds. I will start with the task I routinely dealt with in the laboratory to determine sources of bacterial infections. While working at the U.S. Centers for Disease Control and Prevention, I isolated thousands of bacteria using specific procedures. The challenge was to identify the isolated bacterial cultures. Usually, they belonged to one of the known bacterial species. Occasionally, I was able to discover new bacterial species.

In the next few paragraphs, I will guide you through some of the steps that led to the discovery of new bacterial species and possibly new pathogens for humans or animals. Please be patient! I promise to keep this description as simple as possible.

There are different approaches to identifying bacterial species. For simplicity, let's consider a very effective tool provided to bacteriologists by molecular technology. Essentially, the technique targets a fragment of DNA that matches a specific gene. We can read a sequence of nucleotides that leads us to a taxonomic group to which the bacteria might belong. We can perceive the sequence of nucleotides as a long line consisting of different combinations of four letters arranged in various ways.

In brief, here is the procedure. We begin by growing bacterial colonies on the agar plate (a Petri dish filled with agar and specific nutrition supplements). For example, the morphology of the colonies resembles that of *Bartonella* bacteria, we then select a single colony

---

[7] Quoted from Hyman Schipper, *Did Kabbalah Anticipate the Physics of David Bohm?*

typical of such a bacterium. According to a very detailed protocol, we perform several more microbiological steps. Finally, we obtain the nucleotide sequence (a unique combination of "letters") specific to the isolated bacterium. Let's call it "Strain X."

Now, we have entered the world of Asiyah! Everything is unique: one culture, one colony, and one DNA sample were extracted from one vial. We have obtained a unique combination of "letters" from the nucleotide sequence. We must now pay attention to any detail! First, we need to compare the obtained nucleotide sequence with other sequences in our laboratory to identify it. Some computer programs can quickly help us make such a comparison. The first analysis confirmed that strain X was indeed closer to the bacteria that belonged to *Bartonella*. However, the strain differs genetically from other strains of these bacteria that we obtained previously. The closest match was found with bacterial strains belonging to two species: A and B (Figure 44-A). However, strain X does not belong to any species.

Have you noticed that we have moved to another world? We are still in the same physical space. However, while comparing genetic sequences, we found ourselves in a different world – Yetzirah! In this world, our object is no longer a particular bacterium. Here, we compare the parameters of this strain with the parameters of other strains.

Based on the available data (the unique sequence) and newly acquired information (the distance between this sequence and sequences derived from other known strains), we verify that the newly discovered strain is original. This is verified by the distance between the sequence and sequences derived from other known strains. The discovery of a novel pathogen is tempting for some of my younger colleagues, but based on previous experience, I prefer to obtain additional information. Consequently, I will isolate more bacteria from related sources, sequence more genes, and compare them with a more extensive collection of related species in open databases (e.g., GenBank).

As a result, we discovered more robust evidence that the newly discovered strain differed from other bacterial species. We also found additional strains genetically close to strain X! At the same time, we discovered more strains that *differed* from strain X! Some of them occupied a position between the recently discovered strain X and previously known bacterial species A and B (Figure 44-B).

We then obtained more isolates and used more genetic markers and additional information about isolated bacteria (morphology, ecology, etc.). Then, we compared new data with more information from my colleagues worldwide. This process is my standard research approach. Only by obtaining more data can we continue to pursue new information (Figure 44-C). More data, more information! Note that we are still in the world of Yetzirah and we just generated more information.

One question arose: Did closely related new strains belong to the same species ("pathogen") as Strain X? To answer such a question, we moved to the next hierarchical world of Briah. In this world, we explore the existence of biological species in nature, apply the concept of biological species to bacteria, and define bacterial species according to specific criteria. With my collaborator, Kung-Sik Chan, a Professor of Statistics at the University of Iowa, we formulated the question, "Do new bacterial strains behave as individual species or minor genetic variants?"[8]

Based on our knowledge, we formulated the concept of bacterial species that takes into account not only genetic and morphological data but also reveals information about how to associate these bacteria with animal hosts and some ecological characteristics.[9] In this case, we had to rely on our knowledge and choose criteria to identify the species. In Figure 44-D, data are shown as separate

---

[8] Kung-Sik Chan and Michael Kosoy, *Analysis of multi-strain Bartonella pathogens in natural host population — Do they behave as species or minor genetic variants?* Epidemics, 2010, vol.2, pages 165–172.

[9] Michael Kosoy et al., *Bartonella bacteria in nature: Where does population variability end and a species start?* Infection, Genetics, and Evolution, 2012, vol.12, pages 894-904.

points, information as lines connecting these points, and knowledge as the defined areas according to the images from Figure 17.

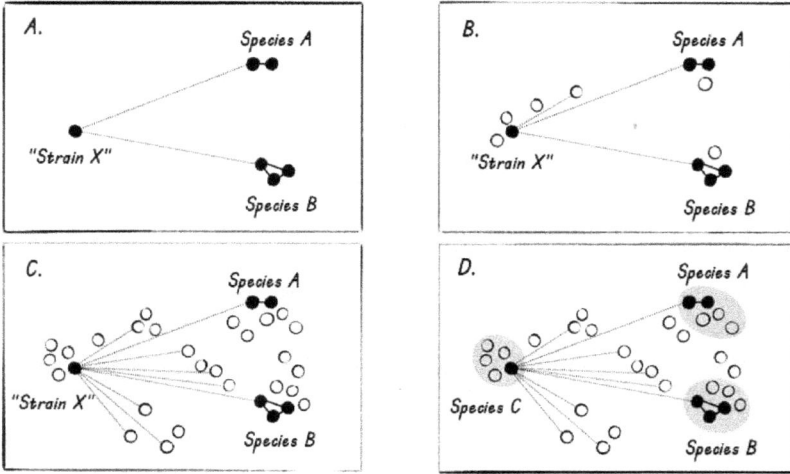

Figure 44. The description of a new bacterial species requires collecting more data (separate points), producing more information (connecting lines), and generating knowledge (defined areas).

In the words of Bateson, knowledge is *multiplied* by knowledge, not added to it! Remember, when it comes to" knowledge," we are not limited by what is "known." The question of limitations in our knowledge is critical. In Chapter 6, I explain how we discriminate between two bacteria species: *Yersinia pestis* and *Yersinia pseudotuberculosis,* by examining their sensitivity to a specific bacterial virus ("bacteriophage"). Would it be enough to discriminate the life-threatening pathogen of bubonic plague (such as *Yersinia pestis*) from the bacterium that lives in soil and causes a mild enteric disease (such as *Yersinia pseudotuberculosis*)? Note that I am talking about the same bacterial species, but my question is different. This is a very practical question, but it requires another level of learning: that of the causality of infectious diseases. To answer such a question, microbiologists and epidemiologists must exchange knowledge.

Plague in the Four Worlds

Talking about plague, here is another example from my professional experience that illustrates the four-tier system of coordinates. In 2016, I was invited to speak at the International Yersinia Symposium held in Tbilisi, Georgia. This symposium is one of the most influential scientific events for plague researchers.

For my presentation, I selected the main questions that have dominated research on plague ecology over the past 100 years. They can be formulated as how plague can persist in specific ecosystems for a long time, and what forces drive the emergence of plague in ecosystems where the infection has not been seen for a long time. There is an enormous number of hypotheses proposed to answer these questions. Plague scientists often argue with each other to prove the advantages of their hypotheses. Thus, I attempted to systematize such hypotheses.[10]

To undertake such a challenging task, I envisioned Olamot as a system of hierarchical worlds. I did not disclose the source of such an orientation system, but instead, I proposed to the Symposium participants that we distinguish the four levels of perception of plague systems.

At the first level, I placed hypotheses based on particular characteristics of system components; these include visible features and genetic elements of the plague, the presence of specific species of rodents and fleas, isolation of the bacterium, the presence of antibodies, and many other characteristics. Here, I stressed the importance of accurate data representation, method selection, thorough records, and repeated observations.

At the second level, I considered hypotheses based on comparing the properties of different components of the plague system. These included the virulence of strains, frequency of mutations, antibiotic sensitivity, antibody level, animal survival time, landscape features,

---

[10] Michael Kosoy, *Are Hypotheses Explaining Plague Activity Mutually Exclusive?*
The 12th International Yersinia Symposium held in Tbilisi, Georgia, in October 25 – 28, 2016,
http://www.gebsa.ge/Yersinia/ConferenceLectureProgramme/

climatic factors, and other external parameters. Designing experiments, applying valid statistical methods, and reporting negative results are important for evaluating hypotheses.

The third level reflects the development of specific concepts. The used examples included the genomic evolution of *Yersinia pestis*, the adaptation of bacteria to specific hosts, transmission mechanisms, the asymptomatic carrying of bacteria by animals, survival of plague bacteria in soil, the metapopulational structure of host populations, and others. I focused on the difference between hypotheses and scientific concepts.

Finally, at the fourth level, I proposed considering plague as a self-regulated ecological system adapted to the environment. At each level of consideration, I identified hypotheses that emphasized either the "persistence" or "emergence" of plague. I also focused on a balance between hypotheses that demonstrated opposing tendencies in plague dynamics.

My conclusion was that although the hypotheses could be seen as competitive and often mutually exclusive, most of them can work synergistically. They can be instrumental in defining a potential framework leading to a more integral picture of plague as a natural phenomenon. After my presentation, the conversations that followed were amiable among the participants, since most of them could see how their own hypotheses fit. The lively discussion lasted until one of my European colleagues noted: "Wait a minute! But that can apply not only to plague but to everything…" "Absolutely!" I responded.

A Mushroom in the Four Worlds

In previous examples, I discussed a biologist's perception of some scientific questions about bacterial species and plague ecology. Here is a more concrete example. Everyone has seen a mushroom. However, only some of us pay attention to the details of the mushrooms we see. When it comes to Asyah's world, it's all about what counts! There are many varieties of mushrooms, and each is a unique creation of its own, meticulously defined by the size, shape, and number of intricate characteristics that define its uniqueness.

Central to mushroom body anatomy are the stem, or stipe, and the cap. The stipe serves as both a supporting structure for the cap and as a conduit connecting the organism to the earth. It manifests in diverse forms, ranging from slender and elongated to robust and stocky. Some mushrooms are vase-shaped. Mushroom caps could be round, conical, bell-shaped, or convex. The cap center may be knobbed or sunken. The cap margin could be unrolled, downcurved, straight, or upturned. It could be wavy, hairy, or smooth. Caps may be dry, moist, sticky, or slimy. They can be raised or made flat. The color of the caps varies greatly, as does the color of the flesh. On the underside of the cap, there might be gills, and plate-like structures. The pattern of these gills can vary from rounded to forked or labyrinthine.

The list could go on and on. Do pay attention to such details if you are considering eating these mushrooms! Studying mushrooms would not involve risking your life if you paid attention to these details. It is imperative to note the habitat and region where the mushroom was found. There are countless records in the world of Asiyah, and each one is unique.

We enter Yetzirah when we compare the collected mushroom with other mushrooms. Some mushrooms look very different from each other, while others resemble each other. With the latter, paying attention to detail could be crucial, even lifesaving. The same mushroom observed on previous and subsequent days can also be compared. As we expected, the mushroom grew taller. Using such information, researchers get updates on the morphology of the mushrooms, their growth rate, and other parameters that characterize them. Importantly, we must learn which mushrooms are edible and which are toxic. Among the approximately 2,000 species, many are delicate and edible, but not all.

There are more surprises in the Yetziratic world: the mushroom body has an underground part, the mycelium, a mass of threadlike hyphae that resemble a root. The analyses of a mushroom and its associated mycelium demonstrate that they are parts of one organism, a fungus. In reality, a mushroom is only the fruit body of a fungus – like apples on a tree, they are barely visible! Through its mycelium, a fungus absorbs nutrients from its environment. The mycelia (plural of mycelium) of two compatible fungi can fuse together, allowing the cells of each fungus to combine and their DNA to mix.

Another fascinating piece of information is the association of fungi (plural of "fungus") with specific tree species. Measurements of the distance between fungi and trees made in Asiyah led to relevant information about the fungi-tree relationship demonstrated in Yetzirah. The discovery of *Armillaria ostoyae*, the world's largest fungus, occurred in 1998. It was heralded as the discovery of an unprecedented world record holder for the title of "World's Largest Organism." Located in Oregon's Blue Mountains, it occupies more than 3.5 square miles.[11] It is estimated that this fungus weighs hundreds of tons and has been around for more than two thousand years. Geographically, the discovery took place in Oregon, but on our 4-tier scale, it was in Yetzirah.

The Yetziratic world brings massive information about these fungi's worldwide distribution. They grow in a wide range of habitats, including extreme environments such as deserts or areas with high salt concentrations. Some can survive intense UV and cosmic radiation encountered during space travel. Around 150,000 fungi species have been described by taxonomists, but the global biodiversity of the fungus kingdom is not fully understood.[12] The amount of information about fungi is increasing rapidly!

At the level of Briah, we come to an understanding of the status of fungi as a separate kingdom, distinct from both plants and animals, from which they appear to have diverged around one billion years ago (around the start of the Neoproterozoic Era). Fungi are not just distinct from plants and animals, they also constitute the highest

---

[11] C.L. Schmitt and M.L. Tatum, *The Malheur National Forest: Location of the world's largest living organism (the Humongous Fungus)*. 2008, Forest Service, US Department of Agriculture.

[12] Gregory Mueller and John Paul Schmit, *Fungal biodiversity: what do we know? What can we predict?* Biodiversity and Conservation, 2006, vol.16, no.1, pages 1–5.

taxonomic rank along with plants, animals, protists, archaebacteria, and eubacteria. This conclusion reflects the level of hierarchical knowledge available, not just accumulated information!

In Briah, knowledge is growing that fungi can closely partner with plant roots through their mycelium. Without the help of fungi in the soil, most plants would not be able to grow and reproduce. Plant roots and fungi create a network called "mycorrhiza" (meaning 'fungus-roots'). Fungi and plants can both benefit from this network.[13] Through photosynthesis, trees and other plants produce sugars and fats, which fungi get from their roots into their mycelium. In turn, fungi help trees absorb water and nutrients from the soil, well beyond their roots and root hairs. We could collect massive information about fungal mycelium and plant root systems, but the realization of the mycorrhizal network is created in Briah!

In Asiyah, we could accumulate observations that some animals, such as reindeer in Siberia, eat the hallucinogenic mushroom *Amanita muscria*. In Yetzirah, we can conduct experiments on measuring the effect of mushrooms' psilocin on memory acquisition in rats.[14] Still, it takes Briah to formulate potentials for fungi in modulating animals' behavior in natural conditions.

In his book, *Entangled Life*, Merlin Sheldrake presents mushrooms in all worlds, but the subtitle refers to Atzilut: "*How fungi shape our worlds, change our minds and shape our future.*" This statement extends beyond current information, data, and existing knowledge. Sheldrake's book uses the hidden kingdom of fungi to question dominant concepts of individuality and intelligence. Such a claim is based on the demonstration that fungi are metabolic masters, earth makers, and key players in most life's processes. It is at that level that Atzilut reaches!

---

[13] Marcel van der Heijden and Thomas Horton, *Socialism in soil? The importance of mycorrhizal fungal networks for facilitation in natural ecosystems*. The Journal of Ecology, 2009, vol.97, no. 6, pages 1139-1150.

[14] Lukas Rambousek et al., *The effect of psilocin on memory acquisition, retrieval, and consolidation in the rat*. Frontiers in Behavioral Neuroscience, 2014, vol.16, no.8, page 180.

A Connectivity Between the Hierarchical Worlds

In Kabbalistic Biology, the Fourth Dimension measures the four planes of reality. There is, therefore, discontinuity in this dimension, as it identifies what level each biological object or process belongs to in relation to the biological process in general. On the other hand, the Fourth Dimension transcends all four planes via the Sefirotic structure hidden in each world: Olam. This dimension measures proximity to the borderline ("screen") that separates the above- and below-existing worlds, unlike the First Dimension, which measures proximity to extreme points, such as Malkhut and Keter.

Each plane is a separate world! Kabbalah teaches that there is a screen or curtain (*massach* in Hebrew) to separate the worlds (olamot). However, the four planes of biological study are not isolated. They are interconnected, with a continuous flow of information and energy between them. Feedback interactions play a crucial role in living systems, accumulating complex information and energy.

Understanding and exploring the interactions between the worlds can lead to a deeper understanding of natural phenomena. The highest worlds interpenetrate the lower worlds. The challenge lies in our habit of linear thinking. Referring to the stages of consciousness, Huston Smith, in *Beyond the Post-Modern Mind*, uses a metaphor that could apply here. He says that the divisions between the stages are like one-way mirrors. If we look up, we see only the reflection of the level we occupy. If we look down, the mirrors are as transparent as glass.[15]

Below, I will introduce some visual models for exploring the connectivity between the worlds of Olamot. All the worlds are joined by Malkhut of the higher world, which overlaps Keter of the world under it. Malkhut of Atzilut overlaps Keter of Briah! In a similar matter, Malkhut of Briah coincides with Keter of Yetzirah, and Malkhut of Yetzirah overlaps with Keter of Asiyah. Overlapping

---

[15] Huston Smith, *Beyond the Post-Modern Mind.* (Cited from David Parrish, *Nothing I See Means Anything.* 2006, page 108).

points are gateways to exchange energy and information between existing worlds above and below (Figure 45).

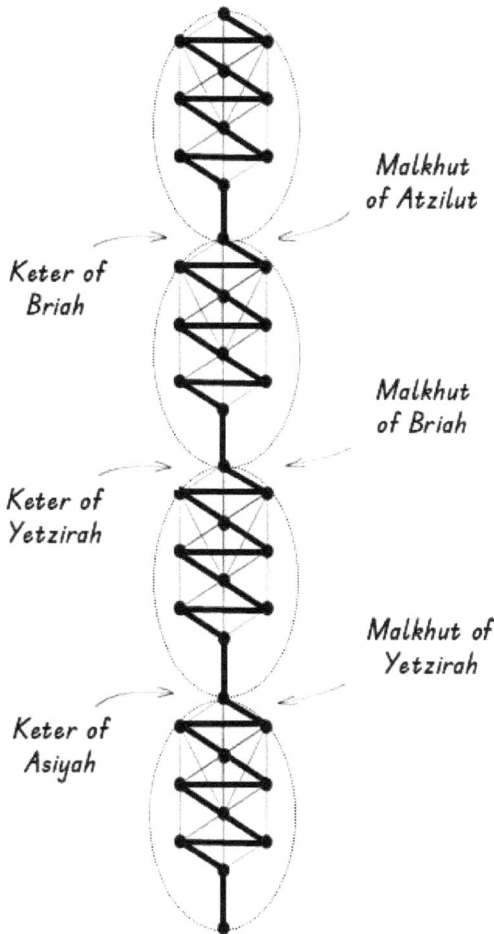

Malkhut of Atzilut

Keter of Briah

Malkhut of Briah

Keter of Yetzirah

Malkhut of Yetzirah

Keter of Asiyah

Figure 45. Malkhut of Atzilut overlaps Keter of Briah! In a similar matter, Malkhut of Briah coincides with Keter of Yetzirah and Malkhut of Yetzirah overlaps with Keter of Asiyah. Overlapping points exchange information and energy between above- and below-existing worlds.

There is a distinct scheme of interactions between the worlds of Olamot. According to Zev Ben Shimon Halevi, an author of books on the Toledano Tradition of Kabbalah, Malkhut of Atzilut encompasses Tiferet of Briah, which in turn corresponds to Keter of

Yetzirah. Similarly, Malkhut of Briah overlaps Tiferet of Yetzirah, which overlaps Keter of Asiyah.[16]

A "Telescopic" Model to Illustrate the Connectivity between the Four Worlds

Now, I invite you to consider the "telescopic" visual model of the four worlds presented in Chapter 12. The flow of energy and information between the worlds remains uninterrupted, despite the divider (screen). The ten Sefirot of the lower world emerge from the ten Sefirot of the higher world. An incompletely extended telescope represents the relationship between Olamot (worlds). Higher rungs overlap the lower rungs, which represent the more basic levels of hierarchical organization. The connection between the Sefirot of the lower world and the world above can be improved. The degree of overlap serves as a measure of how the upper realms influence events within the lower realms.[17]

To illustrate the potential of such a visual model so we can comprehend the challenges of biology, I will refer back to the coupling narratives of ontology and phylogeny– individual development and evolutionary history – we discussed in Chapter 12. To remind you, the "biogenetic law" formulated by Ernst Haeckel is one of only a few theories that project the lofty status of law. Another common name of this law is the "theory of recapitulation" because of its statement that "ontogeny recapitulates phylogeny." According to this statement, the embryonic stages of more advanced species reflect the characteristics of their evolutionary ancestors. However, this concept remains very controversial because of outdated suggestions, such as that a human embryo includes the "fish-like" stage.

A human zygote forms when a sperm penetrates the outer surface of an egg. During mitosis, the zygote divides into two, four, eight, or sixteen-cell stages. Those processes are described in Asiyah – we can

---

[16] Z'ev Ben Shimon, *The Way of Kabbalah.* 1976, page 35.
[17] Hyman Schipper, Kabbalah and the Physics of David Bohm. In *Unified Field Mechanics II: Formulations and Empirical Tests,* 2018, page 361.

follow and describe each zygote division. The cell division composes differentiated tissues and organs via mitosis until the formation of a complete individual. We can still observe the composition of tissues in Asyah, but the formation of individuals can be described in Yetzirah only.

In Yetzirah, we can notice some formative features in human embryos that resemble fish gills at the Yetziratic level. However, here we see the danger of remaining in Yetzirah. Do you remember poor Zoma, who lost his mind? Ernst Haeckel made a similar mistake when he claimed that human embryos include a "fish" stage in their individual development. Yetzirah teaches us to notice "likeness," but not to create concepts!

In Briah, we conclude that despite their diverse forms, mammals share a standard blueprint in their embryonic development. Fundamental processes follow similar patterns across species, from early cell division to organ formation. Therefore, it is only natural that they share common patterns in their embryonic development. Human embryos have a tail, indicating they have common embryonic development patterns. Limb buds arise, a precursor to appendages that could become paws, hooves, or flippers. Neural tubes curve and fold, eventually evolving into the intricate nervous systems that separate each mammalian species.

In Atzilut, the unity in design reflects shared evolutionary history. Early mammalian life unfolds through a series of stages that mirror one another, from the tiniest shrew to the towering elephant. All mammals share this heritage, illustrating the remarkable consistency of Life's evolutionary process. Haeckel's theory of recapitulation emphasizes that all living organisms bear a resonance of their own development and evolutionary history (Fig. 46).

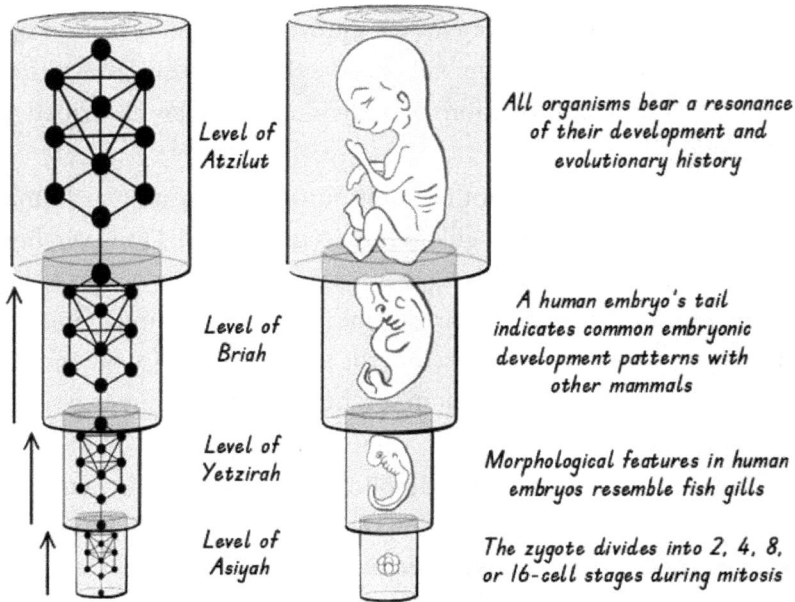

Level of Atzilut	All organisms bear a resonance of their development and evolutionary history
Level of Briah	A human embryo's tail indicates common embryonic development patterns with other mammals
Level of Yetzirah	Morphological features in human embryos resemble fish gills
Level of Asiyah	The zygote divides into 2, 4, 8, or 16-cell stages during mitosis

Figure 46. The "telescope-like" visual model of Olamot – the Four Worlds – allows for comprehending the controversial recapitulation theory.

The "Klein Bottle" Model to Illustrate the Connectivity between Adjacent Worlds

For linear thinkers, it is hard to recognize the connection between the separate four worlds if one is attached to the three-dimensional space model. The following visual model illustrates the possibility of connecting words that exist below or above each other. We learned from the "telescopic model" that one of these worlds is, in fact, inside a higher one.

In Chapter 8, we discussed the "inside-outside" paradox in biology. This paradox refers to the challenge of distinguishing which variables are internal or external to a living system. Kabbalah teaches us not to "solve" a paradox, but to live with it and use it! There, we discussed many examples, such as the microbiome, epigenetic landscape, extended organisms, and biological networks. These demonstrated that we can explore the hidden landscape of biological life.

396

The name of the proposed model, "the Klein Bottle," followed Felix Klein's creation of this image in 1882. Remember the "Mobius Strip" introduced in Chapter 12? The Mobius strip exists in three dimensions but is only a surface. Klein went one step further. He imagined sewing two Mobius strips together to create a single-sided bottle with no boundaries. Its inside is also its outside! It contains itself! A true Klein bottle requires 4-dimensions because the surface must pass through itself without a hole.[18] An ant can walk along the entire surface without ever crossing an edge!

Although it is very difficult to imagine, some masters constructed such a bottle with a stretchable tube. One end of the tube became the bottle's neck, while the other end became the base. The neck was twisted to pass through an opening in its side and then joined to its base. This operation made the tube's neck surface continuous with its own base surface (Fig.47).

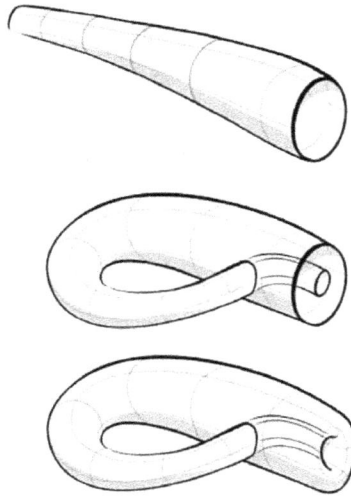

Figure 47. The Klein Bottle model helps to comprehend the "inside-outside" paradox in overcoming the problem of accepting multi-dimensional worlds and maintaining the connection between the Worlds-Olamot. The "bottle of life" has only one side – a face but no real separate and independent backside.

---

[18] https://plus.maths.org/content/imaging-maths-inside-klein-bottle

In his private lessons, kabbalist Joel Bakst repeatedly stressed the importance of the Klein Bottle model to overcome the problem of accepting the reality of multi-dimensional worlds and maintaining the connection between them. Bakst noted that "while traveling along its paradoxical surface ... you are never inside of the tube proper, but rather you are always traveling on its outside even when you are passing through the wall going temporally "inside" before re-emerging from the other "end."[19] Bakst concluded: The "bottle of life" has only one side, a face without a separate and independent backside.

Recursive Hierarchy

Gregory Bateson was an anthropologist, social scientist, linguist, semiotician, and cyberneticist whose work intersected many other fields. In his *Steps to an Ecology of Mind*, Bateson proposed the framework of "levels of learning." Bateson wrote: "If, in the communicational and organizational processes of biological evolution, there is something like *levels* – items, patterns and possibly patterns of patterns – then it is logically possible for the evolutionary system to make something like positive choices."[20]

Bateson's framework can help us understand the possibility of interaction between the different planes of knowledge. He says all living systems are multi-level and recursive. The adjective "recursive" comes from the Latin, *recurrere*. Recursive refers to a procedure or rule that "reoccurs" repeatedly, like funhouse mirrors that are angled to present an infinite number of images.[21] Bateson noticed that informational content in biological systems always assumes a context of interpretation where the term "context" is used in the sense of a higher-order form.[22] Bateson defines context as a "collective term for all those events which tell the organism among what set of

---

[19] Joel David Bakst, *Beyond Kabbalah*. 2012, page 293.

[20] Gregory Bateson, *Steps to an Ecology of Mind:*
Collected Essays in Anthropology, Psychiatry, Evolution, and Epistemology. 2000, page 411.

[21] https://www.vocabulary.com/

[22] Yair Neuman, *Reviving the Living: Meaning Making in Living Systems*. 2008, page 218.

alternatives he must make his next choice."[23] Contexts are always embedded within another context. This embedding also has a dynamic aspect. This dynamic aspect is constituted through feedback loops in which information is fed back and forth between the different levels of the system. We discussed the role of feedback loops in the self-organization of complex systems in Chapter 3. Feedback loops assure each level's stability and constitute the working whole.

Paul Tosey, a senior lecturer at the University of Surrey, and his colleagues arranged learning levels as concentric circles to represent recursive hierarchy.[24] Each successive level extends beyond the previous level boundary. Learning at higher levels means being offered new opportunities with an ever-expanding range of possibilities. Feedback loops connect each level to previous levels and vice versa (Figure 48). Bateson assumes that higher orders of learning are not more desirable than lower orders. The importance of each level cannot be overstated!

---

[23] Gregory Bateson, page 289.

[24] Paul Tosey, Max Visser, and Mark NK Saunders, *The origins and conceptualizations of 'triple-loop' learning: a critical review.* Management Learning, 2011, vol. 43, no. 3, page 299.

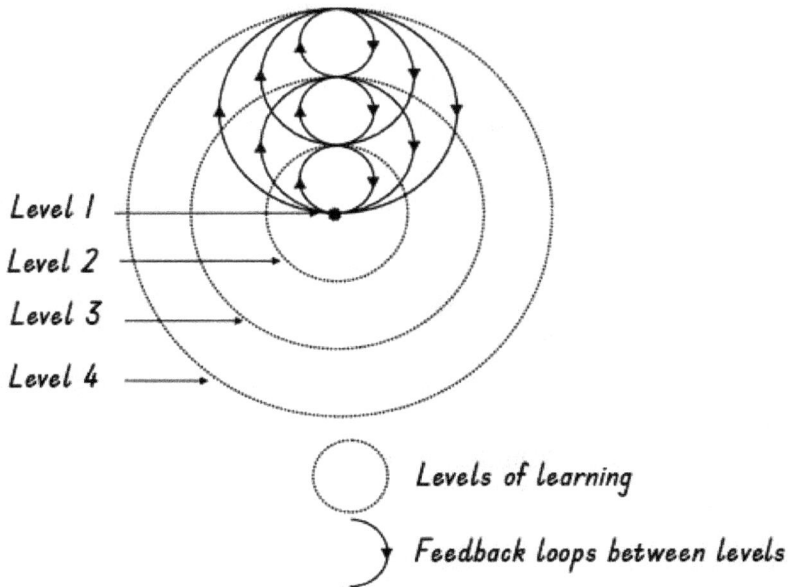

Figure 48. Bateson's levels arranged as a recursive hierarchy (modified from Tosey et al., 2011).

Gary Schwartz, a professor in psychology at the University of Arizona and the director of its Laboratory for Advances in Consciousness and Health, and another psychologist, Linda Russek, wrote a very challenging book, *Living Energy Universe: A Fundamental Discovery that Transforms Science and Medicine*. In this book, Schwartz and Russek propose a four-stage scheme common to all systems. They do not use the Kabbalistic language, but I see some similarities in their description of the proposed stages and Olamot. Schwartz and Russek characterize the stages as follows: stage M – multiple independent laws; stage I – integrative systems of laws; stage L – living systems; stage E – evolving systems of laws.

Regardless of how Schwartz and Russek describe the four stages, it is interesting to observe how they approach the question of connections between them. "Recurrent feedback interaction" is the term they used to describe the constant accumulation of information and energy. They also refer to the process as "circular causality."

Positive feedback systems function, Schwartz and Russek claim, as if they have memory, meaning their history unfolds in complex ways as a function of feedback.[25]

The System of the Implicate Orders

Hyman Schipper, a Kabbalist and neuroscientist, denotes the principle of *Seder Hishtalshelut* ("causal hierarchy") as a top-down chain of command. Considering each Olam-world to be composed of ten Sefirot, dynamic interactions between the Sefirot are necessary for the establishment and proper governance of each world. [26]

Schipper noticed and justified the parallel between the Kabbalistic system of *olamot* - "worlds" and the assertion made by a leading 20th-century scientist, David Bohm, that the reality above us (the explicate order) merely appears to be a surface expression of something much deeper (the implicate order). The implicate (also called the "enfolded") order is seen as a more fundamental order of reality. In contrast, the explicate or "unfolded" order includes abstractions humans perceive. Bohm says an implicate order comprises layer upon layer of "hidden variables" beyond our perception. He viewed each deeper layer as more abstract than that but ultimately responsible for it.[27]

In some limited way, the implicate order contains degrees of enfoldment. In *Bridging Science and Spirit*, Norman Friedman noted that "The whole concept of subsystems and supersystems as part of an infinite yet inseparable totality clarifies one of the major paradoxes of our time."[28] The *focus* of the levels is what separates or specifies them. At the lower level, the focus is more limited than at the higher levels. Even though the higher worlds interpenetrate and sustain the lower levels, this remains true.[29]To be considered as a whole, each biological aspect at a particular level must

---

[25] Gary Schwartz and Linda Russek, *Living Energy Universe: A Fundamental Discovery that Transforms Science and Medicine.* 1999, page 69.

[26] Hyman Schipper, *Kabbalah and the Physics of David Bohm.* 2018, page 363.

[27] Ibid., page 357.

[28] Norman Friedman, *Bridging Science and Spirit.* 1994, page 62.

[29] David Parrish, *Nothing I See Means Anything.* 2006, page 108.

be coordinated with other levels, both below and above. Changes in a biological object's data correlate with changes in the data space that describes the relationship between these objects. They can reflect the conceptual categories that help us define these objects.

Proposing the term "recurrent feedback interaction", Schwartz and Russek refer to Bohm's theory of implicate order. They stressed that Bohm "went on to employ the circular concepts of "re-injection", "re-projection and "recurrent actuality," to explain the origin of memory and wholeness in nature."[30]

David Bohm's profound insights into the role of memory in fostering connections between different realms hold a critical place in our understanding of the interconnectedness of existence. Repeated transitions from one level of learning to another contribute to strengthening the feedback loops that facilitate the exchange of information between worlds. Memory, in its broadest sense, encompasses the enduring effects of past experiences that confer survival benefits and enhance adaptation.

Consider, for instance, the development of antibiotic resistance in pathogenic bacteria – a remarkable manifestation of memory at work. While individual bacteria may not possess cognitive memory, their population benefits from previous exposures to specific antibiotics, akin to learning from past experiences. Point mutations of bacterial genes are recorded in Asyiah. Survival of individual bacteria, which carry the beneficial mutation, is described in Yetzirah. The concept of evolutionary-developed microbial resistance to antibiotics previously lethal to such bacteria belongs to Briah.

The crucial difference between Kabbalistic biology and Bohm's system of implicate orders is the recognition of the four-tier orientation plan correspondent to the four worlds (*olamot*). In the four-tiered dimensional universe, the most profound laws of Nature are hidden inside the four planes of invisibility. They also lead to the four domains of knowledge emphasized by the DIKW pyramid levels - data, information, knowledge, and understanding/wisdom. It is

---

[30] Gary Schwartz and Linda Russek, *Living Energy Universe: A Fundamental Discovery that Transforms Science and Medicine.* 1999, page 125.

important that we ask different questions at each level and look for different approaches to answer those questions.

As demonstrated above, any biological object exists in the four worlds. Each natural phenomenon could be described from these four levels of view. Each level represents a unique world. These worlds are Asiyah, Yetzirah, Briah, and Atzilut.

To summarize, Asyiah is the world of physical objects -- the world in which biological objects are expressed in their elements, forms, and movements.

Yetzirah is the world of perception -- the world of functions and relationships between elements of the biological system that requires definitions and terms.

Briah is the world of conception -- the world of concepts designed to explain biological systems' development and complexity. Atzilut is a world of understanding of life's interconnectedness.

Each of these worlds requires description systems specific and unique for each level of manifestation visibility. Asiah needs measurements of forms and actions of single biological objects; Yetzirah is based on classification of biological objects and their relationship with inside and outside the organism systems and environment; Briah brings conceptualization of biological objects as systems; Atzilut leads to the creation of an image of a biological object as a whole union with the surrounding environment and unbreakable connection between the object and the biologist.

Chapter 17

Fifth Dimension: The Action of Signs in Biology

The intricate tapestry of life extends beyond the confines of physical bodies and transcends the boundaries of individual organisms and cells. The notion of the Fifth Dimension, which navigates the complex realm of meaningful signs, rather than mere physical entities, offers an additional system of orientation within the invisible landscape of organic life.

As we delve into the realm of meaningful signs, we will unearth a hierarchical arrangement of factors that extends beyond the conventional cues of physics and chemistry. This hierarchy traverses realms of biological codes, functional signs, and interpretative symbols, mirroring the profound complexity of living systems.

This chapter introduces the reader to the fascinating merging of biology with Semiotics. Semiosis is any activity, conduct, or process that involves signs. A "sign" is defined as anything that communicates a "meaning" to the sign's interpreter.

An Introduction to Semiotic Biology

The emerging discipline of "Biosemiotics" offers a gateway into this uncharted territory, bridging the domains of biology and semiotics to forge a new understanding of Life. Jesper Hoffmeyer, a professor at the University of Copenhagen Institute of Biology, defined Biosemiotics as "an interdisciplinary scientific project that is based on the recognition that life is fundamentally grounded in semiotic processes."[1] Biosemiotics provides tools to interpret biological systems as a sign-processing study. Through the lens of the Biosemiotics paradigm, the biological world is revealed to be more than its molecular structure and physical bodies; it is also a world that encompasses biological signs and their meanings and highlights cognition as an inherent lifestyle.

Derived from the Greek words, "bios" (life) and *sēmeiōtikos* (observant of signs), Biosemiotics is an interdisciplinary domain that includes "meaning-making" sign interpretation, and communication within the biological landscape. The term, "Biosemiotics," was first used by a German-Israeli psychiatrist and semiotician, Friedrich Solomon Rothschild, in a lecture for the New York Academy of Sciences. Later, he stated that "living systems are constituted from their very beginning as sign systems."[2]

In Chapter 12, we highlighted an interesting parallel between the Kabbalistic hierarchical system (represented by the Tree of Life) and an intellectual approach to Structuralism. Both can be seen as general trends in scientific methodology. Structuralism played a crucial role in developing semiotic thinking. Structuralists emphasize the relational nature of systems, like language. There are more than just elementary units, particles, or entities in a system. The similarities and differences between entities are also critical, as well as their roles, significance, and functions. Like words in languages, individual entities may be substituted or transformed into other entities. Nevertheless, the code defining how the entities relate to each other may remain unchanged. The system of biological species can also be

---

[1] Jesper Hoffmeyer, *Biosemiotics: An Examination into the Signs of Life and the Life of Signs.* 2008, page 3.

[2] Friedrich Solomon Rothschild, *Concepts and Methods of Biosemiotics.* 1968, vol.20, page 174.

viewed as relational in a similar way, based on mutual recognition.[3]

Biosemiotics is a discipline that redefines how we perceive life by positing semiosis: the intricate process of sign interpretation, creation of meaning, and communication as an innate and integral facet of existence. By unraveling the secrets of biological interpretation processes, the production of codes, and the symphony of communication in the natural world, we may find ourselves propelled into uncharted territories of new understanding.

To describe life as a semiotic process, Kalevi Kull, together with several other leading scientists in biosemiotics, made a set of statements. The eight statements (the authors called them "theses") include:[4]

1. The distinction between semiotic and non-semiotic is coextensive with the life–nonlife distinction.
2. Biology is incomplete as a science in the absence of explicit semiotic grounding.
3. The predictive power of biology is embedded in the functional aspect and cannot be based on chemistry alone.
4. Differences in methodology distinguish semiotic biology from non-semiotic biology.
5. Function is intrinsically related to organization, signification, and the concept of an autonomous agent or "self."
6. The grounding of general semiotics has to use specific tools.
7. Semiosis is a central concept for biology that requires more exact definition.
8. Organisms create their environment.

Adding to the descriptions of biological relationships, biosemiotics introduces such hardly analyzed concepts as "function," "information," "code," and "signal." As Kull wrote, biosemiotics "points to the fact that those notions cannot be avoided or fully

---

[3] Kalevi Kull et al. *An Introduction to Our View on the Biology of Life Itself.* In *Towards a Semiotic Biology.* 2011, page 6.

[4] Kalevi Kull et al., *Theses on biosemiotics: Prolegomena to a theoretical biology.* 2009, vol. 4, pages 167-173.

substituted with merely chemical accounts."[5]

Biological Organisms See the World in Different Ways

We must first understand that both individuals and species perceive the "outside" world from very different perspectives. For the organism, it is conditionally perceived as external to the organism. Ecology tells us that organisms and populations are inextricably bound to their environments. From a semiotic perspective, biological signs constitute an organism's environment.

To illustrate this point, look at the pond cartoon in Figure 49. Multiple species of mammals, birds, reptiles, amphibians, fish, and insects live in the same pond. Obviously, the pond has length, width, and depth. Ecologists can also characterize the pond's environment by measuring both abiotic variables (water temperature, water pH, soil properties of pond-bottom sediments, etc.) and biotic variables (richness of animal species, community structure, etc.). These parameters appear objective and are already present in the pond's environment.

Diverse biological species inhabiting the same pond, however, perceive the environment differently. While physically present in the same pond, raccoons, beavers, ducks, herons, frogs, fish, crayfish, turtles, and mosquitoes inhabit distinct "worlds" (environments). Animals perceive the world differently based on the signs they use.

---

[5] Ibid., page 170.

Figure 49. Different animal species perceive the world of a pond in very different ways.

Estonian biologist and semiotician Jakob von Uexküll introduced the concept of *Umwelt* (meaning "environment" or "surroundings" in German). His arguments have greatly contributed to our understanding of how important it is to recognize the different ways living creatures perceive signals. Uexküll published a monograph in 1934 titled, *A Stroll through the Worlds of Animals and Men: A Picture Book of Invisible Worlds.* In it, he revealed how simple animals live in simple worlds, and complex animals live in complex ones. He argued that organisms experience life in terms of subjective reference frames specific to their species.

Uexküll wrote:

> "First blow, in fancy, a soap bubble around each creature to represent its own world, filled with the perceptions which it alone knows. When we ourselves then step into one of these bubbles, the familiar meadow is transformed. Many of its colorful features disappear, others no longer belong together but appear in new relationships. A new world comes into being. Through the bubble, we see the world of the burrowing worm, of the butterfly, or of the field mouse; the world as it appears to the animals themselves, not as it appears to us. This we may call the phenomenal world or the

self-world of the animal… These different worlds… which are as manifold as the animals themselves, present to all nature lovers new lands of such wealth and beauty that a walk through them is well worthwhile, even though they unfold not to the physical but only to the spiritual eye. "[6]

The Umwelt, as described by Jon Deely, the Professor of Philosophy at Loras College, is a "model world" from the point of view of possibility. It is one of the infinite varieties of possible alternatives by which the physical furnishings of the environment can be arranged and incorporated into a superstructure of possible experiences. But from the point of view of its inhabitants, an Umwelt is the actual world they experience as an everyday reality.[7]

A Tick's View of the World

One of the most famous examples that Uexküll used to illustrate his Umwelt concept is his description of the world of a tick. The Umwelt of a tick is different from the Umwelt of a beetle or a honeybee. In his example, a tick is "blind and deaf" with a limited Umwelt. According to Uexküll, the tick's capacity for perception and reaction is limited to only three stimuli, which he calls "effect signs" or "carriers of significance." An adult female tick's whole world is reduced to three carriers of significance: (1) the specific smell of butyric acid released by mammals, (2) the temperature typical for mammals, (3) and the skin structure of mammals. These three signifiers provide the entire orientation of a tick to its surrounding environment (Figure 50).

---

[6] Jakob Johann von Uexküll. *A stroll through the worlds of animals and men: A picture book of invisible worlds.* 1992, vol.89, 319-391.

[7] John Deely, "Basics of Semiotics." Indiana University Press, 1990, page 60.

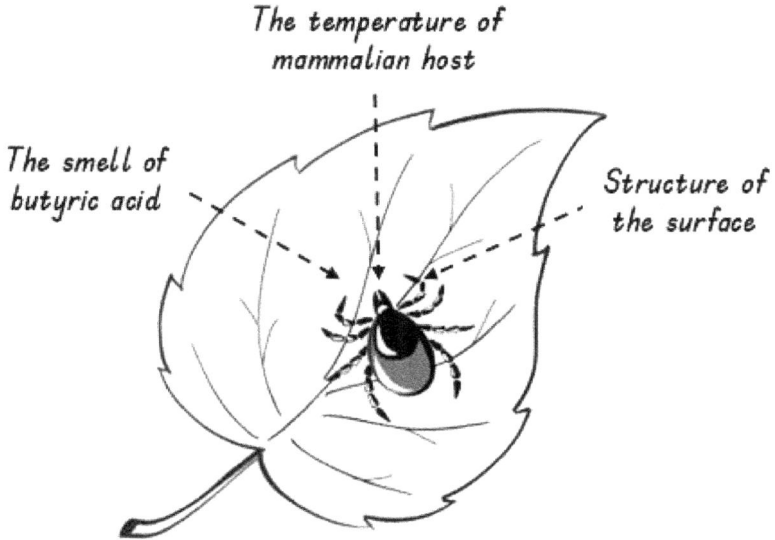

Figure 50. An adult female tick's whole world is reduced to three carriers of signifiers.

However, the sensory responses of a tick include even more stimuli than Uexküll indicated. For example, ticks can also react to gravity, humidity, light, and a sex pheromone.[8,9] There is a crucial point to be made here about Uexküll's insight. His view allows us to decipher the ecological and behavioral significance of certain signs from ticks. First is the orientation of the tick in search of a potential host. Second, is the intensity of each signal that leads the tick to create a uniquely specific behavior pattern. Finding the host, feeding on its blood, and laying eggs allow the tick to continue the survival of its species.

The importance of this conclusion has two aspects. First, some dimensions provide a tick with an orientation in the outside world. Second, identifying those dimensions and measuring them gives investigators tools they can use under environmental conditions. We

---

[8] A.D. Lees, *The Sensory Physiology of the Sheep Tick, Ixodes ricinus L.* Journal of Experimental Biology, 1948, vol. 25, no.2, pages 145–207.

[9] R.S. Berger et al. *Demonstration of a sex pheromone in three species of hard ticks.* 1971, vol.8, pages 84-86.

can even understand why we must develop ways to control the distribution of ticks and their attacks on human beings!

How can the tick at a specific stage see the world around it? Our task is to perceive the world from the viewpoint of the specific animal species being investigated. This enables us to organize investigations and interpret results that shed light on the ecological adaptation and evolutionary patterns of this species.

Meaning-making Signs

Signs can also communicate feelings (which are usually *not* considered "meanings"). Contemporary semiotics is a branch of science that studies "meaning-making" and various types of knowledge.[10] Kalevi Kull, professor of Semiotics at the University of Tartu, defines sign processing (meaning-making) through choice.[11] "Semiosis" is the process of making choices between simultaneously available options. By making choices, semiotic learning leaves traces, which influence further decisions. Such choices form memories.

"Meaning" can be intentional, such as a word uttered with a specific meaning. It may also be unintentional, such as a symptom being a sign of a particular medical condition. A sign can also represent various meanings, depending on its context. Consider the case of someone suffering from a fever, for example. A fever is a body temperature higher than normal. The doctor can be more specific if an adult's temperature is 103°F or higher.

According to physiologists, the hypothalamus in the brain controls body temperature and may raise the body temperature. Fever is usually triggered by an infection, but it can also be caused by other factors. These can include drugs, vaccines, rheumatoid arthritis, stimulation of the immune system, heat exhaustion, and sunburn. Teething in babies can cause a mild, low-grade fever. During the recent COVID-19 pandemic, fever was a possible early

---

[10] C. Campbell, A. Olteanu, and K. Kull, *Learning and knowing as semiosis: Extending the conceptual apparatus of semiotics.* 2019, vol. 47, no.3/4 pages 352–381.

[11] Kalevi Kull, *Choosing and learning: Semiosis means choice.* 2018, vol.46, no.4, pages 452-466.

sign that one might be infected by the virus. Any biological system has a web of linked recognition processes.[12] In the book *Toward a Semiotic Biology,* edited by Claus Emmeche and Kalevi Kull, the authors used the web of recognition processes of a cell as an example. The cell must rely upon energy from its surroundings to maintain its metabolism. Substances containing useful energy must be distinguished from harmful substances. Thus, the cell depends on the recognition processes and is structured as a system of categorization, enabling it to determine whether substances are useful or dangerous.

An Organism Recognizes a Pathogen... as a Pathogen

In order to illustrate the importance of "meaning-making" processes in biology, I will refer to the biology of epidemics and pathogens as examples. Biosemiotics states this paradox: "Interactions at the molecular level cannot fully explain the evolution and transmission of pathogens." In my article in the journal, *Entropy,* I stressed the importance of acknowledging the meaning of "sign" and "signal" for us to understand what a pathogen actually is. [13]

Neuman defined a "signal" as meaningful if it involves communicating something not directly expressed.[14] For example, being an antigen is not an attribute directly expressed by a molecule (a signal). The meaning of "being an antigen" derives from a complex process that is evident in its immune response. Similarly, an infectious agent is not just a microbe with a molecule recognized as "virulent." It is infectious because of its *ability* to create infection in the appropriate host. The "host" is characterized by its *ability* to be infected, not by the presence of a susceptibility gene. The "lock-and-key" model is a popular metaphor for immune recognition; however, living systems at different levels of analysis present a much more flexible interpretation of signals than the "lock-and-key" model

---

[12] Claus Emmeche and Kalevi Kull, *Towards a Semiotic Biology: Life is the Action of Signs.* 2011, pages 15-16.

[13] Michael Kosoy, *Deepening the Conception of Functional Information in the Description of Zoonotic Infectious Diseases.* Entropy, 2013, vol.15, no.5, pages 1929-1962.

[14] Yair Neuman, *Meaning-making in the immune system.* Perspectives in Biology and Medicine, 2004, 47, 317–327.

suggests.[15]

If we were to present the biology of contagious diseases as a stage play, the actors would be infectious agents, their hosts, and vectors. These actors could play different roles. An infectious agent can range from being a neutral symbiont or a beneficial player hiding in the host tissues to a lethal pathogen. The animal host could play the role of an incidental receiver of the infection, or it can be an effective spreader of the infection to other animals, a long-term reservoir host, etc. People can also participate in this play, either as a quiet, passive audience or as a prompter repeating the text. They can also participate as suffering victims or as active participants in the play, assuming different roles during an epidemic. This is not a poetic image, but rather an attempt to illustrate the possibilities of this complex process. DNA and other molecules can provide the text of the script, but the dialogue will be performed between cells, organisms, and populations.

"How," we might ask, "can participants in infectious systems in natural settings recognize each other? How can an infectious agent "recognize" an animal organism as either a "reservoir host" or an "incidental host"? How can a potential host recognize a microorganism as a harmless or beneficial symbiont or a potentially damaging pathogen? To which signs associated with bacteria does an animal organism refer when switching from "resistant" to "susceptible" concerning specific pathogens?

An effective way to analyze the communication between elements of the infectious system on population levels should go beyond studying the interactions between molecules. It should also recognize the complex "images" of lifestyles and histories. A list of specific molecular interactions cannot replace the meaning of "signs" expressed by microbes to animals or by animals to microbes through the two-way and multiple-way communicative processes.

---

[15] M.J.R. Healy, *Paradigms and pragmatism: Approaches to medical statistics.* Annali di Igiene, 2000, vol.12, pages 257–264.

In an article written with my son, Roman, we indicated that an animal organism can recognize a bacterium either as a pathogen or as part of a microbiome. In other words, an animal organism can recognize it as either alien cells ("non-self") or "own" cells ("self").[16] In a related reference, Alexei Turovski wrote, "...viewing the parasite-host relationships ... the most powerful instrument of those dialogical traits appears to be the skillful manipulation of the criteria of 'own-strange-alien'."[17] When describing dynamic host-parasite relations, Horwitz and Wilcox use an analogy with dancing: "...much like a waltz, the 'partners' are constantly adapting to each other's 'moves' in responses to the presence, or potential presence, of each other."[18]

Signification in Biology

According to the teaching of Ferdinand de Saussure, the famous Swiss linguist and philosopher, any sign is composed of the "signifier" (the form which the sign takes) and the "signified" (the concept it represents). In English, the word "sign" has the same root as the word "signification." Merriam-Webster defines "signification" as the act or process of signifying by signs or other semiotic means.

"Signification" is defined as the connection between "signifier" and "the signified." The signifier is the concrete form of the sign that we see, hear, smell, or touch. The signified is the idea expressed by the sign. In the case of text, the signifier is the group of words forming the text. The signified is the interpretation of the message made by the interpreter. As the complexity of a system increases, our ability to make precise, yet significant statements about its behavior diminishes. As we have all learned, precision and the ability to make an urgent decision are often incompatible!

---

[16] Michael Kosoy and Roman Kosoy, *Complexity and biosemiotics in evolutionary ecology of zoonotic infectious agents.* Evolutionary Applications, 2017, vol. 11, no.4, pages 394-403.

[17] Alexei Turovski, *On the parasites association as a vectorizing factor in biosemiotic development.* Semiotica, 2001, vol.134, page 412.

[18] Pierre Horwitz and Bruce Wilcox, *Parasites, ecosystems and sustainability: An ecological and complex system perspective.* International Journal of Parasitology, 2005, vol.35, pages 725–732.

Imagine a situation in the vast African savannah where two gazelles, named Signi and Preci, were grazing peacefully on the grassy plains. Suddenly, they sighted a nearby lion looking straight at them. Signi, who had always reacted quickly to danger, immediately cried out, "Run, Preci, run! The lion is close!" Preci, on the other hand, was more thoughtful and calculating. She wanted to make sure that the situation was indeed urgent: "The lion does not look hungry. I can run at a burst quicker than a lion, and we found some tasty leaves on this bush. Let's keep an eye on the lion and not waste our resources." (Figure 51).

Figure 51. The gazelles must choose between "significance" and "precision."

This story illustrates the tension between the need for precision and the urgency behind critical decision-making. A quick decision may be more important in certain situations than precision and careful consideration. Despite the story's simplicity, real-life situations may require varying approaches to urgent decisions, based on individual circumstances. Our ability to acquire information and take action is affected by the types of reasoning we use. We can either collect more information or take immediate action if the situation threatens one's survival.

A continuous spectrum exists between significance and precision, rather than there being two extremes: "one *or* another." This choice allows us to measure how far an organism's perception of significance departs from physical measurements. Kalevi Kull, professor in biosemiotics at the University of Tartu in Estonia, noted: "Living systems are those that make distinction, or chose." A task of biosemiotics is to demonstrate the multitude, or meaningfulness, of the categories in the living, and particularly, in the invisible worlds of other organisms."[19]

The Triadic Model of a Sign

In contrast to Saussure's model of the sign in the form of a "signifier-signified" pair, Charles Sanders Peirce offered a triadic model. According to philosopher Paul Weiss, Peirce was "the most original and versatile of America's philosophers and America's greatest logician."[20] This version includes: a "Sign-vehicle" (the form of the sign), an "Interpretant" (the sense made of the sign," and an "Object" (to which the sign refers). Semioticians make a distinction between a sign and a "sign-vehicle." Daniel Chandler at the University of Wales, Aberystwyth noted that "the sign *is more* than just a sign-vehicle."[21] This statement emphasizes that a sign extends beyond its surface-level representation. It functions as a dynamic and versatile medium through which meaning, information, and communication are conveyed.

There is more in Pierce's triad, so let us consider it. As a philosopher, Pierce was interested in "phenomenology," a philosophy of experience. For phenomenology, the ultimate source of all meaning and value is the lived experience of human beings. All philosophical systems, scientific theories, or aesthetic judgments are considered abstractions from the ebb and flow of the lived world. The modern founder of phenomenology, the German philosopher

---

[19] Kalevi Kull, *Life is Many*. 2007, page 194.

[20] Paul Weiss, *Peirce, Charles Sanders*. Dictionary of American Biography. Internet Archive, 1934.

[21] Daniel Chandler, *Semiotics for Beginners: Signs*.
https://www.cs.princeton.edu/~chazelle/courses/BIB/semio2.htm

Edmund Husserl (1859–1938), sought to make philosophy "a rigorous science" by returning our attention "to the things themselves." Husserl does not mean that philosophy should become empirical, as if "facts" can be determined objectively and absolutely. Rather, when searching for foundations on which philosophers can attain certain knowledge, Husserl proposes that later reflection should eliminate all unprovable assumptions (about the existence of objects, for example, or about ideal or metaphysical entities) and describe what is given in the experience.

Pierce has come up with a structural classification system for phenomenological elements. He introduces three principal subdivisions of the elements that can be present in the mind, designated "Firstness," "Secondness," and "Thirdness." These phenomenological classes represent a distinction between the sign itself (an example of Firstness), its object (an example of Secondness), and the sense made of the sign (an instance of Thirdness).

Pierce's terminology may sound confusing. The key lesson here is the parallel between Pierce's tripartite structure and the Kabbalistic mode of thinking. The latter is based on three fundamental forces represented by Right, Left, and Central pillars, which we described in Chapters 11 and 15. Pierce particularly liked the number "three" and considered semiosis as a triadic process.

Firstness

In a semiotic process, Firstness is the sign described by Pierce). Pierce defines the feeling of Firstness as "an instance of that kind of consciousness which involves no analysis, comparison or any process whatsoever... it has its own quality which consists of nothing else"[22]

Let's take a look at the idea of "greenness," as described in Chapter 7. The concept of "greenness" does not depend on the presence of a green object. "Green" can be a tree leaf, a grasshopper,

---

[22] Charles Sanders Pierce, *Principles of Philosophy and Elements of Logic*. Collected Papers of Charles Sanders Peirce, 1932, vol.1, page 152.

or a chameleon. It is a general sense of a certain quality or latent potentiality. While "Greenness" is a quality or state of being, it cannot exist by itself.

On the Tree of Life, Firstness belongs to the Right Pillar presented by Hesed (or Hasadim). Pierce wrote: "The idea of Firstness is predominant in freshness, life, freedom. The free is that which has no another behind it."[23] This is a quality of Hesed! "Firstness" belongs to the realm of possibility. It needs an object in which to embody itself. When an object acts as this embodiment, it leads to "Secondness."

Secondness

"Secondness" is the state of being in relation to something else. While "Firstness" is about being, "Secondness" is about existing. In Chapter 7, we also referred to "dog-ness," which cannot be seen, but "dog-like" animals *can* be identified.

In semiosis, Secondness is an object. While Firstness is pure sensation, Secondness is intellectual categorization. This is the level of tangible existence and practical experience. For example, the Firstness of the quality of "green-ness," when attributed to a leaf, becomes Secondness. Even the negation of greenness attributed to a red flower becomes Secondness. Secondness manifests itself when Firstness connects to another object through relation, effect, dependence, independence, negation, or occurrence. If any influential mediation is involved that tends to move the direction of interpretation, it leads to Thirdness.[24]

From the Kabbalistic perspective, Secondness corresponds to Gevurah on the Left Pillar. Pierce says that Secondness acts as a constraint or force on Firstness, which is a characteristic of Gevurah.

---

[23] Ibid., page 148.

[24] https://www.linkedin.com/pulse/understanding-semiotics-firstness-secondness-thirdness-tanvi-gupta/

Thirdness

Thirdness is the mediator through which a First and a Second are brought into a relationship. The Second is the ends, and the Third is the means. Thirdness defines the path taken between the First and the Second. *Thirdness can be seen as the pre-determined or habituated mode of thinking.*

Here, the interpreter becomes a Third element between the sign and the object. The Third is a bridge between the First and the Second. It is a "synthetic consciousness driven by the sense of learning, thought, memory, and habit." In semiosis, Thirdness corresponds to the sense made of the sign (or Interpretant, according to Pierce).

In Kabbalah, Thirdness is represented by Tiferet on the Central Pillar – a pillar of integration. We discussed in Chapter 15 how the Third Point (Tiferet) underscores the importance of each point in the biological process. We defined the central pillar as a sphere of integrity regarding the meaning of biological organization. There, we cited Basarab Nicolescu, who said: "There's certainly a place for meaning: the included middle."[25]

The Semiotics of Invisibility in Kabbalah

Here's the question: "How can Kabbalah help introduce semiotic concepts such as "signs," "signification," and "semiosis" into biological research?" Historically, all these concepts originated in Kabbalah's "humanity domain" – linguistics, human communication, reading texts, etc.

The answer is by going deep into the essence. To explain this, I refer to Irun Cohen, a leading immunologist of our time who received the Robert Koch Prize for his research. Cohen explained the connection between science and the reading of the Torah in his remarkable book, *Rain and Resurrection.* Beyond his scientific research in immunology, Cohen also possesses a deep knowledge of

---

[25] Basaran Nicolescu, *The Hidden Third.* 2016, page 58.

the Written and Oral Torah, including Kabbalah.

Cohen noticed that the word "text" was derived from the Latin *textus*, a fabric or structure. In turn, the word *textus* is derived from *texere* --to weave. Just as a cloth is woven from raw fibers, a text is woven from raw words. The association of written text with a woven cloth is also present in Hebrew. A volume of the Talmud is called a *masekhet*, the Hebrew word which denotes the *warp* of a loom. The *warp* is the fixed foundation of threads of *woof* to fashion a whole cloth. A *masekhet* of Talmud is a *warp* that needs the *woof* of the reader's interpretation to fashion its meaning.[26]

Cohen presents the idea that science, like the Talmud, depends on a skillful interpretation of a received text. Nature is the text of science, and science "reads and writes" it according to pragmatic rules. Many scientists believe that the results of experiments speak for themselves, both among scientists and non-scientists. According to them, scientists who are educated and informed should reach similar conclusions. Alas, this is far from the truth! According to Cohen, "interpretation" can be explained as follows:

> "To interpret is to decipher and explain a hidden meaning. The manifest subject of the interpretation ... is seen to encode some hidden entity – a different object, story, process, or text. To interpret the manifest subject is to break the code and reveal the covert entity. The covert entity is thus *signified* by the manifest entity; to decipher the hidden meaning of the text is to signify its *signification*."[27]

Science is an attempt to read the "text" written by Nature. An observation or experiment made by a scientist is a signifier of a basic truth about the essential nature of the world. The task of the scientist is to interpret just *which* truth the observation signifies! The experimental methodology of science is itself based on signification. An experiment is designed to serve as a signifier of Nature.

---

[26] Irun Cohen, *Rain and Resurrection.* 2010, page 5.

[27] Ibid., page 99.

As Cohen stated, "Nature is too complicated, too uncontrolled, and too noisy to study just by looking at her."[28] Science requires experiments, including thought experiments. An experiment is a signifier, whereas a piece of Nature is signified. To illustrate signification in biological science, Cohen provides the following example. Cells growing in a tissue culture signify something about cells growing in the body. What we learn about artificial cell cultures can later be used to interpret the underlying causes for how cells behave when growing in the body. Thus, observed Nature is both signified and a signifier. Nature is a signifier of hidden causes; Nature is being signified by the experiment. Cohen defines "meaning" as emerging from a process of interaction. Meaning, like causality and signification, refers to a reality hidden behind appearance.[29]

When we first encounter a problem, we are unaware of the hidden option. Rabbi Nilton Bonder wrote: "When we manage to break the bonds of standard solutions, we are surprised to discover these hidden windows."[30] Bonder referred to the Jewish interpretative tradition of thought when a concept may appear in two distinct states: mashal, "signifier," and nimshal, "signified." He illustrated such thinking by using the example of water: "We can say that water is in the hidden form of the apparent when it is solid – that is when it is ice. Ice can be seen as nimshal and water as its marshal. Water reveals certain properties and characteristics of ice that are more evident in the form of water than in the form of ice."[31]

Massimo Leone, a semiotician from the University of Turin, developed the concept known as "cultures of invisibility." This concept, presented in Cultures of Invisibility: The Semiotics of the Veil in Judaism, clearly represents the idea of the hiddenness of the essential nature of any process under a "veil" underlines the imagery

---

[28] Ibid., page 111.
[29] Ibid., page 113.
[30] Nilton Bonder, Yiddishe Kop. 1999, page 2.
[31] Ibid., page 28.

of ancient Kabbalistic texts.[32]

In Leone's view, visual studies from the field of phenomenology to the field of philosophy have been characterized by a general bias. Since the origins of these disciplines emphasize the role of visual representations, these disciplines implicitly support the idea that the best way to know is by giving "an iconic presence to what is absent." In semiotics, iconic representation is the use of visual images to make actions, objects, and concepts in a display easier to find, recognize, learn, and remember. On the contrary, "cultures of invisibility" must be studied from the point of view of what they hide, as well as what they conceal, giving "an iconic absence to what is present." A visual representation of Leone's semiotic dynamics can be found in Figure 52.

---

[32] Massimo Leone, *Cultures of Invisibility: The Semiotics of the Veil in Ancient Judaism.* In *Transmodernity: Managing Global Communication.* Proceeding of the 2nd Congress of the Romanian Association for Semiotics, 2009, pages 189-201.

Absent object

↓

Iconic
presentation

Signs of
visibility

↓

Cultures of
visibility

A.

Present object

↓

Iconic
"absentification"

Signs of
invisibility

↓

Cultures of
invisibility

B.

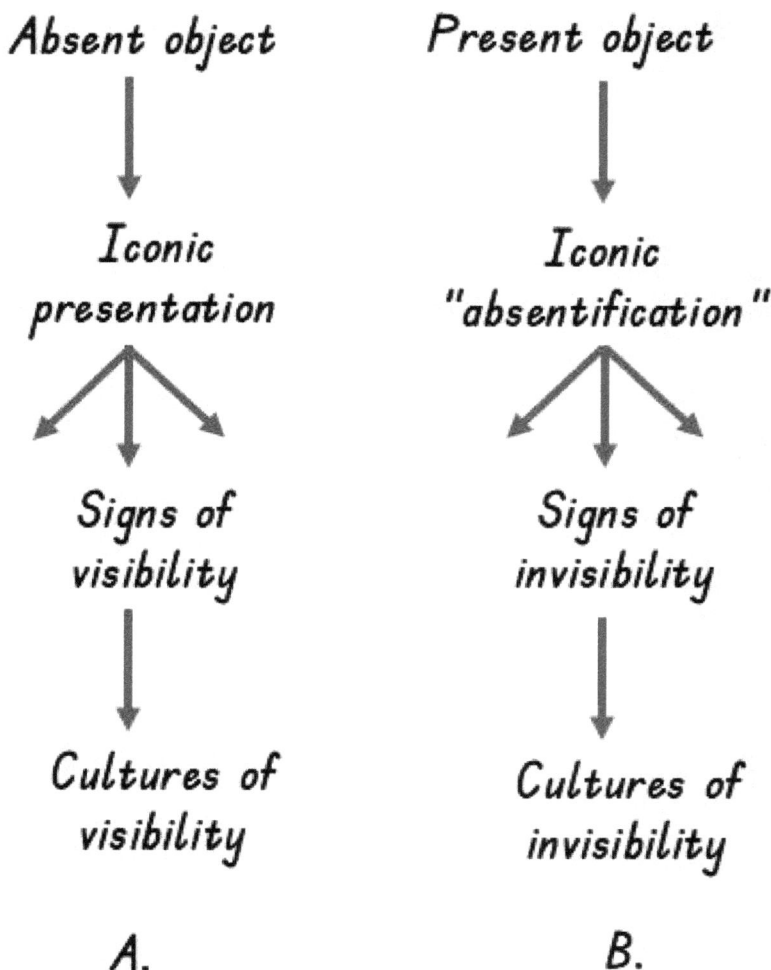

Figure 52. The representation of two semiotics proposed by
Massimo Leone:
A) semiotics of visibility and
B) semiotics of invisibility (modified from Leone, 2008).

Thus, Leone distinguishes two kinds of signs:

1. A "sign of visibility" does not simply aim to present an absent object. On the contrary, it aims to confer an "iconic presence" upon an object.

2. A "sign of invisibility" does not simply aim at "absentifying" (the term used by Leone) a present object.

> For example, hiding an object does not create any discourse of invisibility. On the contrary, a sign of invisibility aims to confer an "iconic absence" on an object.

This comparison leads Leone to a critical conclusion: signs of visibility (or projective signs) are about "*what* can be seen." On the other hand, signs of invisibility (or rejective signs), are about "*who* can see." Signs of visibility answer the questions: What is it possible to see? What is possible to represent, based on what is? What is it possible to see based on what is represented?

Signs of invisibility, on the contrary, answer the questions: Who is allowed to see what is? Who is allowed to represent what is seen? Who is allowed to see what is represented? The same sign can be either projective or rejective, according to the context. In this framework, Leone considers the veil not simply as "dress" but rather as (1) a complex semiotic device, (2) a phenomenological category, and (3) a mechanism of visual enunciation that bestows an iconic absence of certain elements of reality.

Interestingly, Hebrew has different words translated as "sign," depending on the context. Mark Schutzius II, a professor at Union University in Jackson, Tennessee, wrote the book, *The Hebrew Word for "Sign" and Its Impact on Isaiah 7:14.* In the book, Schutzius concentrated on the application of only two words (*ot* and *alma*) that appeared in the Old Testament for a sign. According to him, there is a functional difference between the two words. For example, both words were used to describe the Egyptian plagues. Thus, the impact of the plagues differed based on who was witnessing them. Although both words referred to the same event, their functionality differed. For instance, a plague could be addressed to Pharaoh but it may appear as a wonder for an ordinary Egyptian who was unaware of the purpose behind it.[33]

---

[33] Mark Schutzius II, *The Hebrew Word for "Sign" and Its Impact on Isaiah 7:14.* 2015, page 86.

Visible Signs of Invisible Biology

Leone's idea about two semiotic cultures -- visibility and invisibility -- is powerful. However, Kabbalistic Biology does not stop embracing the duality of the "signs of visibility" and the "signs of invisibility." Rather, it offers the whole spectrum of visibility as an appropriate measuring tool for biological research.

We discussed the variety of levels of visibility in biological objects in Chapter 6. In this context, "visibility" refers to the observable aspects of living life, as well as the factors that underlie these phenomena. "Semiotic reality" refers not to actual objects but to their signs. A biological object may remain invisible, while its sign can be seen. For example, we cannot see wind, but we can observe signs of wind. Similarly, although we cannot observe evolution, we can see "signs of evolution."

Abraham Solomonick, a semiotician and educator in Israel, introduced a theory of general semiotics, a way of understanding how living things communicate.[34] He called it "the visuality of signs," and categorized sign-systems based on their basic signs. These sign systems were structured hierarchically, with each level exhibiting signs of progressively abstract and less directly related character compared to those in the level below. Solomonick uses the term "semiotic reality" as the total of all the signs and sign systems that have been produced by each biological species. Semiotic reality is built from different kinds of signs. These are frequently combined into sign systems of various qualities and coherence.

Solomonick distinguishes different types of visuality: (1) visuality in sign-systems with the weakest sign abstraction, (2) visuality in language sign-systems, and (3) visuality in highly abstract sign-systems.[35] When Solomonick refers to visuality, he is describing the system of "natural signs." These signs represent some part of the phenomenon we are scrutinizing.

---

[34] Abraham Solomonick, *A Theory of General Semiotics: The Science of Signs, Sign-Systems, and Semiotic Reality.* 2015, 415 pages.

[35] Abraham Solomonick, *On Visuality.* https://www.academia.edu/10352019/On_visuality

To explain his point, I will share an experience I mentioned in Chapter 5. Its focus was the fieldwork we did when we investigated animal tracks in winter snow. We could not see the animals themselves, so we created images of animals based on the signs they had left behind. Generally, if the tracks are clear, an experienced zoologist will be able to identify the species and age of the animal from the paw footprint. Whenever the tracks are unclear, or the zoologist is inexperienced, the animal's image has a lower "visuality."

There is a greater challenge when visualizing highly abstract signs. Solomonick calls these signs "symbols" because of their disconnection from the "referent." Besides, these signs can be applied to a variety of objects. Such abstract symbols can be seen in many biological phenomena, such as adaptation, diversity, and evolution. We must explain these signs/symbols thoroughly in the process, which Solomonick calls *the release from the excessive abstraction of signs.*

Living in the Semiotic World

A view of life existing within an ocean of signs rather than within physical boundaries provides a compelling metaphor. Within this semiotic space, communication between organisms and the strength of the signals they emit or receive, play a key role in establishing relationships. These relationships, in turn, assist them in navigating the intricate terrain of this semiotic environment.

Tiit Remm, a scholar from the University of Tartu in Estonia, uses the term "semiotic space" to explain how living entities engage with their surroundings.[36] Remm defines "space" as an assemblage of recognized spatial relationships, i.e., how we perceive our environment. He posits that the spatial element is maintained while a subject-agent is involved in a semiotic situation requiring a degree of separation from both itself and the objective world. This is typically achieved through acts of "distinction."

---

[36] Tiit Remm, *Semiotic space and boundaries – between social constructions and semiotic universals."* Proceedings of the World Congress of the IASS/AIS, 12th WCS, Sofia, 2014.

The concept of space, according to Remm, is tied to the organization of knowledge and, subsequently, the organization of the *object* of knowledge. Remm further suggests that multiple spaces are possible. They differ according to the nature of the relationships involved and the variable organizations thereof. Any space, Remm asserts, possesses definable boundaries. Semiotic boundaries, he contends, depend on being recognized as distinctive by a subject, thus considering interrelated levels of conceptualization and spatiality. The latter is grounded in the semiotic understanding of space as fundamentally connected to relations with the surrounding environment.

Concerning the structure of semiotic space, Almo Farina, affiliated with the University of Urbino in Italy, introduces the term "cognitive (semiotic) landscape." Let's visualize this landscape as a comprehensive map of an ecosystem, with each element representing the needs and functions of various organisms. Farina advocates for a paradigmatic framework comprising species-specific, individual-specific, and function-defined landscapes. This framework offers a systematic way of illustrating how various organisms interact with their environments and each other. It provides a nuanced view of life's communication networks.[37]

Within this world of signs, the focus extends beyond mere molecules and bodies. It encompasses the dynamics of communication, recognition of surroundings, and the construction of cognitive landscapes that foster thriving and adaptation. It also offers a fascinating perspective on life on our planet, where every interaction is a part of the grand symphony of existence.

Levels of Biological Signs Perception

When we recognize that all organisms live within a semiotic world, we must then gauge its vertical layers. In 2005, Andrew Burton-Jones, originally a doctoral student at Georgia State University and now a

---

[37] Almo Farina and Brain Napoletano, Rethinking the Landscape:
New Theoretical Perspectives for a Powerful Agency. Biosemiotics, 2010, no.3, page181.

professor at the University of Queensland in Australia, introduced a system known as "semiotic metrics."[38] This system was developed to assess different levels of ontology. In philosophy, "ontological reality" encompasses concrete objects, facts, and phenomena.

The proposed framework comprises six semiotic levels, each focused on distinct aspects of ontological reality, and each prompting unique questions. Intriguingly, Hong Huang, a scholar from the University of South Florida, has associated Burton-Jones' semiotic levels with three of the four tiers within the DIKW pyramid, an approach we explore in Chapter 5. What's particularly fascinating is that Huang employs this semiotic hierarchy for genome curation, advocating for a "Big Data to Knowledge" perspective.[39]

In this scheme, the first two levels, the "Physical" level (answering the question, "Is it present?") and the "Empirical" level (answering, "Can it be seen?"), align with the realm of "Data." Moving up the hierarchy, we encounter the "Syntactic" level ("Can it be read?") and the "Semantic" level ("Can it be understood?"), which fall under the category of "Information." Finally, the top two levels, "Pragmatics" ("Is it useful?") and "Social" ("Can it be trusted?") correspond to the realm of "Knowledge" (Figure 53).

---

[38] Andrew Burton-Lones et al., *A semiotic metrics suite for assessing the quality of ontologies.* Data & Knowledge Engineering, 2005, vol. 55, pages 84-102.

[39] Hong Huang, "Big Data to Knowledge—Harnessing Semiotic Relationships of Data Quality and Skills in Genome Curation Work." Journal of Information Science, 2017, vol. 44, No.6, pages 785–801.

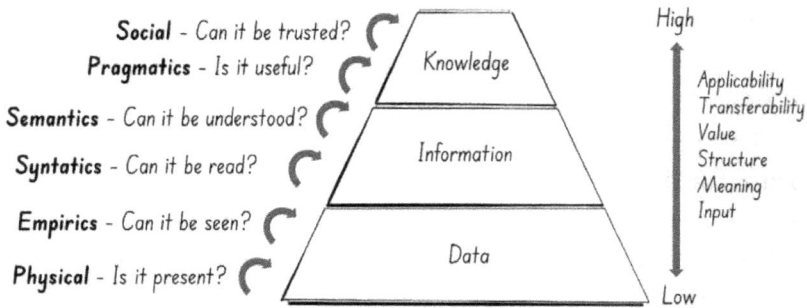

Figure 53. Semiotics levels and DIK pyramid, modified from Huang (2017).

Biosemiotics distinguishes between biological signs and simple clues and signals. While we often think of signals as tangible entities like light, heat, odors, touch, or sound, it's crucial to recognize that signals can encompass a broad range of stimuli. Such cues and signals require no interpretation and lack intrinsic meaning. However, there is a certain degree of ambiguity surrounding the definitions of terms like "signal" and "cue," a common challenge in many scientific disciplines. Some researchers employ these terms interchangeably in their work. Nevertheless, biosemiotics offer a more nuanced distinction. In this nuanced perspective, a "signal" is more adaptable and intentional. It can be activated or deactivated in response to environmental cues.

Cells engage in constant dialogue through signals. Cells don't exist in isolation; they are interconnected and constantly communicating with one another and with the outside environment. Cellular and environmental signals play a variety of roles. They serve as conduits for essential information, triggering events that are pivotal to an organism's life cycle. For instance, they can trigger cell division, where a single cell splits into two. They can also stimulate cell differentiation, where a cell assumes a specialized role in a specific tissue or organ.

Code Biology concentrates on the notion of "code." Here, living organisms are defined as "code-users" or "code-markers." Marcello

Barbieri, an Italian theoretical biologist at the University of Ferrara, defines a code as "a mapping between the objects of two independent worlds that is implemented by the objects of a third world called adaptors."[40] An adaptor is a structure that implements the rules of a code. In the Morse code, for example, the adaptors are the neural circuits that make connections in the brain between letters of the alphabet and groups of dots and dashes. Similarly, in protein synthesis, the adapters are the transfer RNA that serves as a link between the codons (messenger RNA) and the growing chain of amino acids that make up a protein. In this case, a codon (a sequence of three nucleotides) is an organic code, and amino acids provide meaning.

In 1975, Gordon Tomkins, professor of biochemistry at the University of California in San-Francisco, proposed "the Metabolic Code."[41] Various independent discoveries have shown that many organic codes exist in living systems. Among those are the adhesive code, the splicing code, the signal transduction code, the histone code, the neural code, the tubulin code, and others. The distinction between code-based artifacts and meaning-making life is one of the main problems of biosemiotics. Semiosis requires codes, but it cannot be based on a single code. Discussing this problem, Kalevi Kull entitled the paper "Codes: necessary, but not sufficient for meaning-making."[42]

From a semiotic perspective, we earlier indicated the limitations of the "lock-and-key model." However, at the molecular level, many reactions adhere to what is known as the "lock-and-key model." A classic example is the interaction between enzymes and substrates. In this scenario, the enzyme and the substrate possess specific, complementary geometric shapes that precisely fit together. Enzymes exhibit a high degree of specificity, binding to a particular substrate before catalyzing a chemical reaction. No interpretation is

---

[40] Marcello Barbieri, *The Organic Codes: An Introduction to Semantic Biology.* 2002, 316 pages.

[41] Gordon Tomkins, *The Metabolic Code.* Science, 1975, vol. 189, Issue 4205, pages 760-763.

[42] Kalevi Kull, *Codes: necessary, but not sufficient for meaning-making.* Constructivist Foundations, 2020, vol.15, no.2, pages 137-139.

necessary for the enzyme to "recognize" the substrate. Similar examples abound, such as the genetic code, protein-DNA binding, and antigen-antibody reactions. The genetic code, for instance, serves as instructions for a gene, dictating how a cell should produce a specific protein. This lesson explores the genetic code, its role in processes like mitosis and meiosis, and the occurrence of DNA crossing over. Notably, a cell or organism's response to certain environmental cues can be automatic, resulting in a consistent outcome without variation.

One prominent figure in biosemiotics, Jesper Hoffmeyer, asserts that "signs, not molecules, are the basic units in the study of life."[43] The cell is the simplest entity capable of processing genuine semiotic competence. The question arises: Do simple cues and signals hold a place within the semiotic hierarchy? Indeed, they do, as confirmed by Claus Emmeche and his colleagues, using the DNA molecule as an example. They contend that this molecule only holds interest and meaning within its biological context.

To be *biologically* significant, such as in the context of locating genes on a DNA sequence, the chemical findings must be linked to what constitutes significance for a cell or an organism. Therefore, simple cues and signals do occupy a place within the semiotic hierarchy. That place is at the lowest level, where they form the foundation for higher-order biological meaning.

The Semiotic Dimension

The Fifth Dimension is a conceptual tool designed to help us navigate through different levels of awareness of biological signs. These levels provide different perspectives on how organisms interact with their environment and with each other. In this dimension, sign systems are characterized by successively increasing "degrees of abstraction" (using Solomonick's term).[44]

---

[43] Jesper Hoffmeyer, *Biosemiotics: An Examination into the Signs of Life and the Life of Signs.* 2008, 419 pages.

[44] Abraham Solomonick, *Why are signs so different?*
https://www.academia.edu/45162152/Why_are_signs_so_different?email_work_card=view-paper

We can identify five main levels in the hierarchy of biological signs. These are:

Simple cues and signals - At the most basic level, organisms respond to simple cues or signals from their internal and external environment. These cues can include physical stimuli like light, temperature, or chemicals.

Biological codes - These codes determine an organism's traits and functions, playing a pivotal role in life processes orchestration. Genetic information, encoded in DNA, serves as a fundamental set of instructions.

Functional signs - Biological signs represent a web of biological pathways that provide communications between organisms. They allow organisms to interpret, actively react, and build relationships between themselves and the environment. All organisms depend on signs for survival and reproduction.

Interpretative (or variable) symbols - As life becomes more intricate, so does semiotic space. At this level, organisms engage in sophisticated forms of communication. They employ interpretative symbols, such as sophisticated vocalizations, olfactory communication, and complex behavioral patterns to convey information and establish social connections.

Symbols with constant value (archetypes) - Archetypes are deeply ingrained in a species' collective consciousness as universal symbols. Archetypes shape innate behaviors, instincts, and cultural elements, transcending individual experiences.

Understanding these levels within the semiotic space is crucial for those wishing to unravel life's intricate communication tapestry. This multidimensional perspective not only enhances our understanding of the natural world but also provides a framework to explore the fascinating connections between signs and biological processes. By embracing semiosis as a cornerstone of our investigation, we pave the way for a holistic understanding that bridges the gap between living beings and the world of meaning.

Astonishingly, each rung of this semiotic ladder aligns with the configurations of the Sefirot, intricate elements within the mystical Tree of Life. In Chapter 11, we discussed five *partzufim,* meaning "personae" or "faces." If "*olamot*" represents different levels of reality (phenomena), as Solomonick concluded, then a sign never fully reveals its referent. Any object does not completely coincide with the sign designated it. The object of semiosis (a signified) can never become equal to its sign, because "they belong to different spheres of knowledge."

Partzufim represents different levels of perception. The simple cues and signals correspond to the Partzuf of *Nukvah,* the biological codes and signs correspond to the Partzuf of *Zeir Anpin,* the interpretive symbols relate to the Partzuf of *Imma,* and the Archetypes correspond to the Partzuf of *Abba.*

Living organisms are immersed in a world of signs and constantly read their meanings. Acknowledging different levels of biological signs perception, the Fifth Dimension maintains continuity through all levels. As is the case with the Fourth Dimension, continuity illustrates a connection between *partzufim* through the configuration of sefirot.

Chapter 18

Six Dimension: Coordinates within Biological Space

This chapter delves into the intricate formation of biological systems and their intriguing relationship to spatial measurement. The concept of formation is fundamentally spatial in nature. Kabbalistic Biology strongly emphasizes the significance of orientation within the spaces where all biological processes take place. The challenge for researchers is the stark contrast between the orientation needs of biological life and the conventional principles governing physical space. To address this challenge, Kabbalistic Biology introduces a novel concept known as the "six edges" platform. This is specifically designed to measure the unique spatial dimensions within the biological realm.

In the pages that follow, we will explore the practical uses of this unique system by considering real-world applications and examples. By doing so, we can illuminate how the "six edges" platform can help us understand the spatial intricacies that underlie biological life. By so doing, we will uncover the intricacies of biological systems and acquire fresh insights into how they navigate and thrive within their spatial confines.

Why is Physical Space Insufficient for Us to Measure Biological Life?

Franco Moretti, an Italian literary theorist, once pointed out that space is "not a container, but a condition..."[1] The question of whether physical space alone is sufficient for us to measure biological life brings us to the reality of biology: it is inherently invisible. In previous chapters, we explored various levels of invisibility in processes within living systems. Ordinarily, we can only observe living organisms when they take on visible or manifested forms. Can this option provide us with a comprehensive understanding of biological phenomena? Far from it.

The limitations of our perception become strikingly evident when we consider what remains concealed. Biology, as a science, extends beyond the visible to encompass the invisible and its hidden potential. Thus, we cannot directly witness the grand tapestry of evolution unfolding before our eyes. The gradual adaptation of organisms to their environments, a cornerstone of biology, develops beyond the scope of our vision. It is possible to observe certain life forms during their embryonic development, but the dynamic process driving these changes remains hidden. Even within the realm of visible biological creatures, we can only capture a fraction of the organisms we intend to study. Although there are numerous changes taking place, they cannot be measured by the metric of physical space.

Despite this invisibility, every biological process occurs within a specific spatial context. Being able to measure these invisible processes presents a challenge that goes beyond the use of conventional physical measurements. For example, when biologists map the distribution of a species, they consider the organism's physical presence and the range of their potential habitats.

There is more to "biological space" than merely the physical distance between manifested entities; it also includes the invisible realm of potential activity. Thus, the inadequacy of physical space to

---

[1] Cited from Elana Gomel, *Narrative Space and Time*. 2014, page 11.

measure biological life underscores the need for a broader perspective. Biology as a science doesn't confine itself solely to visible organisms or their manifested forms. Instead, it extends its reach to explore not only actual but also potentially possible presences and manifestations. The invisible components of biological systems also occupy space, but not physical space. I call it the "space of variants." To illustrate the challenge of experiencing biological space, let's review some relevant concepts of modern biology.

The Concept of "Niche" in Biology

First, let's look at the so-called "niche concept." The meaning of niche comes from a recess in a wall for a statue that likely derived from the Middle French word *"nicher,"* meaning "to nest." The term has become a mainstream ecological idea since Joseph Grinnell introduced it in the early 20th century in his paper, *The Niche Relationships of the California Thrasher.*[2] Grinnell says the behavior of California thrashers, large songbirds, is consistent with its chaparral habitat. It breeds and feeds in the underbrush and escapes from its predators by shuffling from underbrush to underbrush. Its "niche" is defined by the convenient pairing of the thrasher's behavior and physical traits (camouflaging color, short wings, strong legs) with its habitat.

Chaparral is not only a shrubland plant community prevalent in California; it is also an inevitable component of the thrasher's habitat. We can measure the bird's body like any other body. We can also calculate the distance between the observed birds. Above all, we can estimate the chaparral's territory in California. However, it would be challenging to evaluate chaparral as the habitat, the "niche," in meters or square kilometers. California thrashers can be present in a patch of land covered by bushes, but they can also be absent. I am referring to the space where thrashers *might* exist-- even if no bird is present at a particular time.

In 1957, G. Evelyn Hutchinson, a British ecologist, defined the

---

[2] Arnaud Pocheville, *The Ecological Niche: History and Recent Controversies.* In *Handbook of Evolutionary Thinking in the Sciences.* 2015, pages 547–586.

niche from a different perspective. He defined it as a "multidimensional hypervolume" of environmental conditions that enable an individual or a species to exist.[3] Hutchinson's view is that one can represent an organism's potential environment as an abstract space. Its axes would correspond to environmental factors that affect the organism's performance. In other words, the niche would enable investigators to map population dynamics into this space.[4] Unlike the original Grinnell niche concept, Hutchinson's hypervolume provides a framework to explore interactions among environmental variables that would influence an organism's fitness. Hutchinson called the physical space a "biotope," the physical setting of an ecosystem, such as a lake, forest, or chaparral.

Hutchinson's idea inspired many others to develop models to explain the number and similarity of coexisting species within a given community. This led to the concepts of "niche breadth" (a species' ability to utilize a variety of resources and habitats), "niche partitioning" (differentiation of resources by coexistent species), and "niche overlap" (overlapping uses of a resource by different species).[5] Depending on the niche's breadth along each dimension, we can expect to find organisms in certain environments. In the sense that environmental conditions vary across space to create a visible structure, the environment also has a "texture."

Importantly, an organism experiences an environmental landscape on a scale determined by its size, shape, and mobility.[6] Every organism has "multiple niches" at different stages of its life. As it grows and changes, it requires various resources and conditions in order to survive and thrive. For example, what's suitable for an adult member of a species may be entirely different from what a larva or

---

[3] G.E. Hutchinson, *Concluding remarks*. Cold Spring Harbor Symposia on Quantitative Biology, 1957, vol. 22, no.2, pages 415–427.

[4] Robert Holt, *Bringing the Hutchinsonian niche into the 21st century: Ecological and evolutionary perspectives*. PNAS, 2009, vol.106, pages 19659-19665.

[5] M. Jonathan et al. *Ecological Niches: Linking Classical and Contemporary Approaches*. 2003, page 11.

[6] Michael Angilletta et al. *Fundamental Flaws with the Fundamental Niche*. Integrative and Comparative Biology, 2019, vol.59, no. 4, page 1038.

embryo needs. These changes not only affect where an organism can live but also how it interacts with its environment. For instance, the ability of embryonic lizards to handle various temperatures depends on how much oxygen is in their surroundings. However, this connection doesn't exist for adult lizards from the same group![7]

Moreover, how well a species can cope at a particular life stage is influenced by the conditions it experienced in previous stages of growth or even in previous generations. This adaptability comes from systems in the body that respond to environmental cues, like neural and endocrine systems. When conditions in the environment consistently change in certain ways, it leads to the selection of organisms with genotypes that can adjust their tolerance to match their habitation.

Multiple, sometimes contradictory definitions of the term "niche" create confusion among biologists and even doubts that the niche can be a useful ecological concept. The duality between Grinnell's concept of niche as an aspect of environment and Hutchinson's concept of niche as an aspect of species powerfully illustrates the role of the two levels in the spatial distribution of any biological object or process. Jorge Soberon and Miguel Nakamura refer to the dualistic challenge of defining niches as a "tale of two niches."[8]

The niche as an aspect of the environment represents *visible* and measurable natural resources where *visible* individuals of a particular species are found. Characterizing this point, McInerny and Etienne wrote that "environment is everything except the species."[9] This concept permits species to occupy niches or to leave niches empty if the species becomes extinct. The niche as an aspect of species represents *invisible* "adaptations and traits that determine what

---

[7] Ibid., page 1044.

[8] Jorge Soberon and Miguel Nakamura, *Niches and distributional areas: Concepts, methods, and assumptions.* PNAS, 2009, vol. 106, suppl. 2, page 19644.

[9] Creg McInerny and Rampal Etienne, *Stitch the niche – a practical philosophy and visual schematic for the niche concept.* Journal of Biogeography, 2012, vol.39, page 2103.

biotic and abiotic factors a species interacts with, and how species respond to and affect those variable/factors."[10]

Species Distribution and Ecological Niche Modeling

It is possible to interpret a niche in biology as a factor leading to practical advancements, despite its abstract nature. In the last couple of decades, ecological niche modeling has become a very active field of research. Thanks to this methodological approach, we can estimate not only the *actual* distribution patterns of species but also their *possible* distribution patterns.

The ability to predict species invasions was one of the early advantages of this approach. In 1999, President Clinton signed an executive order dealing with invasive species in the United States.[11] The order was created to "prevent the introduction of invasive species and provide for their control and to minimize the economic, ecological, and human health impacts that invasive species cause." We already know that certain species of animals create problems when they invade areas outside of their native range. Generally, when invasive species are reported, the invasion can be mapped out, and the only question is, "How can we control them?" Different situations exist when invading species are not reported within a particular territory. They are reconsidered after an analysis of environmental variables indicates that alien species could find suitable conditions for invasion.

The challenge of predicting potential species' invasions was solved by focusing on "niche dimensions," not local invasion reports. [12] Species' niche dimensions are affected by geographic factors such as temperature, precipitation, elevation, and vegetation. Ecological niches are divided into "fundamental" and "realized" niches. Fundamental niches represent a species' ecological capacity at its base. Interactions between species have a profound effect on realized

---

[10] Ibid., page 2104.

[11] W.J. Clinton. *Executive Order: Invasive Species*. 1999, http://bluegoose.arw.r9.fws.gov/FICMNEWFiles/eo.html

[12] Townsend Peterson and David Vieglais, *Predicting species invasions using ecological niche modeling: New approaches from bioinformatics attack a pressing problem.* BioScience, 2001, vol. 51, no. 5, pages 363-371.

niches.[13] Townsend Peterson and David Vieglais mapped the distribution of invasive species based on models. Among the predicted invasions were cattle egrets, house finches, Asian longhorn beetles, and Japanese white-spotted citrus longhorn beetles.

Note the crucial difference between the actual distribution of invasive species based on available funding and the potential distribution based on niche dimensions. The potential distribution area does not exist in Nature. Despite this, it is real as far as an abstract space is concerned. Such an abstract space must provide the basis for effective, practical activity. To grasp the essence of biological space, we must appreciate the distinction between the actual and the potential distribution of species.

In my professional experience, I am more familiar with applying ecological niche modeling to infectious disease epidemiology. We cannot see pathogenic microorganisms on endemic territory, so we must detect infected or sick animals and humans to observe the spread of disease. However, we do know that the area in which the disease occurs now can be much wider later. What is the most effective way to predict the distribution of "invisible epidemics?" Such knowledge is crucial for us to be able to prevent and control infectious diseases.

As it happens, most diseases do not spread randomly across landscapes or regions. However, researchers can identify specific environmental factors associated with the occurrence of the disease.[14] Landscape composition, soil typing, and climatic and hydrological variables can provide the basis for predictive maps of the distribution of such zoonotic and vector-borne diseases as anthrax, tularemia, and plague.

There is, however, a fundamental difference between maps that show reported cases of disease and maps that show the potential for the territory to support the spread of the disease. A predictive map is

[13] Robert MacArthur, *Geographical Ecology*. 1984, 288 pages.
[14] Luis Escobar, *Ecological niche modeling: An introduction for veterinarians and epidemiologists*. Frontiers in Veterinary Science, 2020, vol.7.

prepared when the background knowledge, contextual understanding, and inferences from the context are all weighed when assessing expectations for success.

The idea of an ecological niche transcends the mere physicality of an organism's habitat. Instead, it evolves into a conceptual space that accommodates not only the organism's environmental requirements but also its functional role, interactions, and adaptations. This shift from a physical space to a conceptual construct creates a deeper understanding of the intricacy of relationships in living systems.

Space for Invisible Epigenetic Landscape

As the concept of ecological niche examples reveals, one cannot separate the observable qualities of species and individual organisms from their environments. Still, talking about the environment creates an association with geographical space as measured by conventional metrics. In the previous section, we described how such a metric is unable to measure the biological space of species distribution.

Here, we will discuss a biological phenomenon that cannot adequately be estimated by geometrical dimensions alone. I am referring to individual embryonic development. Some may argue that the multiplication and diversification of embryonic cells occur in physical space. The sizes of cells and cell conglomerates can also be measured in micrometers or millimeters. All this is true, when describing visible objects such as cells, tissues, and organs. The challenge is to measure the space where *invisible* processes occur!

The development of organisms is intricately influenced by two key factors: the genetic inheritance they received from their parents and the external ecological conditions in which they exist. This latter pivotal relationship opens up a longstanding debate within the field of biology, commonly referred to as "Nature versus nurture." This debate centers on discerning the relative impact of an organism's genetic makeup (Nature) versus the external environmental conditions it encounters during its development (nurture).

Alas, this debate leads us to a profound question: Where does the interplay of these influences take place? The answer is that it transpires in a dimension that eludes direct observation, an invisible realm beyond the grasp of a simple ruler's measurement. It's within this intangible space that the intricate dance of genetic inheritance and environmental influence shapes the development of an individual organism. This adds layers of complexity to the age-old "Nature versus nurture" discourse. To address this question, I will introduce a concept to help us gain a better understanding of the space of "invisible" biological processes.

During the mid-20th century, Conrad Hall Waddington, Britain's leading theoretical biologist, developed a visionary model. It revealed the possibility that heritable changes in gene expression could be induced without altering the DNA sequence in the underlying genome. This model became popular under the name, "Waddington's Epigenetic Landscape."

In Waddington's era, biology was dominated by the notion that a specific gene dictated the development of a particular trait in an organism. This concept continues to influence the thinking of many laypeople and some biologists, largely because it is simple and convenient. To illustrate Waddington's Epigenetic Landscape, imagine a ball rolling down a landscape with several valleys. As the ball moves through the landscape, it encounters several branching points. Waddington wrote, "Looking down the main valley toward the sea. As the river flows away into the mountains it passes a nagging valley, and then the two branch valleys, on its left bank. In the distance, the sides of the valleys are steeper and more canyon-like."[15]

Each rolling ball travels through a groove on the slope before it comes to a different, relatively stable point. The balls represent particular cells; the valleys represent canal-like pathways of embryonic development named "chreodes" by Waddington. Each final point represents the expressed phenotypes. According to

---

[15] C. H. Waddington, *Organizers & Genes.* 1940, page iv.

Waddington, "canalization" means that up to a certain threshold, neither external factors such as environmental challenges, nor internal factors such as genetic variations, can easily push the ball over the valley's wall into another pass.

Investigations of such transitions are essential to study organismal development, reprogramming, and the initiation of cancer formation (carcinogenesis).[16] Here is an example of how cells transition from one state to another, as defined by Epigenetic Landscape theory. During normal development, a given stem cell rolls downhill through the Epigenetic Landscape. As it rolls, it becomes confined within a particular valley (or lineage), where it is restricted. Eventually, the cell comes to rest in a terminally differentiated state. For a given cell to transition from one cell type to another, it must overcome an "energy wall." An epigenetic landscape can raise or lower the "wall" and either make it more difficult or more accessible for a given cell to transition between states. This low wall is referred to as the "plasticity state." In this state, a cell might move more easily between other states (Figure 54).

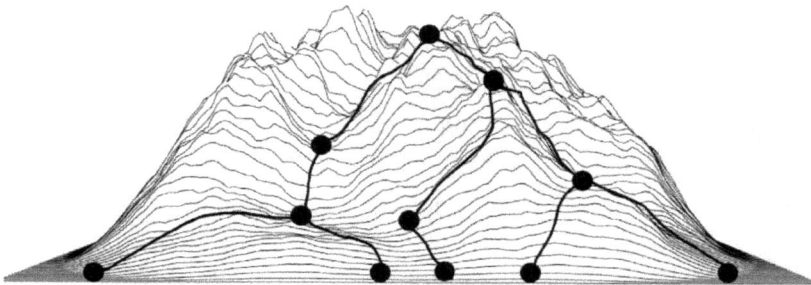

Figure 54. Waddington's depiction of the "epigenetic landscape." The ball represents a cell on the verge of taking a developmental path toward one of a set of alternative states.

I wish to emphasize that the "epigenetic landscape" does exist in invisible space – however, you will never be able to see it in a developing organism. Epigenetic landscapes are virtual forms

[16] William Flavahan et al. *Epigenetic plasticity and the hallmarks of cancer.* Science, 2017, vol. 357, page 6348.

arranged within a space defined by non-physical dimensions.

Waddington's approach to epigenetic landscape imaging revealed potentially adaptive variations that were previously hidden. In addition, it discovered more efficient ways of looking at a dynamic, complex space where the process of individual development occurs. As soon as a protozoan-like spermatozoid reacts with a receptive egg cell, all 75 trillion cells in the human body remain interconnected within an invisible space. Every cell is entangled in this invisible epigenetic landscape!

In the 21st century," epigenetics," the term coined by Waddington, has become one of the world's fastest-growing fields of biology. Several medical applications of epigenetics have already been recognized. The National Institutes of Health, announcing multimillion-dollar funding, noted that epigenetics may explain the mechanisms of aging, human development, and the origins of cancer, heart disease, mental illness, and many other conditions. However, the acceptance of the existence of an epigenetic landscape requires a fundamental change of perspective regarding the space where it exists.

Exploring the Fitness Landscape: A Journey through the Space of Evolution

There is another intriguing concept known as the "fitness landscape" or "adaptive landscape" in the world of evolution. Sewell Wright, an American population geneticist and evolutionist, developed the image of this landscape in 1932. He wanted to illustrate how biological populations might escape the trap of a local peak by imagining what might drive them downhill toward alternative possibilities.

Imagine this concept as a map that helps us understand how different genes (the blueprints within organisms) relate to their ability to reproduce and survive. This map has peaks and valleys, with each peak representing high reproductive success (fitness) and each valley representing low reproductive success. The idea here is that similar genes get close to each other on this map while vastly different genes stay far apart. An organism's fitness is not determined by how many push-ups it can perform but rather by its ability to reproduce. A gene's "fitness" is essentially its ability to assist an organism in reproducing and transmitting its genetic code.

To better understand this process, imagine this fitness landscape as resembling a range of mountains. The landscape consists of high peaks (representing genes with high reproductive success) and low valleys (less successful genes). A landscape with many peaks and deep valleys is referred to as a "rugged" landscape. In contrast, if all the genes have a similar reproductive success, it is considered a "flat" landscape (Figure 55).

In Nature, populations of organisms evolve by making small changes to their genes over time. As they move towards local peaks, they climb uphill. A local peak is like a mountaintop from which every path leads downhill in fitness. However, evolution doesn't always reach the highest peak. Sometimes, it gets stuck on a lower peak. Others find it difficult to reach up because finding the optimal genes can take an incredibly long time. In these harsh landscapes, genes that were once helpful can become harmful, creating a maze-like journey for evolution. In essence, the fitness landscape is like a dynamic map that guides evolution's journey, helping us understand how genes and traits evolve and adapt in the ever-changing world of biology.

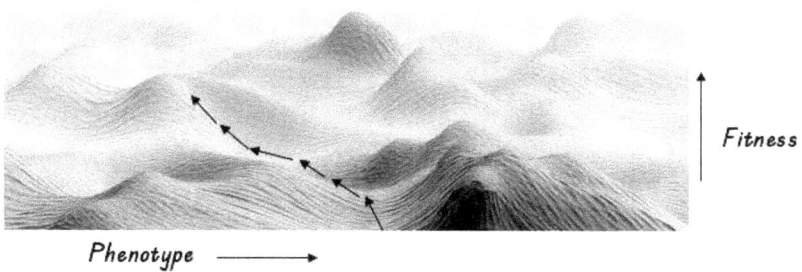

Figure 55. Visualization of two dimensions of a fitness landscape. Interactions among genes can give rise to a rugged fitness landscape. "Adaptive peaks" of high-fitness gene combinations are separated by "adaptive valleys" of low-fitness genotypes.

Furthermore, genetic "shadows" extend the concept of biological space into the realm of genetics and successful adaptation to the environment. These shadows encompass the vast range of possible genotypes and the associated traits that organisms within a species can exhibit according to the ecological context. They represent the potential variations that can emerge within the biological space governed by the Sefirotic orientation points.

Six Coordination Points of Space in Kabbalah

In Kabbalah, space takes on a profound metaphysical dimension. It transcends the boundaries of the physical world and expands into realms that extend beyond our immediate sensory perception. At the heart of Kabbalah's teachings lies its most profound secret, a spatial measurement framework that moves beyond the "visible" aspects of reality and delves into the depths of the "invisible" phenomena that underlie existence.

Within Kabbalah, biological space, like all other aspects of reality, is fundamentally important. It is not merely a void or a backdrop against which life unfolds. Instead, it is a dynamic and intricate tapestry interwoven with divine attributes and energies. Kabbalists often describe this space as defined by six orientation directions. Each one corresponds to one of the six sefirot: Hesed, Tiferet, Gevurah, Netzah, Hod, and Yesod. These six sefirot collectively form a network

447

known as the *middot*, a "limitation" or "measurement." This concept emphasizes that space, in the Kabbalistic understanding, is not an amorphous expanse. Instead, it is a meticulously structured framework where all qualities can be manifested and measured.

As we discussed in Chapter 11, each Sefirah represents a unique aspect of a dynamic, complex system. They interact harmoniously within biological space. Hesed embodies the trend leading to expansion and diversification, while Gevurah represents strength and restraint. In general, form and energy are inversely related. The principle of energy is change, and a form or structure can exist as long as it remains stable and resistant to change.[17] Tiferet balances these forces, radiating harmonious development and sustainability. Netzah and Hod bring tolerance and acceptance, respectively, and Yesod serves as the foundation upon which Life's intricate tapestry is woven. These six sefirot form partzuf Zeir Anpin when they are configured dynamically.

Per this Kabbalistic understanding of biological space, every aspect of life is seen as a complex dance, from an organism's growth to an ecosystem's functionality. These six sefirot – "middots" – are arranged in two triads. They extend a nuanced perspective on biological space far beyond the visible realm. In each triad, there are some sefirot that represent expansion, contraction, and equilibrium at different levels. As we learned in Chapter 11, Hesed, Gevurah, and Tiferet comprise the upper triad "Ha-Ga-T" within Zeir Anpin. Netzah, Hod, and Yesod form the lower triad "Ne-H-Y" within this partzuf.

Spatial relationships between visible individual biological objects and their elements are characterized by the lower Ne-H-Y triad. The upper Ha-Ga-T triad consists of three sefirot. Each represents a biological system as a whole that includes potential manifestations of objects and the variations in their expression. Hence, the six sefirot provide a sense of direction in both the visible and the invisible realms

---

[17] Rupert Sheldrake, *A new Science of Life.* 1995, page 63.

of biology.

The six sefirot of Zeir Anpin represent fundamental forces used as orientational points, as shown in Figure 57. Three sefirot of the lower triad (Ne-H-Y) constitute coordinates to measure relationships between elements of any biological system. Those elements could be cells or organs within an organism, individual animals or plants in a population, territorially separated populations in a species, specific species within biological communities, or entire ecosystems. Each of them can potentially function as a complex system, but the Ne-H-Y emphasizes their actual manifestation as elements of the system above.

Conversely, the upper triad (Ha-Ga-T) of Sefirot serves as coordinates to measure the potential space for entire biological systems, such as organisms, species, ecosystems, and more. Again, we cannot visualize these systems because of their complexity, nor can we analyze all the interconnections between their constituent elements (e.g., cells in an organism, individuals in a species, species in an ecosystem). Ha-Ga-T's sefirot, nevertheless, can provide a map of the potential spaces for the manifestation of these complex systems.

The upper triad of Zeer Anpin plays a crucial role in defining biological space for entire biological systems. Each sefirah represents a distinct aspect:

1. **Hesed** embodies an inclination towards the expansion of the space available for the potential manifestation of entire biological systems. It signifies the inherent drive for growth and the exploration of possibilities within these systems.

2. **Gevurah** delineates the boundaries of space, distinguishing between the potentialities and the constraints that apply to whole biological systems. It acts as a regulator, identifying the limits and conditions within which these systems can operate effectively.

3. **Tiferet** acts as a harmonizing force, shaping the space that

449

will allow biological systems to develop in harmony. To optimize their growth and evolution, they must both consider both the expansive tendencies represented by Hesed, as well as the constraints established by Gevurah, to find a balance and an optimal path.

The lower triad of Zeir Anpin, on the other hand, defines the biological space for individual elements that make up biological systems in their entirety:

1. **Netzah** signifies the drive to expand the space available for potential elements within biological systems. It reflects the inherent tendency for diversity and the exploration of different components that can contribute to these systems.

2. **Hod** outlines the space of constraints imposed on elements participating in biological systems. It sets limitations and rules governing these components' behavior and interactions.

3. **Yesod** plays a pivotal role in forming the space for the development of biological systems' constituent elements. Similar to Tiferet, it seeks to strike a balance between Netzah's acceptance and tolerance versus the selective power defined by Hod. This fosters the optimal conditions for the individual elements' functionality and survival within larger biological systems.

The two triads of Zeer Anpin, therefore, provide a comprehensive framework for us to understand and navigate the intricate biological spaces that encompass both entire biological systems and the elements that make up those systems.

Gevurah ⟷ Hesed

Space of feasible and infeasible constraints for biological systems

Expansion of space for potential manifestations of biological systems

Tiferet

The formation of a space for the development of biological systems with both trends of expansion and constraints

Hod ⟷ Netzah

Space of constraints of elements of biological systems

Expansion of space for potential elements of biological systems

Yesod

Foundation for developing individual entities within biological systems

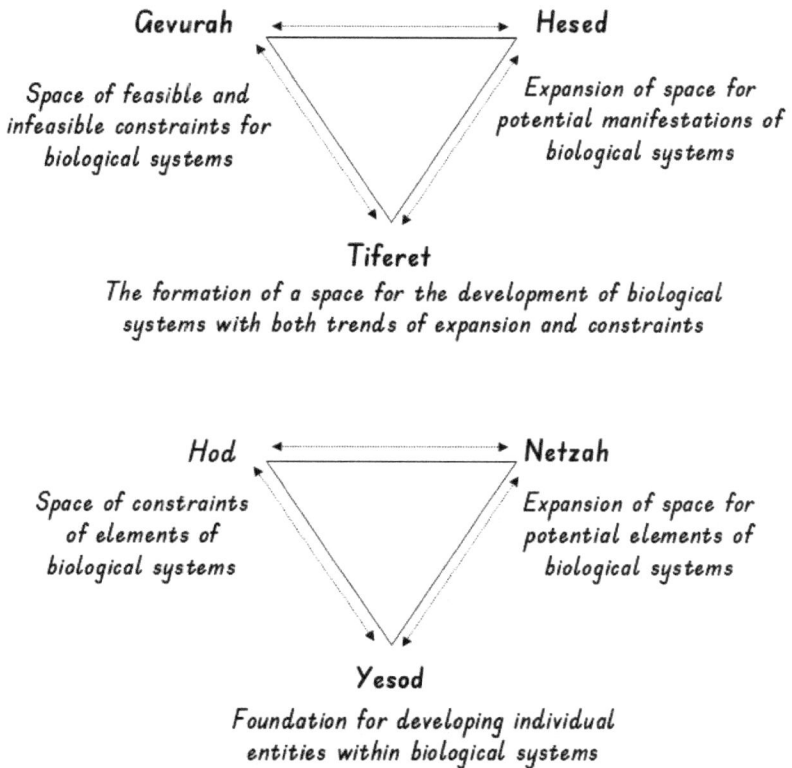

Figure 56. The two triads of Zeer Anpin provide a framework to navigate the biological spaces that encompass entire biological systems and the elements comprising those systems.

Bottom-Up and Top-Down Directions in Biological Space

Generally, the term "causality" refers to the relationship between an occurrence and what causes it. Sometimes, "causation" is used to refer to the process that connects causes to their effects. As for the terms used here, I will use them interchangeably. Certain events or factors can influence biological outcomes. If a plant receives sunlight ("cause"), it will undergo photosynthesis and grow ("effect"). The concept of causality helps us understand how actions, conditions, and interactions affect the behavior, development, and functions of living things. This is a fundamental concept in biology that helps scientists explain and predict how organisms will respond to their environment and internal changes.

In the intricate web of biological systems, causation takes on a variety of forms. Two fundamental modes stand out: "top-down causation" and "bottom-up causation." When one begins to understand the dynamics of life based on these concepts, it has profound implications for progress, from observing molecular interactions within cells to exploring the complexity of ecosystems. Let's explore the differences between these two causal relationships and their roles in biology.

The concept of "top-down causation," first introduced by philosopher and social scientist Donald Campbell in 1974, describes how higher hierarchical levels influence lower levels by creating a context for them.[18] Through system structuring, top-down causation occurs when high-level variables impact lower-level dynamics. Therefore, the outcome depends only on the high-level structural boundaries and initial conditions.

A key feature of top-down causation in biological systems is its inextricable connection with the environment. Danko Nikolić, a neuroscientist from the Max Planck Institute, emphasizes that downward causation in biology involves environmental factors.[19] Rather than direct causal effects coming from higher up to affect below-level levels, top-down causality occurs indirectly. When mechanisms at higher levels fail to fulfill lower-level requirements, environmental cues signal lower-level mechanisms to initiate action. When a species experiences downward pressure to adapt to novel circumstances, it may alter its behavior and genetic expression as a result.

More often, biologists use a bottom-up approach instead of top-down causation. Thus, certain actions taken at a lower-level lead to higher-level biological functions. For instance, physics and chemistry

---

[18] Donald Campbell, *Downward causation in hierarchically organized biological systems.* In *Studies in the Philosophy of Biology: Reduction and related problems,* 1974, pages 179–186.

[19] Danko Nikolić, *Practopoiesis: Or how life fosters a mind.* Journal of Theoretical Biology, 2015, vol. 373, pages 40-61.

principles underlie biochemistry, cell biology, etc. It's like building a complex structure from the ground up.

When the fundamental rules and major players in modern biology are not yet fully understood, bottom-up causation is the most effective remedy. It is common for researchers to construct a theory based on a limited set of data or observations. The underlying rules governing the system are gradually revealed as more data become available, refining and expanding the theory. When a phenomenon is well understood, and its foundational rules are clear, top-down approaches are preferred. The bottom-up approach, on the other hand, is used when the rules and major components are not yet fully understood.

The bottom-up analysis begins with a small set of data and gradually builds a theory around it. As a result of this approach, one must keep an open mind, since fundamental rules may not become apparent until deeper exploration is conducted; bottom-up analysis is similar to primary research since it involves observing, experimenting, and collecting data. By contrast, secondary research is often used in top-down studies, in which researchers build on existing knowledge.

The distinction between top-down and bottom-up causation is crucial to understanding the intricate web of life in biological systems. In contrast to top-down causation, bottom-up causation thrives in an uncertain and exploratory environment. Both approaches provide valuable insights, and their choice depends on the depth of understanding and complexity of the biological system under investigation. Both modes of causation provide complementary perspectives that enrich our understanding of the natural world in the ever-evolving field of biology.

The conventional 3-D spatial platform, often used to represent biological systems, has inherent limitations when seeking to understand the intricate relationships between cause and effect within them. This platform primarily serves as a tool for visualizing the static, physical arrangement of elements within a biological system. It also offers a snapshot of the system at a particular point in time. However, it has difficulty depicting the dynamic and nuanced interactions that underlie the functioning of biological entities.

Biological systems are characterized by complex cause-and-effect relationships. Cellular, organismal, and ecosystem events trigger cascades of reactions that are often interconnected. Conventional perspectives struggle to represent these intricate causal networks effectively, often oversimplifying them or linearly presenting them. This doesn't capture their true complexity.

The holistic study of biological systems, known as "systems biology," requires a deep understanding of how components interact to produce system-level behaviors. 3-D spatial models are not well-suited to capturing the holistic, systems-level view necessary to advance in this field. Besides, the conventional spatial platform cannot illustrate dynamic processes within biological systems. These systems are constantly changing, with elements interacting and responding to one another over time.

Kabbalistic Biology introduces a novel technique for reflecting on the relationships between distinct elements and whole biological systems, employing a conceptual framework called the "six-edges model." This model involves rearranging the six Sefirot of Zeir Anpin, resulting in the formation of two unique types of triads, each representing specific aspects of these relationships.

1. Top-Down Triad

Within the first triad, Hesed represents the expansion of space within the holistic biological system, and Gevurah signifies the constraints imposed on it. These two aspects are interconnected with Yesod, which serves as the foundation for the development of individual biological units. This triad represents a top-down relationship, where overarching factors such as system-wide expansion and constraints influence the development of individual components within the system.

2. Bottom-Up Triad

The second triad consists of Netzah, which represents the multiplication of individual units, and Hod, which defines the selection of the units. Tiferet maintains the unity of parts while developing space for the entire biological system. In this triad, factors like expanding potential space and defining constraints guide the coherent development of separate biological entities.

Importantly, these two triads are not independent. Instead, they intersect to form a distinctive pattern resembling a six-pointed star, commonly known as *Mogen David*, literally meaning "shield of David." By rearranging the Sefirot of Zeir Anpin and forming these triads, this interconnected framework provides a conceptual tool to construct the space that governs the development and interactions of elements within the biological realm (Figure 57).

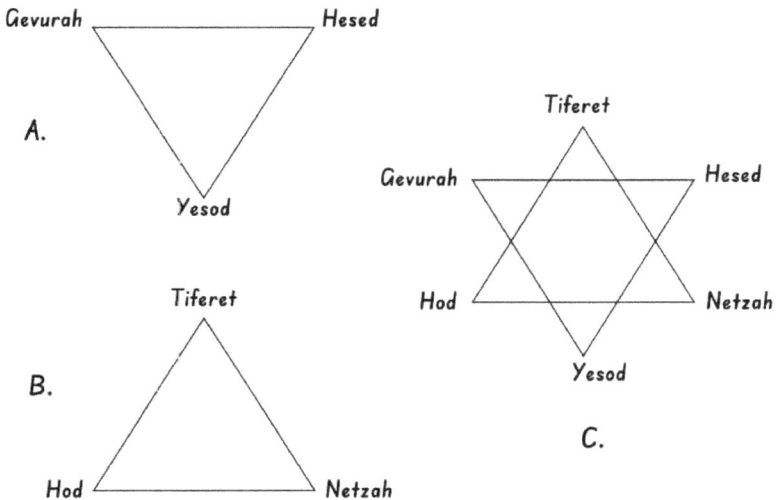

Figure 57. The interconnected framework of six sefirot of Zeir Anpin to construct the space that governs the development and interactions of elements within the biological realm. A - Top-Down Triad; B - Bottom-Up Triad; C - The Six-Points framework to map and navigate biological spaces.

Application of Six Spatial Dimensions in Species Distribution

A Kabbalistic "middot" system manifests itself in every biological system. The following examples will illustrate how the six sefirot are used in measuring biological space. One good example is the distribution of species in a given area. In Colorado, where I live, black-tailed prairie dogs are the main carriers of bubonic plague. To organize plague control in this area, I need to understand the animals' population distribution and density.

The most direct way to learn about prairie dog distribution is to collect data about where this particular species of animal lives. There are several ways to collect data, such as visual observations or traces of prairie dog activity reported by experts. Analysis of such data leads to gathering information about prairie dogs' geographic distribution. Depending on available resources and the organization of the research, we can obtain more or less information about the actual presence of prairie dogs within a defined range.

However, we are limited to searching for available observation areas. Even with abundant resources and excellent organization, we will be limited to a tiny portion of the species' population. Some territories will remain completely devoid of investigation. Animals might also be inactive during observation times. Diseases, predators, or poisoning may have wiped animals out. The main point is that a predominant portion of the population will remain "invisible."

To overcome such a limitation, we can use information about the habitats typical of the species. Black-tailed prairie dogs inhabit shortgrass prairie, mixed-grass prairie, sagebrush steppe, and desert grassland. Habitat preferences for these animals are influenced by vegetation type, slope, soil type, and the amount of rainfall. Their foraging and burrowing activities influence environmental heterogeneity, hydrology, nutrient cycling, biodiversity, landscape architecture, and plant succession in grassland habitats. As we discussed above, such data and their analysis can help model the ecological niche. Such niches are considered possible habitats for organisms and species that could help them survive.

It is imperative to understand that the information obtained from actual animal observation and the information obtained from ecological niche modeling have fundamental differences. The prairie dogs' ecological niche can be measured in size, structure, and dynamics. *However, the niche itself is an abstract category.* The parameters that define the ecological niche, such as vegetation and landscape, are visible, but the ecological niche is not! Experts in ecological niche modeling can produce maps outlining niche borders. Nevertheless, these maps are virtual. An ecological niche defines a potential existence, not an actual existence.

Thus, we have two levels of occurrence of prairie dogs: actual and potential. At each level, the phenomenon is driven by three forces: expansion, constraint, and balance, mirroring the inherent qualities of the Sefirotic triad. Considering the situation of finding the biological space occupied by prairie dogs, the lower triad of Zeer Anpin defines a space for the actual presence of these animals. Within the scope of visible manifestations, it is relatively easy to determine whether the number of animals has increased or decreased. In addition, the diversity of competitive species may also have increased or decreased; food resources may have expanded or diminished, and more.

Netzah signifies the intensity of the birth rate of prairie dogs, the increase in genetic diversity within the population, the dispersion of young animals outside the colony, the occupation of adjacent territories, the exploration of new food resources, and so forth. In contrast, Hod outlines the space of constraints that limit the growth of the population: decline of density, unfavorable climatic conditions, crash of food resources, death from diseases, predators, competitive species, reduced genetic diversity, etc.

A space is defined by Yesod for each prairie dog colony based on local conditions, seasonal conditions, and optimal conditions for colony functionality and the survival of the animals within larger biological systems for that colony.

However, if these three trends (expansion, constraint, and balance) are viewed from a potentiality perspective, then it would also be possible to estimate the growth and decline of populations and species as a whole without having to refer to any particular individuals. The upper triad plays a crucial role in defining space for potential development of the prairie dog populations as a whole system.

Hesed represents an inclination towards the potential expansion of space for prairie dogs. Related parameters could include the growth of grasses, the diversity of plant communities, conservation practices, new behavior patterns, etc. In the opposite direction, Gevurah delineates potential boundaries of space for prairie dog distribution. Those could include both natural (geography, climate, other species) and social parameters (pest control, reintroduction of black-footed ferrets, residential development across the foothills of the Rocky Mountains). In order to allow the prairie dog population to function as a self-regulating system, Tiferet acts as a harmonizing force, shaping the space that ensures its functioning (Figure 58).

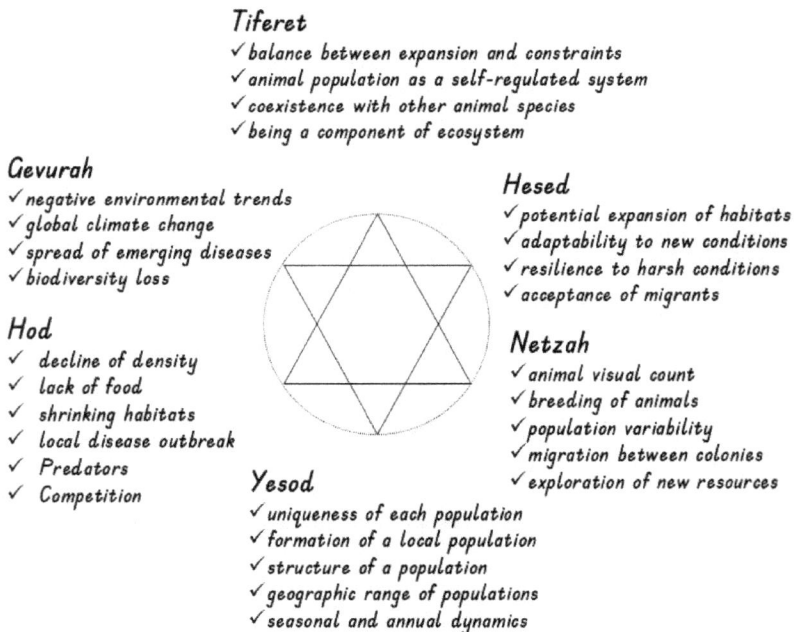

Tiferet
✓ balance between expansion and constraints
✓ animal population as a self-regulated system
✓ coexistence with other animal species
✓ being a component of ecosystem

Gevurah
✓ negative environmental trends
✓ global climate change
✓ spread of emerging diseases
✓ biodiversity loss

Hesed
✓ potential expansion of habitats
✓ adaptability to new conditions
✓ resilience to harsh conditions
✓ acceptance of migrants

Hod
✓ decline of density
✓ lack of food
✓ shrinking habitats
✓ local disease outbreak
✓ Predators
✓ Competition

Netzah
✓ animal visual count
✓ breeding of animals
✓ population variability
✓ migration between colonies
✓ exploration of new resources

Yesod
✓ uniqueness of each population
✓ formation of a local population
✓ structure of a population
✓ geographic range of populations
✓ seasonal and annual dynamics

Figure 58. Biological space of an animal population distribution based on the six-sefirot orientation model.

The two triads provide a comprehensive framework for understanding and navigating space for both entire species and individual animals. Remember Figure 22 in Chapter 7, where the external observer can see dogs, but not "dog-ness?" In this situation, figuratively speaking, we define a space for "visible" prairie dogs and "invisible prairie dog-ness."

Navigating the Space of Evolution

Disputes about evolution and the role of specific mechanisms in the evolutionary processes extend far beyond scientific communities. Yet, the majority of biologists do not question the role of evolution as the core of the entire science of biology. As Theodosius Dobzhansky eloquently put it in his 1973 essay, "*Nothing in biology makes sense except in the light of evolution.*" Still, within the scientific community, a significant dispute rages on, particularly regarding the role of natural selection in shaping evolution's course.

Instead of engaging in these debates, let us consider a different challenge, one that often goes unnoticed. In many cases, miscommunication among scientists can be attributed to the absence of a unified framework for understanding the complex space in which evolution takes place. The evolutionary space is vast, encompassing changes on scales that range from the minute, such as shifts in gene frequencies within bird populations. It also includes the monumental, like the evolution of entirely new taxonomic lineages above species. These two extremes represent microevolution and macroevolution, respectively.

Microevolution focuses on changes within a single population. Scientists delve into gene frequencies or the emergence of visible traits, known as "phenotypes." On the other hand, macroevolution zooms out, transcending single-species boundaries. Here, we witness the emergence of entirely novel features, like the evolution of complex structures such as the eye. Another example is the development of entirely revolutionary life strategies, like bird flight. Recognizing these two distinct scales, microevolution and macroevolution, helps us avoid fruitless debates. However, it introduces a duality. Scientists on each side may have different objectives, employ different methodologies, and think critically in differing ways.

Kabbalistic Biology approaches this duality not as an obstacle but as a powerful tool. It introduces the concept that microevolution and macroevolution are governed by different aspects of the framework represented by the two Sefirot triads. In this framework, microevolution is guided by the Sefirot in the lower triads: Hod, Netzah, and Yesod. Macroevolution, on the other hand, is shaped by the Sefirot in the higher triad: Hesed, Gevurah, and Tiferet.

Some evolutionary mechanisms, such as gene mutations, gene recombination, horizontal gene transfer by viruses, animal migration, and others, lead to increased variability inside populations, a diversity of species, the appearance of new features, and an expansion of carriers of new characteristics. Hesed symbolizes such changes at the level of macroevolution, whereas Netzah is at the level of microevolution. On the other hand, some evolutionary patterns will lead to the elimination of some genes or manifesting features.

There is a phenomenon called "genetic drift." Genetic drift refers to direct variation in the relative frequency of different genotypes in a small population. For example, in a rabbit population with brown and white fur, white fur becomes dominant over time. However, due to genetic drift, only the brown population might remain, with all the white ones eliminated. Such a change refers to Sefirah of Hod.

Natural selection, a cornerstone of evolutionary theory, can be understood differently depending on the observation scale. Hod's influence leads to the microevolutionary elimination of specific genes and traits. Gevurah comes into play at the macroevolutionary level, eliminating organisms and populations ill-suited to their environments or incapable of adapting to significant changes.

In evolution, we often see opposing trends—diversification and uniformity, change and stability. Kabbalistic Biology introduces central points as manifestations of Tiferet in the upper triad and

Yesod in the lower triad. These points embody attributes like harmony, optimality, directedness, and sensitivity to circumstances. (Figure 62).

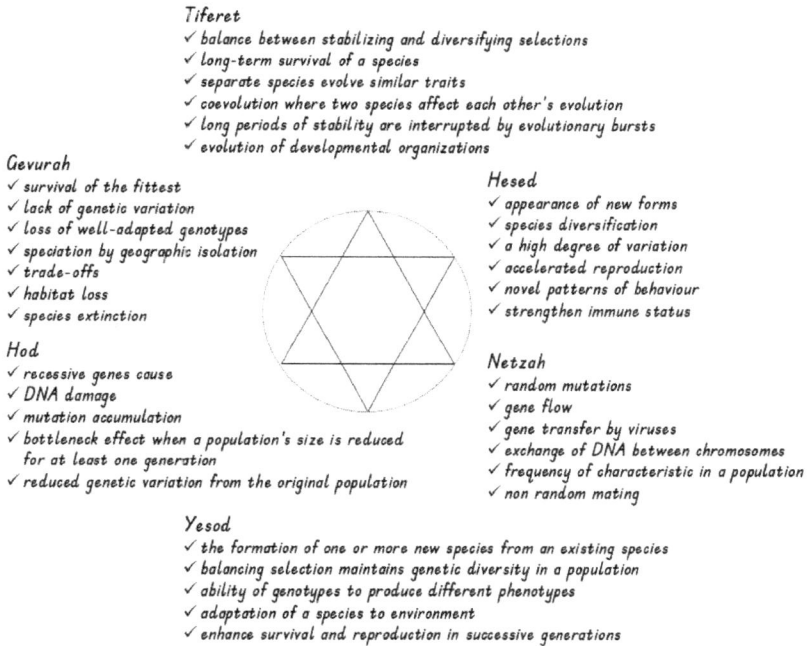

Tiferet
✓ balance between stabilizing and diversifying selections
✓ long-term survival of a species
✓ separate species evolve similar traits
✓ coevolution where two species affect each other's evolution
✓ long periods of stability are interrupted by evolutionary bursts
✓ evolution of developmental organizations

Gevurah
✓ survival of the fittest
✓ lack of genetic variation
✓ loss of well-adapted genotypes
✓ speciation by geographic isolation
✓ trade-offs
✓ habitat loss
✓ species extinction

Hesed
✓ appearance of new forms
✓ species diversification
✓ a high degree of variation
✓ accelerated reproduction
✓ novel patterns of behaviour
✓ strengthen immune status

Hod
✓ recessive genes cause
✓ DNA damage
✓ mutation accumulation
✓ bottleneck effect when a population's size is reduced for at least one generation
✓ reduced genetic variation from the original population

Netzah
✓ random mutations
✓ gene flow
✓ gene transfer by viruses
✓ exchange of DNA between chromosomes
✓ frequency of characteristic in a population
✓ non random mating

Yesod
✓ the formation of one or more new species from an existing species
✓ balancing selection maintains genetic diversity in a population
✓ ability of genotypes to produce different phenotypes
✓ adaptation of a species to environment
✓ enhance survival and reproduction in successive generations

Figure 59. Biological space of evolution based on the six-sefirot orientation model.

How can we avoid the duality created by the separation between microevolution and macroevolution, each symbolized by lower or higher sefirotic triads? We turn to the conceptual map represented by the intersection of these two triads, a six-pointed star. This intersection forms a space that allows us to navigate and discuss evolutionary processes without losing sight of the interconnection between the micro and macro scales.

Six Sefirot as the Centers of Space

Sefirot should not be viewed as static points representing a fixed shape or structure of biological space. Instead, they can be more aptly understood as indicators of trends or dynamic tendencies in the development of space and its underlying structure within biological systems. What sets this Kabbalistic interpretation of biological space apart is its unique approach to measurement. Instead of relying on conventional Euclidean geometry, it adopts the rules of topology. Proximity, in essence, aligns more with topology and is intimately connected with the notion of "neighborhood." When we navigate within a space endowed with metrics, we can envision proximity as a sphere characterized by a specific radius. It's imperative to note that topological space doesn't require metrics, offering a more abstract and flexible perspective.

While the concept of "distance" inherently involves units of measurement providing a quantifiable means of assessing spatial relationships, proximity shares common ground with the idea of "vicinity" or "locality." Proximity thrives in relativity, finding its value in the dimensions and context of other components within a system. Crucially, the determination of whether a particular distance is deemed "near" or "far" hinges on the dimensions and scale of the system in question. If we lack a comprehensive understanding of a system's dimensions, we remain uncertain about the categorization of a specific distance as close or distant. Without a grasp of these dimensions, distance categorization remains elusive.

A position within a proposed biological space is determined by its relative proximity to all six sefirot of Zeir Anpin. Instead of relying on notions of absolute distance between objects, this approach employs relational proximity to specific orientation points. Samuel Avital referred to these orientation points as "six edges of space."

Furthermore, it emphasizes that the manipulation of biological space adheres to the principles of topology as opposed to those of conventional geometry. Traditional geometric measurement methods focus on physical distances and Euclidean space. In contrast, the proposed framework adopts topological principles. Topology is concerned with properties preserved under continuous deformations, making it particularly suitable for capturing biological systems' flexibility and dynamic nature. It allows for a more abstract and flexible representation of spatial relationships.

The sefirot does not operate independently but rather has a cumulative effect on biological systems' spatial arrangement. The cumulative effect of the six sefirot highlights their interdependence. Each sefirah represents a different aspect or quality, such as expansion, constraint, or balance. Their combined influence shapes the overall dispositions of biological systems. It's not merely the presence of one or a few sefirot that matters, but the synergistic impact of all six.

David Sheinkin, an eminent psychiatrist, lecturer, and student of mysticism, wrote:

"A Sefirah never appears separately, acting alone. It is similar to the situation of the human body when one organ never acts independently of the others. There is a fundamental balance that exists. Even if one organ is overactive, somehow, it is connected to everything else in the body that is occurring. Certainly, the organ does not act totally by itself. Each of the Sefirot can combine in different combinations with other Sefirot. For an analogy, think of the genetic structure of human beings: from a relatively small number of genes, billions of different people are created."[20]

This approach aligns with systems thinking, which emphasizes the interconnectedness of components within a system. It highlights the fact that a holistic understanding of biological space requires an awareness of all the relevant factors represented by the sefirot, rather than isolating individual aspects. This concept can be used in various biomedical disciplines, from ecology to genetics. It encourages researchers to consider the combined effects of different factors when studying and modeling biological systems. This leads to more comprehensive and accurate insights.

It is productive to refer to sefirot as "centers of power," borrowing words from psychologists Joseph Berke and Stanley Schneider.[21]

---

[20] David Sheinkin, *Path of the Kabbalah*. 1998, page 83.
[21] Joseph Berke and Stanley Schneider, *Centers of Power*. 2008.

When considering the sefirot as centers, I wish to refer to Christopher Alexander, a Professor of Architecture at the University of California, Berkley, and Director of the Center for Environmental Structure. In the four-volume book, *The Nature of Order*, Alexander proposes a scientific view of the world in which all space-matter has a perceptible degree of life. He defines space as filled with certain centers, "all helping each other, all created by other centers, but all field-like, all radiating centeredness."[22] Armed with the idea that each center is a multi-leveled field-like phenomenon, Alexander states that each center has a distinguishable degree of life.

Proposing the following unusual way of thinking, Alexander made five assertions:[23]

1. Centers arise in space.

2. Each center is created by a configuration of other centers.

3. Each center has a certain life of intensity... The life of any one center depends on the life of other centers. This life or intensity is not inherent in the center by itself but is a function of the whole configuration in which the center occurs.

4. The life or intensity of one center is increased or decreased according to the position and intensity of other nearby centers.

5. The centers are the fundamental elements of wholeness, and the degree of life in any given part of space depends entirely on the presence and structure of the centers there.

Alexander's view is very similar to the Kabbalistic interpretation of space. Crucially, this perspective redefines our understanding of biological space. It ceases to be a mere container for living processes and becomes a set of dynamic conditions, much like the Kabbalistic sefirot themselves. Life within this framework is not confined to static spatial dimensions; it thrives within a web of relationships and interactions that mirror the intricate interplay of divine attributes in

---

[22] Christopher Alexander, *Nature of Order*. Book One: *The Phenomenon of Life*, 2002, page 120.

[23] Ibid., page 122.

Kabbalistic thought.

In conclusion, this chapter invites us to accept the gift of a new lens through which we can explore the complexity of biological systems. By adopting this spatial measurement framework, we gain fresh insights into the intricate dance of life. Here, ecological niche and genetic potential intersect with the profound principles of expansion, contraction, and balance a harmonious symphony of existence that transcends the limits of conventional space.

Chapter 19

Seventh Dimension: Biological Time as a Transformation

Time is often conceptualized as a dimension, just like the three spatial dimensions: length, width, and height. In physics, this idea is formalized as "spacetime," where time and space are intricately linked. Albert Einstein's Theory of Relativity explains this connection, showing how mass and energy affect both time and space. However, it was Hermann Minkowski, one of Einstein's teachers at the Zürich Polytechnic, who brilliantly demonstrated the inseparable nature of space and time. He explained that moving through space always entails moving through time; you can't instantly move from one location to another. Instead, you must progress through both space and time to reach your destination. This insight has profoundly altered our understanding of space/time interplay.

Kabbalistic biology, however, introduces biological time as a distinct dimension. This innovative concept challenges our conventional understanding of time and offers a fresh lens through which to comprehend organic life processes. At the heart of this perspective lies the fundamental premise that biological time is not merely an external measure or a passive backdrop. Instead, it is an integral and intrinsic aspect of living systems. Time is considered an inseparable component deeply intertwined with the Life processes of

467

development and transformation.

In this paradigm, time ceases to be a detached entity, and instead becomes an active participant in the unfolding drama of biological existence. This unique perspective challenges us to view time as an ever-present, dynamic dimension deeply embedded within Life. Kabbalistic Biology invites us to explore the profound interplay between time and biology.

The Challenge of Measuring Time

The nature of time has been a subject of contemplation and discourse throughout human history. The question of time's nature is a fundamental one and has been the subject of philosophical and scientific debates for centuries. It remains a topic of profound interest and inquiry, reflecting our deep-seated curiosity about natural phenomena. Two pivotal questions emerge before we embark on discussions concerning time in the realms of both biology and Kabbalah.

The first question concerns choosing between linear and nonlinear conceptions of time, which questions the very structure of reality and our place within it. Are we bound by unidirectional flow, forever moving forward along a predetermined path? Or does our existence unfold within a more intricate tapestry, where the past informs the present, and the future influences the past? This suggests a dynamic interplay of possibilities!

In the conventional approach, we perceive the progression of events when time flows in a straight line from the past to the present and then to the future. They call such a perception "linear time." It operates in a single direction, like a continuous thread, with one timeline that represents the order of occurrences. Accordingly, events are arranged in a "chronological" order based on the time they occurred. This linear concept of time aligns with our everyday experience, where we measure time in minutes, hours, and years. It also implies that we can't move backward in time. Hours, days, and years are some of the signifiers used in chronological order.

However, there is a compelling alternative to linear time: a vision of time as a multidimensional, interconnected web where the past, present, and future coexist and interact. According to such a view, time moves in various directions simultaneously, like a web with multiple interconnected paths. Nonlinear time is not consistent with the traditional concept of time as a unidirectional flow. Normally, we watch one point in time followed by the next one. To connect one point to another, a step must precede or follow that point in the sequence in which it occurs. According to a nonlinear perspective, each point is connected by the nature and texture of the *event* it contains!

The choice between linear and nonlinear perspectives on time hinges on how we perceive the interconnectedness of events. In a linear perspective, time unfolds as a series of distinct and separate links, akin to a chain of cause and effect. Each event is viewed in isolation, and the connection between them is seen as a linear progression.

On the other hand, a nonlinear perspective views events as "stages" within a broader transformational process. A continuous influence of the past is exerted on the present, bridging the temporal gap. Based on this perspective, events do not follow a linear sequence but are bound together across time, reflecting their intrinsic connection within a larger, holistic process. Time is no longer a linear progression or a one-dimensional entity; rather, it is a multidimensional force that plays a vital role in shaping Life's intricate tapestry. From the growth of a single cell to the evolution of an entire species, biological time is the unseen thread that weaves these processes together (Figure 60).

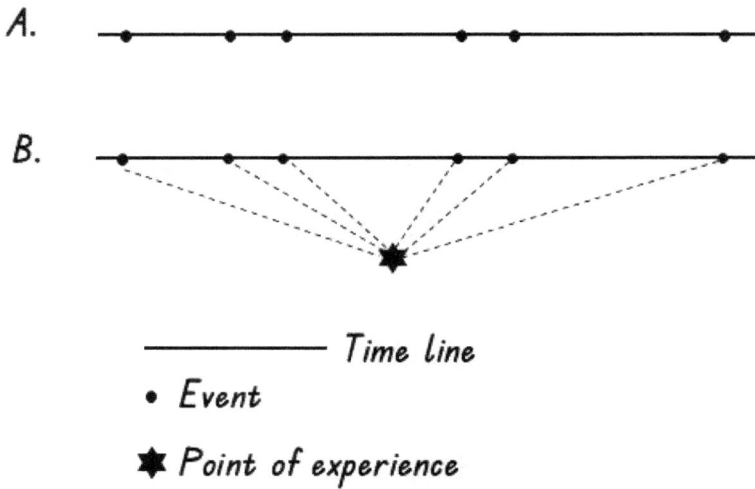

Figure 60. The choice between linear and nonlinear perspectives on time hinges on how we perceive events' interconnectedness. A). In a linear perspective, time unfolds as a series of separate, distinct links. B). In a nonlinear perspective, events are bound together across time, reflecting their link to the point of our experience.

The second crucial question leads to the distinction between time in nature ("biological") and chronological ("physical" or "astronomical") time. An example of this dichotomy is the historic debate between Albert Einstein and Henri Bergson over the very nature of time. It occurred in Paris on April 6, 1922. In her book, *The Physicist and the Philosopher*, Jimena Canales tells the remarkable story of how this explosive debate transformed our understanding of time and created a rift between science and the humanities that continues today.

Einstein's perspective boils down to the theory that only two types of time exist: physical time, measured by clocks, and psychological time, which is subjective and experiential. He sought simplicity and consistency in his understanding of time. However, Bergson took a different stance, emphasizing time's intricacies and complexities. He contended that time is not an entity distinct from the individuals who perceive it, but is intimately intertwined with their perceptions. In essence, Bergson's argument revolved around the idea that time cannot be separated from the observer's experience.

Bergson described time as a "duration" (*durée*) that cannot be reduced to a succession of distinct, instantaneous states. This historic debate illuminates the fundamental dichotomy between these two perspectives on time. Einstein pursued objective, quantifiable time, while Bergson recognized the inseparable connection between time and individual perception.[1]

Bergson's idea resonates with the groundbreaking ideas of the great scientist, Vladimir Vernadsky, the founder and first president of the Ukrainian Academy of Sciences. He proposed that time manifests itself as "the development of biological material." Vernadsky, in his philosophical musings, categorizes time into three different processes: individual existence, generational switching without changes in life forms, and evolutionary time accompanied by changes in life forms and generational shifts. Each of these processes is underpinned by different *aspects* of time.[2]

Physicists typically regard time as an independent variable. However, biological systems appear to be influenced by *external* factors. This is because interactions between systems can have an impact on their unique notions of time. This makes it challenging to describe changes as "simple trajectories in a phase space." Instead, system evolution appears to be a *combination* of potential trajectories.

Time as a "Flowing Process" in Kabbalah

Time, as perceived by the Western mind, often appears as a linear progression with a discernible beginning and end. However, the Kabbalistic perspective challenges this conventional notion, presenting time as a dynamic, multidimensional entity. Thorleif Boman, an influential Norwegian scholar, believes that European societies are dominated by "static time "thinking because it emphasizes rhythmic recurrence. Alternatively, "Hebrew thought"

---

[1] Jimena Canales. *The Physicist and the Philosopher: Einstein, Bergson, and the Debate that Changed Our Understanding of Time.* 2016, 488 pages.

[2] Vladimir Vernadsky, *Philosophical Thoughts of the Naturalist.* 1988, page 82.

(as Boman calls it) emphasizes the uniqueness of events at non-repeatable intervals, which makes them dynamic.[3] Boman wrote in his book, *Hebrew Thought Compared with Greek*:

> "For the Hebrews who have their existence in the temporal, the content of time plays the same role as the content of space plays for the Greeks. As the Greeks gave attention to the peculiarity of things, so the Hebrews minded the peculiarity of events ... The Semitic concept of time is closely coincident with that of its contents without which time would be quite impossible. The quantity of duration completely recedes behind the characteristic feature that enters with time or advances in it."[4]

Kabbalistic thought recognizes two distinct versions of time. First, there is "local time," measured by conventional clocks and calendars. The second version introduces a more profound dimension: the "unified time flow." [5] Elliot Wolfson, Professor of Hebrew and Judaic Studies at New York University, distinguishes between two vectors of time according to Kabbalistic teaching. One applies to the physical universe – the world of discriminate beings-- and the other applies to the unfolding of the enfolded force of the Higher Realm – the world of integration.[6] This overarching temporal force orchestrates Life's evolution, transcending the boundaries of local time. It is this unified time flow that causes the emergence, manifestation, and actualization of "latent potentials."

In the Kabbalistic worldview, existence is not viewed as a linear process culminating in a finite ending. Instead, it is likened to a continuous rebirth, pulsating instantaneously into and out of existence. In general, Jewish tradition does not perceive time as a

---

[3] Bruce Chilton, *Redeeming Time.* 2002, page 9.

[4] Thorleif Boman, *Hebrew Thought Compared with Greek.* 1970, page 138.

[5] Herman Branover and Ruvin Ferber, *The Concept of Absolute Time in Science and Jewish Thought.* 1999, https://www.ldolphin.org/jtime.html

[6] Elliot Wolfson, *Alef, Mem, Tau: Kabbalistic Musing on Time, Truth, and Death.* 2006, page 81

relentless march toward death; rather, it acknowledges an open universe, receptive to infinite sources of renewal. Each moment is imbued with new potential, representing a fresh choice and an opportunity for perfection. In this context, the Kabbalistic measure of time lies in the degree to which Life's purpose has been achieved. This transcends the limitations of the physical body. Allen Afterman, a poet who wrote a remarkable book on Kabbalah, wrote: "The true measure of time is the degree to which life purpose has been achieved."[7]

Kabbalistic thought introduces a profound paradox in the understanding of time, the convergence of linear and circular time concepts. Wolfson speaks of "linear circularity" or "circular linearity." He stresses a notion of time that is circular in its linearity and linear in its circularity. This concept suggests that time possesses the qualities of both linearity and circularity simultaneously, challenging our traditional distinctions between the two.[8] It emphasizes the idea of "flowing presence," where repetition and novelty coexist harmoniously. In this framework, what recurs is not a mere replication of the past; rather, it is a unique occurrence, blending the familiar with the unprecedented.

Within the Kabbalistic tradition, the sefirot, representing ten attributes, shapes time perception. Wolfson describes sefirot as "time without time," a dimension of time that transcends mundane temporal constraints. This realm of sefirotic potential is envisioned as a continuum of events, manifesting either as an externalization of the internal or as an internalization of the external. Time, in this context, is seen as inner space, and space is external time. Wolfson wrote: "The dimensionality of each point charted in the time-space of a sefirotic graph can be demarcated as the duration of a blink or the extension of a swerve."[9] Referring to Yehuda Leib Ashlag's *Talmud Eser Sefirot* ("Ten Luminous Emanations"), Philip Berg wrote,

---

[7] Allen Afterman, Kabbalah and Consciousness. 2005.
[8] Elliot Wolfson, *Alef, Mem, Tau: Kabbalistic Musing on Time, Truth, and Death.* 2006, page 82
[9] Ibid., page 87.

"Time is, in fact, a sequence of events, a distinct number of connected phases branching from each other in their order of cause and effect."[10]

Kabbalistic wisdom introduces another profound concept—the eternal recurrence of the ephemeral. It highlights the repetition of what is repeatedly different and the perpetual coming-to-be of that which has never seemingly come to pass. Judah Loew of Prague (known as Maharal), the famous Kabbalistic mystic of the 16th-17th centuries, wrote that the time that was "before" is the order of time [*seder zemannim*], not a particular time [*zeman meyuhad*].[11]

Yaacov Agam, an Israeli sculptor and experimental artist, noted:

> "The concept of Time in Judaism differs in essence from that of other civilizations, and constitutes ...one reality of its fundamentals. Other civilizations freeze Time (eternity)...As opposed to this the Jew finds himself in a world of reality which cannot be reconstructed... All is transient... My aim is to show what can be seen within the limits of possibility which exists in the midst of coming into being."[12]

C.H. Waddington, the prominent British biologist, commented on Agam's point in his book *Behind Appearance*: "Time is an essential and not merely a contingent aspect of reality — the world processes, not things."[13] This perspective invites us to embrace the coexistence of time and eternity, not as logical antinomies but as ontic variations held together in the identity of nonidentity. In this intricate dance between time and eternity, the timeless is concealed within the eternal, and the transcendent is manifestly concealed within time.

Time as a Function in Kabbalah
In delving deeper into Kabbalah's traditional perception of time, we find that it encompasses various dimensions, like a multi-layered

---

[10] Cited from Philip Berg, *The Energy of Hebrew Letters*. 2010, page 270.

[11] Ibid., page 222.

[12] Jasia Reichardt, *Yaacov Agam (Art in progress)*. 1966 (cited from C.H. Waddington).

[13] C.H. Waddington, *Behind Appearance*. 1969, page 220.

tapestry. I would like to draw attention here to an approach to measuring time deeply rooted in Kabbalah as interpreted by Nilton Bonder, a Brazilian rabbi who has authored several Kabbalah books.

According to Bonder, time is a function of purpose, representing the speed of transformation of form in the pursuit of a specific course. It's a constant flow, characterizing the passage of time as it serves the purpose of existence. Bonder provides a compelling perspective: time is a function—a mere direction. It is something that can only be measured by observing the transformation of form. Bonder's insights highlight the interplay between form and time, and state that "*Form is just another way of representing time.*" [14]

Nilton Bonder's exploration of time in his book, *The Kabbalah of Time,* delves into the multifaceted nature of how we experience time. He divides time into four distinct dimensions or worlds, each with its unique attributes and significance.

1. *Present Tense-Time (Physical Time)*: This dimension is characterized by concrete experiences, actions, and physical measurements. It is the realm where sensations of pleasure and pain mark existence. In this dimension, life is lived in the "now" and represents the time when a separate question arises: "How?'

2. *Past (Inclusive Tense-Time):* The past is a shared temporal dimension where collective experiences and memories form. It operates in the first-person plural, representing a reference point for the collective "us." It raises questions about origins and sources as individuals collectively shape criteria to both measure and interpret results.

3. *Future (Virtual):* The future exists as a virtual realm devoid of real form, much like a blank canvas waiting to be painted. It lacks tangible substance but is filled with expectations, projections, and possibilities. People often extrapolate future scenarios based on past experiences, generating

---

[14] Nilton Bonder, *The Kabbalah of Time.* 2009, page 100.

models and potential outcomes by exercising their intellect.

4. *Everlasting Time*: This is a timeless backdrop against which all other temporal dimensions unfold. It's a realm of essence and purpose. Here, the question, "Why?" takes precedence over those of "How?" or "When?"' The "everlasting time" lies beyond ordinary perception, a constant that defies our understanding of past, present, and future.

Bonder makes a crucial distinction between "the past" and "before." The past is a construct influenced by individual and collective perceptions, giving rise to notions of identity, morbidity, and loneliness. These notions are illusions that arise during the course of our existence. In contrast, "before" represents a living experience that seamlessly leads into the present. It forms an integral component of the "before-now-after" continuum.

Furthermore, Bonder emphasizes that each moment in the present (now) represents a pulsing disturbance of the "everlasting time," creating a dynamic temporal experience. He also challenges the notion of a linear chronology, particularly within Torah, where time's purpose is found more in its form than in its content. The future, Bonder asserts, exists as demands that influence the present— a tool for projecting potential outcomes. This perspective aligns with the idea that sequential time is not a rigid matrix, but an experiential result of existence, constantly shaped by changing forms.[15]

Ultimately, Bonder proposes that time is not merely a quantity but a determinant of quality measured through the transformation of form. He draws a parallel between the loss of Life's organic relations and the development of diseases like cancer, where cells prioritize quantity over quality. Bonder wrote:

> The disappearance of life's 'organic' (ecocosmic) relations constitutes a kind of disease similar to cancer. What is a cancer if not cells that do not want to yield to the body's

---

[15] Ibid., page 108.

discipline, that replace qualitative guidelines with quantitative ones? The cells' 'ego' goes crazy, out of control. In their zeal to devour everything in front of them, they quit caring about the organism as a whole. These cells trade parameters of quality for parameters of quantity. In structural terms, immortality undoubtedly resembles a carcinogenic manifestation.[16]

This process represents a certain expanse to be covered in a course and time stands for a function of the speed of transformation of form. The course (purpose) is covered by the constant transformation of all forms, generating the perception of a flow characterizing the passage of time. Time, as Bonder stated, "is nothing more than a direction, left behind as a trail and imagined as the continuation of a purpose."[17]

Time is a Measure of Changes in Biology

The notion that biological processes are inherently linked to the passage of time has fascinated many thinkers throughout history. Soviet paleontologist Sergey Meyen was among the early visionaries who grasped the intimate connection between biological processes and the unfolding of time. Importantly, his concept is very much in line with Bonder's view of time as a function of purpose, presented in Kabbalistic terms.

Meyen delved into an intriguing inquiry during his studies of fossil plants: "What is time?" In paleobotany, the temporal aspect ("when?") and the nature of events ("what?") often seem intertwined. For instance, if there was an evolutionary occurrence in the Triassic period, the precise physical or astronomical time becomes less relevant. "The Triassic period" serves as the answer to both temporal and categorical inquiries. Meyen's perspective posits that time is not an abstract backdrop against which other changes occur. Instead, it

---

[16] Ibid., page 135.
[17] Ibid., page 141.

embodies variability or change within real entities, such as living organisms and geological formations.[18]

Today, the prevailing evolutionary theories posit the development of living systems in an external, independent, and absolute physical space-time (akin to Newton's conception). In sharp contrast, Meyen proposes that time is relational, contending that all biological systems possess unique temporal dimensions. The processes of embryonic development and evolutionary changes require the interaction and continual rescaling of these individual times. Mayen's conclusion about biological time matches Bonder's point that "time is a function of purpose."

The structure of living systems is hierarchical, with each component or subsystem operating under its own dynamics and temporal dimension. This hierarchical structure of biological entities results in a complex biological time system. Various subsystems distributed across different hierarchy levels exhibit distinct temporal and spatial boundaries. They also adhere to specific change rules. For instance, change rules differ between cells, organisms, and populations. Meyen expressed this viewpoint, stating, "From the observer's point of view, time is the variability of each object (individual) in the environment... We have to consider general features in the variability of some set of individuals because we are interested in receiving information concerning not individuals but classes of individuals." [19]

Meyen introduced the concept of "typological time" in biology as a set of phases that encapsulates change, variability, and development. Importantly, Time isn't just a sequence chosen for convenience--it mirrors Nature's real processes. It's a way to quantify the ever-shifting nature of biological entities. Meyen succinctly encapsulates the concept of time in biology as a "type" that reflects a

---

[18] Alexei Sharov, *Analysis of Meyen's typological concept of time*. In *On the Way to Understanding the Time Phenomenon: The Constructions of Time in Natural Science*, 1995, pages 57–67.

[19] Sergey Meyen, *Methodology of the Study of Temporal Relations in Geology: Development of Time Concept in Geology*. Naukova Dumka, Kiev, 1982, pages 361–381 (in Russian).

specific phase, form, or cycle—a variability intrinsic to any living phenomenon. The crux of the matter is that any biological process can be understood through a distinct temporal scale, unique to that process. The concept of typological time in biology suggests that this temporal scale consists of phases of change, development, and variation, each ordered by nature. Time, then, is not a mere byproduct; it's the essence of biological life itself.

The typological concept of time can be used for historical reconstructions (e.g., in geology or paleontology).[20] Patterns of variation or change in one kind of system (or its parts) can be extrapolated to another similar or related kind of system (or parts). In some cases, a structural series of variations (e.g., annual rings on fossil trees) can be used to reconstruct tree growth dynamics that are not readily visible. Fossil records provide information on various stages of organism development and growth and can be used to reconstruct a whole life cycle.

In practical terms, it is significant that a transformation or alteration within a system or process imparts discernible "footprints" that can be harnessed for subsequent reconstruction or analysis. These distinctive tracks are valuable markers or indicators, offering insights and data that ease the processes of informed decision-making and problem-solving. By doing so, these markers help create a deeper understanding of the changes that have occurred.[21]

In this hierarchical structure, changes in parts can be accommodated across a broad range of time. These range from the rapid changes during embryonic development to the protracted alterations that define evolution. However, this hierarchy also emphasizes the importance of synchronicity between components when they collaborate and serve a common purpose. Thus, the overall time frame of a biological system cannot be reduced to the

---

[20] Sergey Meyen, *The Principles of Historic Reconstructions in Biology: Systemness and Evolution.* Nauka; Moscow: 1984, pages 7–32 (in Russian).

[21] Alexei Sharov, Analysis of Meyen's typological concept of time. In *On the Way to Understanding the Time Phenomenon: The Constructions of Time in Natural Science,* 1995. pages 57–67.

time of its constituent components; rather, it constitutes a higher-level coordinating structure.

"Synchronization" is a prevalent type of interaction. Here, time adjusts to changes in another system, whether living or non-living. For instance, plants and animals adapt their diurnal and seasonal cycles to match their environment's rhythms. Similarly, parasites align their seasonal cycles with those of their hosts. When the individual times of system components coordinate in this manner, it results in the emergence of an integrated time for the entire system. This integrated time can be quantified as one or more principal components within the coordinate system of individual processes.

Consider the example of growth or aging, which doesn't occur uniformly across all parts of the body. As a result, the main components can integrate multiple aspects of these processes. By projecting individual times onto these principal components, we can predict various processes with reasonable accuracy. Principal components serving as generalizations of individual events and reducing the number of dimensions involved, help us distinguish between the process itself and the concept of time. This allows us to plot each process against time, using projections onto the subspace formed by these principal components.

Time as Transformation from One Life Stage to Another

Change becomes a part of time only when it is reproducible. The main reproducible element of any living organism is its life cycle. Throughout evolution, organisms provided cyclic processes such as cell cycles, circadian rhythms, and photoperiods.[22]

To illustrate the concept of time as a set of phases of change, development, and transformation, we can use an image of early motion-picture film technology. There is a photograph of a famous film director, Sergei Eisenstein, which was frequently reproduced in specialized books and articles (Figure 61). Eisenstein is considered a

---

[22] Alexei Sharov, *Time in living systems*. Presentation at the ArcheTime: Cross-Disciplinary Conference, 2014.

pioneer in the theory and practice of "montage" for his seminal silent film *Battleship Potemkin* back in 1925. In the photograph, he is editing one of his films. Film montage involves combining short pieces of film to condense space, time, and information. It is the timing of the cuts that creates a meaningful connection.

Figure 61. Sergei Eisenstein editing one of his films.
https://www.russianarchives.com/samples/photos/sergei-eisenstein/

Notably, this picture was selected for the cover of a book written by Sergey Dolgopolski about the Talmud, a part of the Jewish oral tradition along with Kabbalah. In his book, *The Open Past*, Dolgopolski explains that in a film--and by analogy in Talmud--"...there always are two other lines of time — the time of the representational content of the film ... and another line of time pertaining to connecting the film's cuts together, otherwise called, 'the film montage.'"[23]

Let us now use the thought experiment with the "cross-cutting" technique to evaluate the time of biological development and

---

[23] Sergey Dolgopolski, *The Open Past: Subjectivity and Remembering in the Talmud.* 2013, page 219.

transformation. Our examples will be two biological processes presented in Chapter 2: first, an embryonic development, and second, a butterfly's metamorphosis.

How, then, does one measure individual development from a single cell to an embryo? Employing the "cross-cutting" technique, we can see a series of images in rapid succession as being in "continuous motion," with each image representing a single stage in the early embryonic development of an animal (Figure 62).

Figure 62. *The continuous motion of images represents stages of the early embryonic development of animals.*

Let's keep the description of the complex embryonic process as simple as possible. The point is to measure the functional importance of the entire process at each stage. We can observe this by watching one image of embryonic transformation after another. First, an ovum (egg cell) enters the scene. The successive motion starts with fertilization when the spermatozoon (a motile sperm cell) successfully enters the ovum, and the two sets of genetic material carried by the gametes fuse. Later, the fertilized zygote (a single diploid cell) divides through mitosis into two cells. This mitosis continues as the first two cells divide into four cells, then into eight cells, and so on. At some point, a single-layered hollow sphere (blastula) forms an inner fluid-filled cavity.

At the stage of gastrulation, the blastula is reorganized into a multilayered structure known as the "gastrula."

Let's watch another "thought film" about butterfly metamorphosis time. One of the most impressive illustrations is this insect's complete metamorphosis from a young larva to an individual adult differentiated from other butterflies by their form and lifestyle. Remember the example of the Western North American Monarch in Chapter 2? We can observe images from this insect's life cycle using the old motion picture film technology that we applied to observe animal embryonic development (Figure 63).

Figure 63. The continuous motion of images represents stages of the Monarch butterfly's life cycle.

This description is presented cinematographically to emphasize a single aspect: the sequential changes between the forms and lifestyles of one insect species. Butterflies lay their eggs. Then, caterpillars emerge from eggs, feed, and grow. Afterward, the caterpillar turns into a motionless pupa attached to a plant. Finally, an adult butterfly emerges from the pupae, and once again, the cycle begins. Each sequence represents a life stage from eggs to a fully developed butterfly.

C.H. Waddington wrote in *Behind Appearance*:

"The biologist finds that his material presents to him some processes of rather different type. He watches an egg developing and differentiating into an adult organism. This can hardly be regarded as a mere succession of repeating processes; something or other has definitely undergone changes which can be referred to as 'development' or even 'progress'; the system has become more complicated, what were at first only potentialities have become realized."[24]

Both listed processes--embryonic development and insect metamorphosis--can be measured on two temporal scales. One is external or astronomical, estimated in hours, days, etc. A second is internal, or biological, and is determined by the succession within a whole process of development and transformation.

For example, "gastrulation" is one of the most critical steps of development. It takes place roughly between days 14 and 21 after fertilization and lasts about a week. From the point of biological time, gastrulation is a stage during which immature and single-layered cells of the blastula are transformed into a multi-layered structure known as the "gastrula." It is not about how long it takes the transformation from one stage to another. More critical is the *order* of switching stages.

In the same way, entomologists estimate the length of the Monarch's life cycle in days: egg-3 to 4 days; caterpillar-10 to 14 days; pupa (chrysalis) - 10 to 14 days until the transformation from pupa to adult is complete. However, a biological time scale is not determined by counting days, but by observing cardinal changes in the insect's morphology, physiology, and lifestyle as parameters of internal development.

One notable manifestation of biological time is the phenomenon of synchronicity between various biological processes. Unlike the straightforward cause-and-effect relationships that we often rely on for explanations, synchronicity represents a different kind of connection— one that might not always yield a simple causal narrative. However, it undeniably points to a profound

---

[24] C.H. Waddington, *Behind Appearance*. 1969, page 106.

interconnection in Nature.

Consider the synchronized behavior of migrating birds as they collectively alter their migration routes. This coordinated shift is a testament to the intricate web of timing and communication among these avian creatures. Similarly, when an ant colony functions as a "superorganism," displaying almost hive mind-like behavior, we witness the orchestration of countless individual actions into a synchronized whole.

Even when faced with events that appear mysterious and lack an obvious cause, biological time remains relevant. For instance, the rapid and seemingly unexplained decline of an entire lemming population challenges our understanding. Lemmings don't die by running blindly off cliffs—that's a myth. But lemming populations in their Arctic tundra home can rise and fall dramatically in just a few years.

The Timing and Phases of Epidemics

To understand the essence of biological time, consider the question, "When does a plague epidemic begin?" The question is significant. In conventional terms, an epidemic indicates the rapid spread of an infectious disease among a large number of people in a given population. The plague (commonly called "bubonic plague") is a classical model for the study of infectious diseases. Not only has plague had a strong role in influencing human history, it has also been researched extensively. My selection of plague is also influenced by my personal experiences while investigating this disease on five continents.

In epidemiology, a plague epidemic can refer to as few as ten cases per year in New Mexico since plague cases in the United States are usually low in number. In some African countries, for example, Tanzania, the Democratic Republic of the Congo, and Madagascar, several cases might not be called an epidemic. However, this would not have been the case during a past pandemic. Whatever criteria are used to define an elevated number of sick or dead people, the more fundamental questions are: 'When and how did this epidemic begin?"

Here is an example to illustrate the challenge. In Georgia, the country sandwiched between Russia and Turkey, we found an active plague epidemic (die-off) among wild rodents. It was the first evidence of plague after 49 years of inactivity, though plague was widespread in this territory for centuries. Presumably, our report of plague in rodents prevented its transmission to people who lived in or visited this area. The question is, "When did the plague begin to appear here?" And in this case, human cases did not appear--but plague did!

It began with a plague outbreak in animals, but the plague had circulated in animal communities before without being noticed. We can delve deeper: the plague pathogen could have survived in some rodent ectoparasites, soil protozoans (single-celled microscopic organisms), or rodent nests before infecting animals. Many hypotheses still remain possible, based on available scientific information and current views about the "invisible" phases of the plague pathogen between "visible" epidemics.

So, when does a plague start? Here, we come to the question of time. A plague epidemic, like any other epidemic, is a process, not a static state. Epidemiologists commonly register human or animal cases per period (day, week, month, or year). Then, they try to identify the first case to determine the beginning of an epidemic (an "index case"). However, the registered case is not the beginning of the process.

Epidemiology-—the study of disease spread and dynamics in populations can be approached from a new perspective that transcends the conventional external chronological framework. While chronological time remains a valuable reference, epidemiology faces an equally critical dimension: the intrinsic time frame of life stages unique to each disease. These stages can manifest in various ways, some readily observable, others remaining concealed. The observable stages might include tangible markers such as the registration of cases, deaths, or illnesses.

Moreover, these stages can extend beyond the human scale and

486

into the microscopic world. Visualizing the presence of bacteria responsible for a disease or detecting specific antibodies as a response to a pathogen shows how epidemiological time scales can be rooted in biology's microcosm. Yet, not all stages are easily discernible. Some crucial phases of a disease may remain hidden from plain sight, lurking as undetected pathogens in the environment. Pathogens may also lie dormant as latent infections within animals and humans. Identifying these hidden stages requires more than just empirical observation; it requires science's artistry, honed by professional expertise and informed by a deep reservoir of knowledge from previous investigations.

In essence, epidemiological processes are multidimensional endeavors where external chronological time converges with disease-specific life stages. This holistic approach enables epidemiologists to unravel the intricate tapestry of disease dynamics, whether manifest or concealed. It ultimately enhances our understanding of how diseases evolve, spread, and impact people and populations. Holism's blend of scientific rigor, experience, and intuition characterizes the field of epidemiology.

The ability to measure the biological timing of epidemics goes beyond curiosity. I cannot overstate the importance of this challenge. For instance, tracking COVID-19 cases or monitoring mortality rates due to a particular disease constitutes a visible timeline epidemiologists often work with. However, the registration of COVID-19 cases and the discovery of new variants of a viral protein were still not enough to prevent the disease during the pandemic of 2019-2020-2021. Nor did it contribute to our ability to determine the source of the outbreak.

There is an evident lack of understanding of the essential stages of the COVID epidemic, such as the host-pathogen relations in the circulation of related viruses in bat populations in Yunnan, maintenance of isolated viral strain in the laboratory, spillover of the virus to a new animal reservoir, and transformation of an enzootic strain into the pandemic one. We must recognize that the failure to prevent the pandemic was not solely due to a lack of data and information. It was due to our inability to change our mental model of biological time!

Phases of Egyptian Plagues in the Exodus

Life stages are unique for each biological process, including specific diseases. Therefore, I will refer to a highly metaphoric situation — the biblical story commonly called "The Ten Plagues." This reference in "Exodus" highlights the principle of "time" as a series of transformations. The narrative of the Ten Plagues of Egypt offers us meaningful principles with which to understand the evolution and manifestation of epidemics of zoonotic infections, such as plague.

When we talk about the plague, a critical issue is epidemics. A bacterium or one human case can be managed, but plague epidemics cannot. The fear of plague is deep in the human mind, due to the experience of sudden and unexpected epidemics that resulted in the extinction of large numbers of people. As we know, plague pandemics (huge epidemics spread worldwide) likely altered mankind's history.

But was it really about the plague? Among the proposed interpretations are some that appear curious and represent a major opportunity for the human imagination to consider. As a plague researcher, I have my own opinion about various interpretations made from reading certain Bible portions. I fully realize that the epidemics described as "Egyptian plagues" may or may not represent the disease we now call "the plague." Some events described in Exodus, such as frogs, hail, etc., might not appear to be related to any infectious diseases!

On the one hand, people tend to ignore the links between passages in the Torah passages and concrete physical manifestations. On the other hand, there have always been doubts that historical documentation of plague during ancient and medieval epochs corresponded to a specific disease currently called "plague."

Many attempts have been made to find a rational interpretation of Exodus's described events. John Marr, who served as the director of the Bureau of Preventable Diseases in the New York City Department of Public Health, and his co-author, Curtis Malloy, a research associate with the Medical and Health Research Association of New York City, wrote: "The Ten Plagues of Egypt described in the Book of Exodus are the first example in a historical written record of what today might be described as emerging infections."[25]

During my presentation, "The Ten Plagues from the Point of Evolutionary Biology of Infectious Diseases," at the International Conference on Torah and Science in 2019, I posited that each plague represents a specific stage in developing epidemic manifestations:

Plague 1 - Dramatic environmental disruption.

Plague 2 - A sudden change of ecological niche for a principal animal host species.

Plague 3 - The presence of arthropod vectors for infection transmission.

Plague 4 - A critical level of biological diversity.

Plague 5 - Epizooty among animals.

Plague 6 - Microbial evolution leading to specific pathogenesis.

Plague 7 - Rapid climate change.

Plague 8 - An invasion of alien animal species.

Plague 9 - Social disturbance.

Plague 10 - A high incidence of fatal cases manifested in a susceptible population.

The Torah addresses fundamental questions and speaks in "essential" language, not "conventional" language. My interpretation of each Egyptian plague was presented to the conference as a transitory stage in the development of an epidemic. This was intended to support such a radical view. Below is my summary of

---

[25] J. Marr and C. Malloy, C. *An epidemiological analysis of the Ten Plagues of Egypt.* Caduceus, 1996, vol. 12, no. 1, pages 7-24.

these points:

Plague 1 "Bloody waters" – *dam* in Hebrew (Exodus 7:19–24).

The emergence of diseases can be attributed to various factors. Still, the source creating this emergence is almost always a dramatic change in the environment on a global and/or regional scale. The leading role of ecological catastrophe in emerging epidemics is still seeking acceptance, but it was clearly articulated in Exodus. Scientists can rely on biochemical analysis to confirm ("yes, this is blood") and even conduct a blood analysis. Instead, the Torah speaks about blood's "likeness." To illustrate this point, I showed a photograph of a river in Siberia that turned into "bloody water."[26] Surely, it was not blood—the body fluid that delivers nutrients and oxygen to people and animals. Two major facts related to the phenomenon. The first is that the red color comes from the large quantity of iron that occurs naturally in the ground in that region. The second is that it came from a chemical leak from a nearby metallurgical plant. The first plague was an environmental catastrophe: the river turned red, the water became undrinkable, fish died, and people suffered severely.

Ecological disaster is not just a factor affecting the epidemic--this phase is essential to plague development and represents its first stage. This is not about gradual changes in the environment but rather radical changes through the creation of novel ecological situations. The result may be the emergence of infectious diseases caused by mutated variants of bacterial species or their radical transformation. The formation of microbial variants with properties that potentially lead to massive epidemics can result from the dramatic shuffling of all components of the environment.

Plague 2 "Frogs" — *tzefardea* in Hebrew (Exodus 8:2–10).

In Exodus 8:2, "*the frog-infestation ascended and covered the*

---

[26] National Public Radio,
https://www.npr.org/sections/thetwo-way/2016/09/08/493139519/a-siberian-river-has-mysteriously-turned-blood-red.

land..." What role do frogs play in plague or other infectious diseases? Ehrenkranz and Sampson argued that increased water temperature and toxic river environment forced "frogs to flee to land," using the words from the Bible.[27] This phase represents the emergence of a newly developed parasitic system as a result of a change in the ecological niche. During its life cycle, an organism's role in the environment (its niche) may change. In Chapter 18, I discussed the importance of ecological niche modeling in the prediction of infectious diseases. Pathogens co-evolve with their main animal hosts, and when those animals occupy a new ecological niche, it can lead to the formation of new relations between pathogens and their hosts. When an ecological niche is suddenly expanded, it can create a strong risk of emerging pathogens spreading.[28]

If the first "Egyptian plague" led to the development of the property that we define as "pathogenicity," the frog story can teach us about changes that accelerated the formation of an "elementary parasitic system" with a primary host (frogs in this case) as a result of co-niche construction. Frogs are the model by which we can demonstrate this idea. Human activity, including anthropogenic modification of the environment, leads to the construction of new ecological niches.

Plague 3 "Vermin" — *kinim* in Hebrew (Exodus 8:13–14).

The third plague introduced the appearance of arthropods required to transmit vector-borne diseases such as plague. The land was hit by an infestation of insects described commonly as "lice." "*The lice infestation was on man and beast; all the dust of the land became lice.*" However, those can also be identified as fleas or other small insects; it is not easy to distinguish lice from fleas for a non-specialist.

The main point we can learn from this is that one of the essential

---

[27] N. J. Ehrenkranz and D.A. Sampson, *Origin of the Old Testament plagues: explications and implications.* Yale Journal of Biology and Medicine, 2008, vol.81, no.1, page 35.

[28] E. Fodor, *Ecological niche of plant pathogens.* Annals of Forest Research, 2011, vol.54, no.1, pages 3-21.

parts of a plague phenomenon is vector transmission. Soon after the identification of the etiological agent of plague, Paul-Louis Simond, a French physician, demonstrated that rat fleas played the most important role as a vector for plague transmission between animals. Thus, the first Egyptian plague describes the origin of microorganisms' potential pathogenic properties. The second plague describes the construction of an elementary parasitic system. The third plague teaches us the necessity of transmission mechanisms in epidemic evolution.

Plague 4 "Mixture of wild beasts or flies" — arov in Hebrew (Exodus 8:20–27).

The word, arov, is generally used to signify a "mixture." Many interpret it as a mixture of various living creatures. While some authors define it as a mixture of wild beasts (including Rashi, a medieval French rabbi who is considered the most comprehensive commentator on the Torah), others describe it as a mixture of flies.

Ibn Ezra, a Torah commentator who lived in Spain in the twelfth century CE, said it signified beasts "mixed together," such as lions, wolves, bears, and leopards. This "plague" may have consisted of more than one kind of animal, and as some suggest, insects (animal ectoparasites) were possibly part of the mix.

There are many scholarly publications dedicated to the role of biological diversity in infectious diseases' emergence.[29] For example, in the U.S. Southwest, all native rodent species are very sensitive to plague infection and quickly die from it. A certain level of diversity of rodent communities is required for the continuous transfer of plague from one species to another.[30] Biological diversity plays a critical role in the emergence of infectious diseases, as indicated by the fourth plague.

---

[29] J.C. Maillard and J.P. Gonzalez, *Biodiversity and emerging diseases.* Annals of the N. Y. Academy of Sciences, 2006, vol.1081, pages 1-16.

[30] Kenneth Gage and Michael Kosoy, *Natural history of plague: perspectives from more than a century of research.* Annual review of entomology, 2005, vol.50, pages 505–528.

Plague 5 "Pestilence" — *dever* in Hebrew (Exodus 9:1–7).

In the fifth plague, animals suffered massive epizooties. Plagues may persist unnoticed in their natural environment for long periods but suddenly manifest in mass animal die-offs. As said in Exodus 9:6: "*all the livestock of Egypt died.*" Plague can cause the death of many domestic animals, e.g., camels, goats, and pigs, but also wild animals. The devastating impact of the plague on wild animal populations has raised conservation concerns over the past few decades.

Plague 6 "*Boils*" — *shkhin* in Hebrew (Exodus 9:8–12).

The sixth plague indicates the stage when microbial evolution leads to the development of pathogenic properties that cause specific clinical symptoms. In the Exodus story, it was the first plague that caused harm to the bodies of the Egyptians; "*it became boils and blisters, erupting on man and beast.*" *Shkhin* has also been translated as "cutaneous disorders characterized by marked inflammation."[31] In most cases, a bite by a plague-infected flea leads to its inoculation into a person's skin. The bacteria migrate through cutaneous lymphatics to regional lymph nodes, where they rapidly multiply, causing the destruction and necrosis of the lymph node and subsequent septicemia. This can quickly lead to shock, fever, chills, and the development of acutely swollen, tender lymph nodes, or "bubos" (this is why it is called "bubonic plague").

Plague 7 "Hail" — *barad* in Hebrew (Exodus 9:13–26).

The seventh plague symbolizes rapid climate change. Exodus 9:17 says, "...*rain and a very heavy hail, such as had never been in Egypt.*" The influence of climate change on infectious diseases has become a very significant topic recently. There has been a special focus on the role of climate change as a driving force in vector-borne diseases. Ehrenkranz and Sampson (2008) explicitly analyzed the role of local

---

[31] L. Hoenig, *The plague called "shechin" in the Bible.* American Journal of Dermatopathology, 1985, vol.7, no.6, pages 547-548.

climate change in plague as described in the Torah.[32] They proposed that the root cause was an aberrant El Niño–Southern Oscillation connection that brought progressive climate warming along the ancient Mediterranean littoral, including biblical Egypt. This, in turn, initiated the serial catastrophes in the biblical sequence—in particular, arthropod-borne and arthropod-caused diseases.

Plague 8 *"Locusts"* — *arbeh* in Hebrew (Exodus 10:3–20).

The eighth plague symbolizes the stage characterized by alien species invasion. *Arbeh* in Hebrew also means "many," and the locusts came in great numbers. "The locust-swarm ascended over the entire land of Egypt, and it rested in the entire border of Egypt" (Exodus 10:14). One of the important questions when investigating ecology and the emergence of infectious diseases is whether there have been invasions of alien species of mammals, birds, and insects into a new geographical area. Overall, invasions of locusts commonly lead to devastating natural disasters that have been feared throughout history.

Another notorious example in most countries is an invasion of rats. Rats of the genus *Rattus* originated in Southeast and Central Asia, but these animals traveled by ships to most countries[33] A little more than a hundred years ago, rats (*Rattus rattus* and *Rattus norvegicus*) arrived in some US seaports (San Francisco, Los Angeles, Seattle, Galveston, and New Orleans) and brought the plague with them. Currently, plague still presents a risk in some countries through the movement of infected rats and fleas from rural areas to cities. Animal migrations are expected to enhance the global spread of pathogens. The potential for animal migrations to disperse pathogens across large geographic territories has resulted in a growing number of investigations of the interactions between

---

[32] N.J. Ehrenkranz and D.A. Sampson, D. A. *Origin of the Old Testament plagues: explications and implications.* Yale Journal of Biology and Medicine, 2008, vol.81, pages 31-42.

[33] Michael Kosoy et al., *Aboriginal and invasive rats of genus Rattus as hosts of infectious agents.* Vector Borne and Zoonotic Diseases, 2015, vol.15, no.1, pages 3–12.

migration and infections.[34]

Plague 9 *"Darkness"* — khoshech in Hebrew (Exodus 10: 21-29).

Social distortions are associated with the ninth stage of plague development. It was written in Exodus 10:22: "A thick darkness throughout the land of Egypt…" The word *khoshech* is commonly translated as "darkness," but can also be understood as misery, destruction, death, ignorance, sorrow, or wickedness. The darkness described in the Torah could be understood not to refer to physical darkness but rather to a mass psychogenic disorder or epidemic hysteria. This clearly relates to "social darkness," when everything "goes black" and "No man could see his brother" (Exodus 10:22). Social darkness is accompanied by wars, public distortions, and societal collapse. Egyptians could not see each other; their vision was blocked. Modern epidemiology of plague and other zoonotic diseases (carried by animals) has not ignored social processes in epidemics. The crucial role of wars, famine, evacuation of people, international trade, the structure of urban populations, governmental policies, and many other societal factors is obvious in many cases.

Plague 10 "Death of the firstborn" — makat bechorot in Hebrew (Exodus, 12:29–30.)

Only during the tenth and last phase was the plague manifested as an epidemic in the common sense of this word: "There was not a house where there was no corpse" (Exodus 12:30). The literal reading of the word, *beachfront,* as "firstborn," can be misunderstood by limiting this definition to the special status of the firstborn. From an epidemiological perspective, it would be more constructive to emphasize a different definition of those who appear first, e.g., pathfinders, pioneers, and early explorers.

These are the first people exposed to danger, jeopardized, and at risk; they are vulnerable to infections. From the vantage point of modern epidemiology, we can assume that those people are

---

[34] S. Altizer et al. *Animal migration and infectious disease risk.* Science, 2011, vol.21, no. 331, pages 296-302.

"immunologically naïve," as they have never been exposed to the newly introduced pathogen. Therefore, the death of the firstborn could mean a higher rate of plague deaths among individuals susceptible to the infection.

Interestingly, Robert Haralick, a professor at the City University of New York, argued that the sequence of the ten plagues corresponds to the sequence of the ten Sefirot. This is starting at the bottom of the tree, at Malkhut, and moving up to the top of the tree, Keter.[35]

The narrative of the Ten Plagues in Exodus is a foundational story, not a textbook on epidemiology. Nevertheless, an epidemiologist can learn from it on the condition that the reading is not limited to its literal meaning. The Bible cannot replace scientific books. If people study infectious diseases, they must learn from many books on general biology, ecology, microbiology, immunology, and epidemiology.

Here are two key lessons from the "Ten Plagues" story. First, time measured by the development and transformation of epidemic phases may provide meaningful principles for comprehending infectious diseases' epidemiology, evolution, and ecology. Second, most of the listed phases are invisible from the point of conventional epidemiological practice. Plague, like other diseases, is not just a collection of individual cases of disease caused by a specific bacterium. There are multiple and various processes that precede massive epidemics; those processes should not be ignored and can be used for early warning. A plague is a chain of developing "invisible stages" that leads to massive destruction of human populations, known as epidemics.

For some reason, modern epidemiology started counting and analyzing cases of infectious diseases and deaths related to them from the point when these cases were first reported. Epidemiologists may

---

[35] Robert Haralick, *The 10 Plagues and the 10 Sefirot.*
http://www.kabbalah.torah-code.org/kabbalah/plague_sephirot.pdf

or may not refer to some natural and social circumstances that happened before an epidemic, but they do not consider them components of the epidemic. The crisis described in Exodus clearly falls under the category of "epidemic," although all previous stages described in the preceding lines and chapters remain beyond a conventional definition of epidemics.

"Footprints" of Time

In Kabbalistic Biology, the Sefirot plays a pivotal role in shaping the temporal dynamics of existence. In this exploration, we embark on a journey to understand how various activities of life are intrinsically linked to when specific Sefirot manifests at different stages of time. Temporal changes relate to the "variation of emanations" represented by the seven lower Sefirot. Each Sefirah embodies distinct qualities and energies, and their emergence at different times imbues life processes with unique characteristics.

This encompasses Nature's ordered phases of change, variability, and development. By identifying signs ("footprints") that signify these changes, we can measure and understand time in the biological world. Time is viewed as both a function and a direction.

While studying biological systems, we cannot abandon conventional time based on external astronomical criteria. We use timers in the laboratory for diagnostic tests. We use clocks to measure the duration of animal interactions and calendars for estimating population dynamics and making recommendations. The point is that such external time is not sufficient to measure complex and often hidden stages in the functioning of organisms, species, and ecosystems.

When biologists discuss biological time, they do not refer exclusively to clocks ticking or the passing of days on the calendar. Instead, they dive into a profound aspect of existence, one that underpins the very essence of life itself. Biological processes can be analyzed not only according to a system of external astronomical time. However, it can also be used to define an internal time scale specific to each biological process.

Chapter 20

Eighth Dimension: Invisible Pathways of Hebrew Letters

There is a profound secret hidden within the ancient Hebrew alphabet. This secret is a connection that runs very close to the manifold and intricate structure of organic life. Each of the 22 letters of the Hebrew alphabet possesses a distinct vibrational frequency and symbolic significance. These letters are not arbitrary symbols; they are energetic blueprints that resonate with the very core of existence. When combined in various sequences, they unveil the hidden secrets of both visible and concealed patterns in organic life. A connoisseur of the Hebrew language, Samuel Avital, wrote in his impressive book, *The Invisible Stairway:* "As a practical system, [Kabbalah] is based on 22 sacred letters which hide ... invisible cosmic powers and various manifestations of the Universe Laws." [1]

To avoid venturing into universal scales and philosophy, we shall concentrate on specific junctures in biological investigation. This focus will require using Hebrew letters as symbols that define specific connections between the investigator and the biological subject under scrutiny. These symbols will serve as indispensable signposts in our quest to unravel the mysteries of life's intricate tapestry. Such

---

[1] Samuel Avital, *The Invisible Stairway: Kabbalistic Meditations on the Hebrew Letters.* 2003, page 44.

knowledge will ultimately aid us in forging a deeper connection with the living systems we seek to comprehend.

The Alphabet of Life

One might wonder why biologists (myself included) who do not speak Hebrew, would delve into the significance of Hebrew letters. How can letters reveal an independent dimension?

This is a reasonable question. Fortunately, during the second half of the 20th century, biology provided a compelling argument for the notion that letters can offer entry to an independent dimension. I refer to the discovery of the "Genetic Code," the set of rules used by living cells to translate the information encoded within genetic material into proteins. Since that discovery, nobody can deny that the order and arrangement of DNA or RNA sequences of nucleotide triplets, or *codons*, serve as the blueprint, the very essence, of an organism's internal nature.

Imagine for a moment the entirety of biological diversity, ranging from the simplest bacteria to the majestic elephant. Each of these life forms carries a "genetic code," a sequence of letters unique to their species. This code defines their characteristics, functions, and existence. It is a testament to the universality and precision of the genetic code that such vast complexity can be distilled into a language of just four nucleotide bases: adenine (A), cytosine (C), guanine (G), and thymine (T) in DNA, or uracil (U) in RNA. In our laboratory jargon, we refer to each nucleotide base as one of these letters.

The discovery of the genetic code unveiled a realm where letters—the nucleotide bases—transcend mere symbols. Indeed, only four letters become the construction blocks of both biological diversity and the existence of Life itself! In this sense, the genetic code is a testament to the power of letters to provide an independent dimension. In this biological makeup, every word, sentence, and paragraph is a unique genetic sequence that defines and shapes the living world. The internal structure of a biological organization can be measured by analyzing genetic sequences (aka letters). For example, by comparing DNA sequences of several bacteria, we can talk about their genetic similarity, common traits, and even evolutionary adaptations to the environment.

In this chapter, we go even further. In fact, instead of four letters of the genetic code, we can use twenty-two letters of the Hebrew alphabet! What relevance do these 22 letters hold for our work? The answer lies in the fact that these letters serve as the basic tools for us to comprehend life as a biological system. They offer guidance right from the outset, addressing how to approach a biological subject.

The letters also address which questions to pose when exploring this intricate natural system, and how to position the biologist in the pursuit of understanding. The combination of these letters can help us comprehend both visible life patterns coded by nucleotides and invisible patterns that we have discussed throughout this book. "Genetic determinism" (an extreme interpretation of genetics) suggests that all living organisms express themselves implicitly through their DNA, as early geneticists suggested. Following the discovery of epigenetics, the scientific platform discussed in Chapter 18, most biologists would feel uncomfortable with this reductionist perspective. The genetic code, with its four "letters," cannot describe invisible connections between genetic makeup and environmental conditions in which organisms live. However, we have such an opportunity with the "twenty-two letters" as a code!

In the most ancient Kabbalistic book, *Sefer Yetzirah,* we read, "Two stones build two houses; three stones build six houses; four stones build 24 houses; five stones build 120 houses; six stones build 720 houses; seven stones build 5,040 houses; from here on go and calculate that which the mouth cannot speak and the ear cannot hear."[2] According to Aryeh Kaplan, a stone is a letter, while a house is a word -- a particular combination of letters. Two stone letters build two words, and six letters make 720 words. Each letter permutation becomes a house – for a house is a container that holds meaning.

---

[2] Aryeh Kaplan, "Sefer Yetzirah." 4:16, page 190.

Hebrew letters build and enliven reality the same way DNA-encoded letters build and define the qualities of a living body. Rabbi Yitzchak Ginsburgh noted that "the twenty-two letters of the Hebrew alphabet are reflected in the number of chromosomes in a human seed."[3]

Nosson Scherman, the general editor of Mesorah Publications, presents an analogy to clarify how endless combinations of letters create a diversity of life. He likens this concept to the way chemical elements combine to form various compounds, each with distinct properties. Just as the combination of hydrogen and oxygen yields water, each with unique properties, the combination of letters in different sequences creates Life's myriad forms and functions.

This analogy highlights the complexity and elegance of Life's "chemical formulae," where even subtle variations in the "elements" (letters) can lead to vastly different outcomes, much like in chemistry.[4] We can compare letters to the elements of the Periodic Table--another very small set of fundamental elements from which the entire universe is created by combining and arranging them in different combinations.

The Essence of Letters Beyond Communication

The letters of the Hebrew alphabet have been a focus of Jewish mysticism for millennia. Samuel Avital warned, "The study of Kabbalah demands a perfect knowledge of what these letters represent in the essence of being... There is no Knowing of the Kabbalah without the knowing of the Hebrew letters."[5]

To begin our exploration, it is essential to grasp the unique characteristics that set Hebrew apart as a language. Every language, regardless of its origin, assigns specific terms to various objects and concepts. These terms are constructed by using combinations of letters. Take, for instance, the English language. In the past, it was

---

[3] Yitzchak Ginsburgh, "The Hebrew Letters." Gal Einai Publications, Jerusalem, 1990, page 6.
[4] Cited from Michael Munk, The Wisdom of the Hebrew Alphabet. 1986, 240 pages.
[5] Samuel Avital, The Invisible Stairway. 1982, page 120.

universally agreed to refer to a common domestic rodent as a "rat." In written or spoken form, this term is crafted by arranging the letters "R'" "A'" and "T" in a particular sequence. The scientific community reached a consensus to employ the Latin name, *Rattus,* as the scientific equivalent.

As a biologist, my responsibility involves categorizing organisms, species, and functions. This often sparks debate among scientists, but the biologist's primary concern is to determine how effective these labels will be in simplifying communication among colleagues. This matter leans more toward linguistics than biology, and it does not fall within our current discussion.

Kabbalistic biology uses letters differently. Here, letters represent entities that bear a close resemblance to the subjects of biological investigations. These can be invaluable tools for biologists to use in their daily scientific pursuits. Still, it is crucial to understand that the nature of the relationship between an investigator and a biological object might remain enigmatic.

Jason Shulman, the founder of "A Society of Souls," a group dedicated to the awakening of the human spirit through the work of Integrated Kabbalistic Healing, emphasizes that each Hebrew letter is intrinsically linked to a particular condition or quality.[6] In Hebrew, unlike many languages classified as "conventional" by Samuel Avital, each letter carries a specific energy, force, meaning, and direction. When these letters are arranged in a particular sequence, they form a word. The meaning of each word is inherently intertwined with its constituent letters. Consequently, Avital describes Hebrew as "an essential language."

In the fundamental book, *The Hebrew Letters*, Rabbi Yitzchak Ginsburgh explains that the letters "appear at each of these three levels: as the energy building-blocks of all of reality, as the manifestation of the inner life-pulse permeating the universe as a whole and each of its individual creatures ... and as the channels

---

[6] Jason Shulman, *Kabbalistic Healing: A Path to an Awakened Soul.* 2004, 192 pages.

which direct the influx."[7] Menachem Mendel Schneerson, the Lubavitcher Rebbe, believed that the letters of the Hebrew alphabet were the "building blocks of Creation." These are the generative units that comprise the names of creatures, objects, concepts, and so forth.

In Kabbalah, the Hebrew letters are both units of energy and channels for the "Life Force" as both a force and a container of the force. This situation is akin to Quantum Theory, where elementary physical particles can be seen as both particles or waves, depending on how one is assessing them. According to Kabbalah, the fact that words possess the same root letters, even if in different combinations, indicates that they are inherently linked.[8]

The Power of Triads: Three-Letter Roots

American physicist and biophysicist George Gamow postulated that sets of three bases (triplets) must be used to encode the 20 standard amino acids used by living cells to build proteins. This would allow a maximum of 64 amino acids. Gamow named this DNA/protein interaction (the original genetic code) the "Diamond Code" because it depended on the diamond-shaped cavity formed between four nucleotide bases in DNA. An amino acid would fit snugly into a diamond in a stereospecific manner. The code relies on amino acids and DNA interaction. It is a triplet code: only three bases are significant since the fourth is automatically specified by base-pairing rules.[9]

A lasting gift from working on the coding problem was that it made the concept of information *central* to biology. Interestingly, Gamow possessed great enthusiasm that touched everyone and made the breaking of the code a uniquely social enterprise. He founded the so-called "RNA tie club," with twenty full members (one for each amino acid) and four associate members (one for each base). Club

---

[7] Yitzchak Ginsburgh, *The Hebrew Letters*. 1990, page 3.

[8] Cited from Joseph Burke and Stanley Schneider. *Centers of Power*, 2008, pages 152-155.

[9] Brian Hayes, *Computing Science: The Invention of the Genetic Code*. American Scientist, 1998, vol. 86, no.1, pages 8–14.

members were invited to communicate with each other about progress on the coding problem.[10]

The triplet-based structure of the genetic code draws a parallel with the concept of three-letter roots found in Hebrew words. These three-letter roots serve as profound metaphors to help us understand biological processes, which themselves often unfold as triads. Typically, each root of the Hebrew word consists of three letters. It can be less or more, but regardless, it is distinguished as having three parts—usually separate letters. Each letter within this triadic structure holds unique significance and contributes to the overall meaning and understanding of natural phenomena. Understanding these elemental traits represented by the letters is crucial for us to be able to understand and describe the complex journey through our living world.

Samuel Avital teaches that the structure of each Hebrew word consists of three components, which define stages in the exploration of the word's meaning. He calls such an exploration, "a journey." I consider myself lucky to have taken such journeys under Avital's guidance each week for fifteen years!

The first letter indicates the departure of the journey (origin) or object. The middle letter indicates the way (process) or the traveler. The last letter represents the destination (goal) or the expression. The same structure can be used to describe any biological system as a process and, equally important, as a description of the research project attributed to the investigation of this system. The letters within each word relate to each other in a certain way, and so do elements of each biological system:

1. **Origin** - Just as a Hebrew word's meaning derives from its three-letter root, life processes trace their genesis back to a source, an origin from which they sprang forth. This origin, represented by the first letter of the root, signifies the beginning of a journey.

---

[10] Vidyanand Nanjundiah, *George Gamow and the Genetic Code.* 2004, page 48.

2. **Traveler** - Life is a continuous voyage and the traveler, indicated by the second letter of the root, represents the dynamic journey itself. This traveler navigates the complexities of existence, manifesting the growth, transformation, and adaptation inherent in all living organisms.

3. **Destination**- Every journey has a purpose. The destination, symbolized by the third letter of the root, signifies the culmination or result of a biological process. It represents the fulfillment of a purpose or the realization of a particular state or condition.

Thus, Hebrew letters represent elementary blocks that can be intelligently used to construct our search as scientists and to comprehend the results we obtain.

Letters as Semiotic Signifiers

Beyond mere linguistic characters, Hebrew letters are more than a collection of individual symbols. In Kabbalah, these letters are seen as fundamental building blocks of the world! They serve as semiotic signifiers that convey profound meanings. The Hebrew word for "letter" is *"ot,"* which, as we learned in Chapter 17, also means "sign." Here, we will explore the idea that Hebrew letters are not only visual and auditory symbols but are also powerful carriers of meaning.

In *Sefer Yetzirah*, it is said that the entire world is constructed from the 22 foundational letters, *otiyot yesod,* in Hebrew.[11] These letters are believed to be the building blocks of Life, with each having a unique role and significance. In this context, each letter is a "signifier" that represents a specific aspect of reality, and "signified" is the meaning it conveys.

Rabbi Yitzchak Ginsburgh, a renowned Kabbalist, provides insights into this concept. Ginsburgh says each letter has distinct

---

[11] Arieh Kaplan, *Sefer Yetzirah.* 1997, page 27.

parameters, such as form, name, and number.[12] Each letter's form or visual structure represents its manifestation in Asyah, the physical realm. The name of each letter is its representation of the world of Yetzirah, the world of formation. Each letter is assigned a numerical value, which is significant in the world of Briah, the world of Creation. The numerical values of words and phrases reveal even deeper connections and insights.

Each of the 22 Hebrew letters possesses three distinct creative powers that traditionally represent three levels: "energy," "life," and "light." According to Ginsburgh, the Alef-Beit letters appear at each of these three levels: [13]

1. As the energy building blocks of all reality.
2. As the manifestation of the inner life pulse permeates the universe as a whole and each of its creatures ("pulsing" through every created being, instantaneously, into and out of existence).
3. As the channels that direct the influx of reveales energy into created consciousness.

The Meaning of Hebrew Letters

To understand the full scope of how Hebrew letters function as semiotic signifiers, we must consider the various parameters through which they convey meaning. Each letter of the Hebrew alphabet is believed to carry inherent meaning. When combined to form words, they offer profound insights into the nature of the concepts they represent. This perspective asserts that the "true meaning" of a Hebrew word can be deciphered from its constituent letters' meanings.

The tradition of interpreting the meaning of Hebrew letters finds its origins in the *Sefer Yetzirah*, the most ancient Kabbalistic book. Commenting on the text of that book, Arieh Kaplan pointed out that

---

[12] Yitzchak Ginsburgh, *The Alef-Beit: Jewish Thought Revealed through the Hebrew Letters.* 1992, page 10.

[13] Ibid., page 3.

while ordinary thought is composed of words, and these words are made up of letters, "These are not physical letters, but mental, conceptual letters. These conceptual letters, however, are built out of "Voice, Breath, Speech."[14]

The *Sefer Yetzirah* divides the letters into three categories: three Mothers, seven Doubles, and twelve Elementals. The three letters: "Alef," "Mem," and "Shin" are called "Mothers" because they are considered to be primary letters, derived from Understanding (Binah). Alef is the first letter in the Hebrew Alphabet, while Mem is the middle letter, and Shin is the second from the last. The *Sefer Yetzirah* states that Mem represents water, Shin represents fire, and Alef represents air. The *Sefer Yetzirah* also explains that the Mothers are manifestations of the stages in a full cycle ("Year"). It says, "Three mothers (Alef-Mem-Shin) in the Year are the hot, the cold, and the temperate. The hot is created from fire, the cold is created from water, and the temperate from Breath (air) decides between them."[15]

Kaplan explains that the last letter, Tav, is not used because it is one of the Doubles. The Doubles are letters that can express two sounds: Beit, Gimel, Dalet, Kav, Peh, Resh, and Tov. This property depends on whether they are written with or without a dot, called *dagesh*. The *Sefer Yetzirah* characterizes two possible sounds as "a structure of soft and hard, strong and weak."[16] It's possible to connect the seven double letters to the seven lower sefirot.

The Elementals, or "simple," are the twelve remaining letters, which have a single sound. Those are Hei, Vav, Zain, Heit, Teit, Yud, Lamed, Nun, Samekh, Ayin, Tzaddi, and Quf. The *Sefer Yetzirah* provides particular characteristics to each of the letters: "Their foundation is speech, thought, motion, sight, hearing, action, coition, smell, sleep, anger, taste, laughter."[17]

Kabbalistic scholars and mystics have delved deeply into the

---

[14] Arieh Kaplan, *Sefer Yetzirah*. 1997, page 90.
[15] Ibid., page 148.
[16] Ibid., page 159.
[17] Ibid., page 197.

meanings of Hebrew letters. Their teachings have yielded a rich tapestry of spiritual and metaphysical insights. Works such as Avital's *The Invisible Stairway* and Ginsburgh's *The Hebrew Letter* are invaluable resources for those seeking to understand the profound meanings embedded within each letter.

In Kabbalistic Biology, Hebrew letters are used as an autonomous dimension with a twofold objective. First, we try to define meanings pertinent to biological systems' structural and operational aspects. Second, we seek to clarify how these meanings can inform and guide biological research methods.

In subsequent sections, we will delve into specific activities that are significant in biology, with each activity intricately linked to a particular Hebrew letter. To ease this exploration, we will begin this journey with Beit, the second letter of the Hebrew alphabet. This order is chosen for a specific reason that will be revealed later in the discussion.

Beit - ב

The letter "Beit (ב)" is integral to biology's inception and symbolizes containment. Biology, whether manifested through unicellular or multicellular organisms, hinges on the fundamental principle that life unfolds within an enclosed space. This space acts as a sanctuary, defining a clear boundary between the internal and external realms. Beit symbolizes the cardinal property of segregation between "inside" and "outside." Whether in the form of skin for mammals, chitin for insects, or cellular membranes for cells, this boundary serves as a protective barrier containing the innermost components of a living organism.

Nevertheless, this enclosure is *not* hermetically sealed: it is an interface through which exchanges between the internal and external worlds can occur. This semi-openness is pivotal, enabling the survival, maintenance, and evolution of biological entities, even in the face of the Second Law of Thermodynamics. By its form, the letter

Beit represents an internal space that is closed from three sides but open from one side.

Beit embodies containment and structural organization within living systems. Take, for instance, the microscopic world of an animal cell. Within this seemingly minute unit of life, we find not only an assemblage of membranes, cytoplasm, nuclei, mitochondria, and endoplasmic reticula; deeper within the cell exists an intangible essence that enables it to function as a unified whole. This is where all parts harmonize as a space of the particular biological system.

Beit transcends the cellular realm, affecting every tier of biological organization. Whether examining an individual organism's place within a population, its membership in a particular species, or its role within an ecosystem, the common thread is Beit's differentiation between the internal and external realms. This differentiation, the hallmark of Beit, marks the separation of a biological entity's constituent parts.

Consider, for example, the study of a drosophila (fruit fly). Within this single organism lie distinct organs and numerous cells, each engaged in intricate interactions and specialized functions. Biologists can investigate these components separately or as interconnected contributors to the fly's overall functionality. In this context, the "individual drosophila" serves as a biological container, encapsulating all attributes relevant to this specific organism. When the lens shifts to a broader biological perspective, such as the study of *Drosophila melanogaster* as a species, the container expands even further. The species, defined by its Latin name, represents a collective entity. Here, the species itself functions as a container housing distinctive characteristics inherent to all members of the species.

Gimel - ﬡ

The Hebrew letter, "Gimel (ﬡ)" is the quintessential force behind the biological systems' movement, development, and dynamic functionality. It represents the vital relationship of "giving and receiving" that acts as a linchpin in every biological system. Just as Beit symbolizes containment and structural organization, Gimel defines the dynamic interplay within space and time. Gimel is the conduit through which biological systems exist, evolve, and function.

According to French writer and Kabbalah author, Carlo Suares, Gimel stands as the archetype of all movement.[18] This movement is intrinsic to Life itself, and to understand it is to comprehend space and time parameters within a biological context. Just as Beit delineates the constituent parts of a biological system and their structural relationships, Gimel delineates the way these components interact dynamically across space and time.

Gimel is intimately associated with development. At the cellular level, it transforms nutrients into organic energy. Consider the cell as an example of this: within its confines (Beit), every component at every level of organization, from organelles to molecules, engages in movements, functional activities, and energy production. Gimel orchestrates these processes, propelling them in various potentially viable directions.

Suares defines Gimel as "a double movement." The first facet of Gimel encompasses the unbridled outward flow of life, a force driven by the inherent vitality of biological entities. It encapsulates Life's perpetual dynamism. Conversely, Gimel's second facet delves into inner, controlled equilibrium within biological systems. Inner control is the unceasing response of biological systems to maintain stability amid ever-changing environmental conditions and internal processes.

While Gimel governs the observable movement and activity of a biological entity, it also governs the potential for uncontrolled movement. For instance, Gimel transforms inorganic substances into organic energy within a cell. In this context, the cell remains open to various substrates that may serve as nutrition.

---

[18] Carlo Suares, *Cipher of Genesis*. 2005, 232 pages

Dalet - ד

"Dalet (ד)" signifies the force that defines resistance, stability, and responsiveness within biological systems. Dalet's essence lies in its ability to stabilize biological functions. It restrains and binds the ceaseless movement represented by Gimel, ultimately contributing to the formation of specific cellular organizations. Dalet signifies the "stability of biological functions," a vital aspect of Life's intricate dance.

Dalet signifies both the common inertia of organic structures and their capacity to react to external stimuli—a duality inherent in the living world. Dalet manifests itself as the interaction between structural components of biological matter engaged in states of motion. It serves as the equilibrium point, resisting Gimel's momentum, which represents movement and activity within biological systems. It is a stabilizing force that maintains the integrity and form of these structures, even amid life's continuous dynamism. Dalet functions as the guardian of structural stability, ensuring that the various components within a biological entity maintain their relative positions and functions.

Dalet embodies the biological response to external stimuli. It is the element that enables organisms to react to changes in their environment. It is the force that allows organisms to adapt, thrive, and maintain their functionality within the ever-shifting context of their surroundings. Its role in biology extends to defining the relationship between biological objects and their environments. Dalet encapsulates the external factors that shape the biological entity, exerting control over Gimel's movement and dynamics. In this regard, the concept of "niche construction" within ecosystems acts as a compelling illustration of Dalet's role.

Hey - ה

The letter "Hey (ה)" acts as the agent of organization, the vivifying

512

force, the link between inner and outer structures, and the coordinator of Life's activities. Hey is the expression of Life itself, reminding us that the intricate web of existence is sustained by the harmonious interplay of its various components. To truly grasp the significance of Hey in biology, we need to recognize that it is the agent responsible for optimally organizing, managing, and transmitting the primary energies and information that underlie the universe. Within living organisms, Hey ensures that every element plays its role in harmony with the whole.

Hey is the outward-moving force that breathes life into an enclosed or protected form. In the intricate dance of existence, Hey is the vital link between the inner structure of a biological object and its external environment. It is the conduit through which an organism adapts to the natural cycles and rhythms of the world around it. Hey reminds us that nothing in the biological realm exists in isolation; every living entity is intimately connected to the broader ecosystem and must navigate the cycles of birth, growth, reproduction, decline, and death. Samuel Avital teaches that "Hey is formed by itself. No other letters could cause its existence. It is life and breath. To become self, it needs to be itself."[19]

One of Hey's most remarkable attributes is its role as the coordinator of all Life activities. It ensures that the various elements within an organism work in unison to perform the entire life cycle, from birth and formation to death and disintegration. While Hey coordinates these activities, it simultaneously signifies the separateness of each element by assigning distinct functions within the cycle. In this way, Hey completes the circle of attributes that describe any manifestation of Life, emphasizing the interconnectedness of all living things. In cellular biology, Hey is the messenger that facilitates communication among cell components, allowing them to function as a coherent whole.

---

[19] Samuel Avital, *The Invisible Stairways*. 2005, page 101.

Vav - ו

In biological systems, the letter "Vav (ו)" emerges as a symbol of the fundamental concept of "connection." Vav means the conjunction "and," which is used to connect words of the same part of speech. In biology, Vav links, integrates, and harmonizes life's diverse components. It reminds us that within biological systems, unity emerges through connection. It represents the power of connection that unites all the disparate parts of a biological system.

Vav manifests both external and internal connections within biological systems. The external force of Vav plays a pivotal role in creating order within an organic object, establishing hierarchical structures that differentiate and separate its various aspects. It is the force that assigns each component its place within the broader system, creating a sense of organization and structure. Its exterior aspect delineates the dimensions of a living entity, distinguishing between the influences of sefirot.

Conversely, the internal force of Vav reveals the inherent interconnectedness of the various aspects of the biological object. It represents the idea that these components are not isolated entities but are intricately interwoven, existing hierarchically. Vav's internal aspect brings together all the various elements, forming a unified organic whole. It signifies the internal organization that allows components of each biological system to function harmoniously.

Edward Hoffman, professor of psychology at Yeshiva University, adds another layer of depth to its symbolism. He suggests that Vav conveys space, mass, and physical wholeness. In this view, every complete, self-contained object encompasses six dimensions, including above-below, right-left, and front-back.[20] Vav, therefore, expresses the idea of a biological space as an independent spatial entity. This quality ensures the wholeness inherent in living systems.

Fred Wolf, a world-renowned physicist who researches the

---

[20] Edward Hoffman, *The Hebrew Alphabet: A Mystical Journey.* 1998, 96 pages.

relationship of quantum physics to consciousness, extends the symbolism of Vav by associating it with the role of a fertilizing agent.[21] It represents "male" energy, the primary force behind fertilization and reproduction. This concept aligns with the ability of biological systems, such as cells, to reproduce themselves. Vav signifies the power of copulative reproduction, seed generation, and multiplication.

Zayin - ז

The letter, "Zayin (ז)" represents the very essence of possibilities and the fundamental notion that choices and movements can initiate new realities. Zayin signifies movement and possibilities. It stands as a testament to the idea that Life is in constant motion, and from this movement arises a myriad of potential outcomes. In this sense, Zayin is the spark that ignites change and transformation. It embodies the concept that within every system, evolution and adaptation are possible.

One intriguing perspective on Zayin is its association with indeterminacy, a principle at the heart of quantum physics. Fred Wolf suggests that if the preceding Hebrew letter, Vav, represents the discrete particles of possibility, then Zayin can be likened to the expansive wave of possibilities that emerge from these particles. These are, in essence, the forces of uncertainty and potentiality. Just as Heisenberg's principle asserts that we cannot precisely know both the position and momentum of a particle, Zayin reminds us that Life is inherently unpredictable.

Carlo Suares stressed that Zayin is the life-generating force.[22] It is the moment of vital transformation when the field of all possible possibilities is unlocked. The word Zayin means "weapon" in Hebrew, but it derives from the word meaning "nourishment." This concept resonates with the idea of fertilization, where the union of

---

[21] Fred Alan Wolf, *Mind into Matter: A New Alchemy of Science and Spirit.* 2000, 188 pages.
[22] Carlo Suares, *Cipher of Genesis.* 2005.

elements leads to something new. Zayin, in this context, is the gateway to infinite potential, and it represents the act of the origin of life itself.

Het - ח

The letter "Het (ח)" symbolizes life dynamics in biological systems. Het is the first letter of the Hebrew word for Life – Haim (Chayyim is an alternative spelling). This letter represents the remarkable ability of living organisms to make distinctions and choices among various trends or components within their systems. This ability is fundamental to survival and adaptation, allowing organisms to differentiate between what is living and what is not, what is beneficial and what is harmful.

Het is closely associated with the distribution and storage of energy within biological systems. It serves as a reservoir for substances still unstructured or in the early stages of their evolution. It represents the raw potential of energy, in contrast with the realized possibilities symbolized by the preceding letter, Zayin.

To understand Het more deeply, we can examine the two levels of Life distinguished by Yitzchak Ginsburgh. The first level, known as "essential Life," represents the core vitality inherent in all living entities. The second level, which encompasses Life as we commonly perceive it, involves pulsation and the cyclical process of "run and return."[23] This cycle sustains and nourishes living creatures while allowing them independence to grow and develop. The dynamic pulsation of Life is akin to the constant ebb and flow of energy, characterized by change with repetition.

In the realm of worlds, Het mirrors phenomena where a biological entity, akin to a subatomic elementary particle in quantum mechanics, ascends to a higher energy level. It then returns to its previous stable state, enriched with its new experiences. Het is the embodiment of dynamic life in motion, born from the interaction of

---

[23] Yitzchak Ginsburgh, The Hebrew Letter. 1990, page 122.

Hey and Gimel. It signifies the synthesis of energy in its most primordial form, akin to a seed that holds the sum of possibilities represented by Zayin.

This is the power of choice, allowing organisms to manifest their specific states of being. To illustrate this concept, consider the example of the beetle, which can transition between the larval and adult stages, but cannot exist in both simultaneously. Het finds resonance with the notion of individuality within a collective system. For example, an ant colony is a living system because individuals work together to preserve the colony as a whole. Biophysicist Bernard Korzeniewski said, "*An ant is alive in the manner, say, a liver or a heart is - only as a part of some bigger system.*" [24] *Het,* in this sense, encases the interconnectedness of individual existence and collective survival.

Tet - ט

"Tet (ט)" functions as a balancing point that harmonizes the interplay between force and form. It is guided by the principles represented by the preceding letters, Zayin, Hey, and Gimel. Tet represents the hidden potential that gives rise to the diverse forms of life we observe in the biological world. At its core, Tet represents individuation, the process by which entities become distinct and separate from one another. It bestows the remarkable ability to adapt to unique circumstances, demonstrating a quality of centeredness. This allows for the establishment of a focal point from which all biological activities originate.

Shirley Chambers explains that Tet serves as a connecting bridge between the catabolic and anabolic processes in an animal organism. To maintain an animal's body properly, the metabolic triad strength must grow in proportion to the body's development. [25] Carlo Suares describes Tet as the archetype of primeval female energy, drawing life

---

[24] Bernard Korzeniewski, *From Neurons to Self-Consciousness: How the Brain Generates the Mind.* 2010, 193 pages.

[25] Shirley Chambers, *Kabalistic Healing.* 2000, 216 pages.

from Het and gradually shaping it into structures. It signifies a profound step toward existence, forming the primary cell and acting as a focal point for energy to converge. One of its illustrative aspects is its connection to pregnancy and the power of the female to carry a fetus through gestation.

Tet encapsulates both tendencies in biology: the creation of possibilities and the selection of forms. It unites force and form, serving as the "and" between them. This letter offers a solution to the classic biological dilemma of the "chicken and egg" paradox by embracing the concept of a simultaneous and interconnected existence.

Yud - י

"Yud (י)" is the smallest letter; it is a small "formed" point. Yitzchak Ginsburgh explains: "The secret of this point is the power of the Infinite to contain finite phenomena ... and express them to apparent external reality. Finite manifestation begins from a zero-dimensional point, thereafter developing into a one-dimensional line and two-dimensional surface."[26] Ginsburgh illustrates this thought through the full spelling of the letter's name – Yud (point) – Vav (line) – Dalet (surface).

Samuel Avital says *Yud* is the point where "enough is enough," meaning that a cell becomes a whole cell, an organism becomes an individual, and a biological species emerges as unmistakably different from others. In developmental biology or evolution, Yud serves as the phase switch or bifurcation point. This is a critical juncture that defines a biological entity's trajectory.

In its symbolic role as a seed, Yud symbolizes the growth and transformation of the individual. The onset of a biological organization or process signifies the beginning of the process. During it, a distinct biological entity emerges, representing a pivotal moment

---

[26] Yitzchak Ginsburgh, *The Hebrew Letters*. 1990, page 154.

in the development of an organism when it becomes complete--whole.

In quantum biology, Yud is likened to the smallest discrete quantity of a property involved in an interaction or that represents an entire biological system. It is the initial manifestation of the quantum principle that measurement can alter the qualities being measured. Yud represents the starting point of a biological journey, the spark of existence that sets the stage for growth and transformation.

Kaf - כ

"Kaf (כ)" represents the containment of potential and its transformation into actuality. It embodies the idea that to create something from nothing, it must first be contained. This containment process creates a complete "something" with the potential for various possibilities. While Beit provides containment initially, Kaf takes this potential one step further by initiating a cycle of life and death within the enclosed space. This dynamic power enables hidden potentials to be actualized within this contained system.

Kaf makes a system alive, functional, and capable of development. However, there is one crucial condition: the presence of Yud, the turning point where potentiality becomes reality. This transformation can only occur within a contained space that behaves as a unified system. Kaf doesn't contain separate parts; instead, it encompasses an entire system with a complete cycle. Kaf, in essence, serves as an environment for biological systems, whether organisms or genomes. In this environment, a specific space is created and maintained through discrete moments of actualization by Yud.

Two critical parameters define the function of Kaf: the volume of the container and the frequency of Yud within it. In this context, the container is not just a physical space, but a conceptual one that holds potential for biological functions. Yitzchak Ginsburgh highlights the power of Kaf to continuously actualize potential, a fundamental aspect of functional biological systems.[27] For a biological system to exist and function, it must perpetually maintain and actualize its potential. This concept underscores the idea that potential is ever-present within the actualized system, with Kaf serving as the vessel for this continuous process.

Samuel Avital shares a similar perspective, describing Kaf as a container that follows the function of Beit in the realm of manifested life. This is where possibilities are manifested and contained. Thus, Kaf represents the containment and continuous actualization of potential within biological systems, enabling their existence and functionality.

Lamed - ל

"Lamed (ל)" represents the actual, organic manifestation of biological systems and the controlled, animating motion within them. If Gimel represents the archetype of all movements, Lamed is the embodiment of this concept within biological entities. It encapsulates the motion of biological objects, the cellular activities within organisms, and the dynamic processes of all biological systems. While Gimel embodies potential movement possible in all directions, Lamed signifies directed and purposeful motion.

Lamed refers to the representation of directives within biological systems. In contrast to *Gimel*, it unifies separate components within a biological object by creating a cohesive, singular motion. This is akin to the guidance provided by a nucleus that contains the fundamental elements of organic life, including nucleic acids. *Lamed* serves as the force that combines the beginning, intermediate phases, and completion of biological processes, emphasizing the unity and purpose of these processes.

Haim Shore, a professor at the Ben-Gurion University of the Negev, explains *Lamed* as "highlighting the real intention or direction

---

[27] Ibid., page 168.

an organism or biological object takes."[28] It encapsulates the true purpose of Life within biological systems. It indicates the specific direction that an object or organism follows to achieve its goals.

Furthermore, the letter's direct meaning of "learning" has special significance for biology as a science. In this capacity, Lamed symbolizes the process through which a biologist becomes a biologist! It encompasses the journey of acquiring knowledge, understanding, and expertise in the field of biology. It also reflects the broader theme of purpose and direction in the life sciences.

Mem - מ

"Mem (מ)" symbolizes the circulation of vital elements within and outside biological systems. It represents the duration of biological cycles and emphasizes the interplay between internal and external factors in organic life. This connection operates via a system of positive and negative feedback, ultimately achieving completion. One of its vital roles is to represent the duration required for a biological cycle. It serves as a symbol of the time for life processes to come full circle.

Mem lacks resistance and hardness, unlike Dalet. It is highly responsive to forces acting upon it, much like water's fluidity in fluid mechanics. This responsiveness reflects the vitality of Life, which is enabled by the circulation of organic elements and energy within and outside biological systems. This holistic approach unites the external and internal aspects of a biological entity, emphasizing their inseparable connection. This concept aligns with the awareness of the unity and interdependence of all entities and events, fostering a deeper understanding of the intricate web of Life. Yitzchak Ginsburgh highlights how the embryo "swims" within its nurturing aquatic environment, emphasizing the confinement of the "law of conservation of energy/matter."[29] Everything necessary for an

---

[28] Haim Shore, *Coincidences in the Bible and in Biblical Hebrew*. 2008, page 101.
[29] Yitzchak Ginsburgh, The Hebrew Letters, 1990, page 194.

organism's growth and development comes from its environment in a continuing cycle of nourishment.

Carlo Suares adds depth to this letter's meaning, noting its twofold manifestation: the existence of the container and the Life that is contained. This contrast underscores the importance of the biosphere as the optimal environment for Life's evolution. Organisms actively modify their environment, as discussed in Chapter 18 as the "niche construction," representing the interplay between internal and external cycles.

Within an organism, *Mem* is mirrored in the intricate network of cell interactions. This includes plasma's role in facilitating blood cell circulation and the endoplasmic reticulum's function in protein transport. Genes, although not entirely discrete due to mobile elements, are still part of the "fluid genome" that undergoes constant transformation.

Nun - נ

"Nun (נ)" encapsulates the concept of "continuity" in biology. It signifies a continuum that spans from the point where a biological container holds nothing but the potential for Life, to the point where it transforms into a fully manifested biological system.

At its initial stage, Nun represents simplicity and limitations. However, it is characterized by its inherent readiness to receive and fulfill a specific function. It stands as the point of preparation for full realization, demonstrating its eagerness to embrace its role in life's complexity. This notion resonates strongly with the concept of "pre-adaptation" in biology, where an organism exhibits traits or features that may serve a future purpose, further emphasizing Nun's role in the evolutionary process.

Laibl Wolf beautifully describes the letter's twofold significance as "regeneration and disintegration."[30] Nun embodies the eternal

---

[30] Laibl Wolf, *Practical Kabbalah*. 1999, 272 pages.

cycles of life, mirroring the constant rhythm of birth, growth, blossoming, peaking, and decline that permeates the biological world. In essence, Nun teaches us that our true identity transcends these cycles. This offers a profound perspective on the enduring essence that exists beyond biological temporal shifts.

Samekh - ס

"Samekh (ס)" serves a vital role in biology—it is the indicator of growth and the support system for any seed poised to unleash its full biological program. Samekh acts as a fertile ground for fertilization, nurturing the eternal circle that cradles the seed and transforms it into a fully realized female actuality. It responds to the call of regeneration.

In essence, Samekh represents the nurturing embrace that fuels seed growth, multiplication, fruitfulness, propagation, and fertility. It does so by enveloping the seed in a protective "seal," safeguarding this phase of development. This cycle can either lead to expansion or condensation, depending on the biological context. Samekh, in its role, also underscores the omnipresence of support and protection in the biological realm. This cycle of Samekh's influence is evident in both the seasonal changes of the external environment and the internal rhythms within biological organisms.

Carlo Suares portrays Samekh as the embodiment of female sexual energy in action. This can manifest as the proliferation of biological cells or as pathogenesis, reflecting Samekh's capacity to concentrate biological resistance within physical supports. If Yod symbolizes the seed, Samekh can be likened to the oocyte, an immature egg. In the impregnated cell, Yod represents the condensed Samekh, while Samekh embodies the expanded Yod. Furthermore, its imagery mirrors the process of mammalian fertilization. This is where the interaction between the oocyte and spermatozoa culminates in the creation of a single diploid cell, the zygote. This process initiates the development of an embryo, marking the inception of the organism's existence.

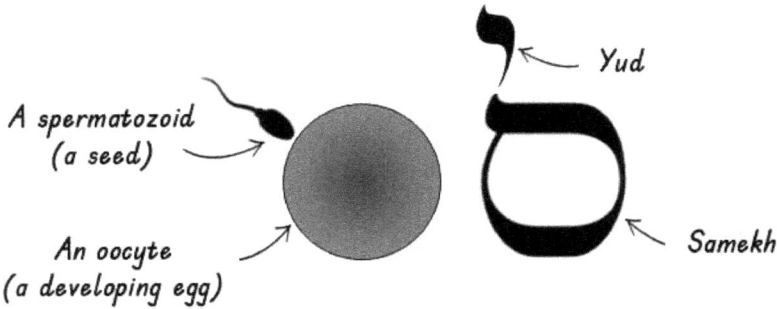

A spermatozoid
(a seed)

An oocyte
(a developing egg)

Yud

Samekh

Figure 64. The letter Yud symbolizes the seed, while Samekh can be
likened to the oocyte, an egg before maturation.

In another biological context, Samekh is reminiscent of the ribosome, often dubbed as the cell's "energy manufacturer." Ribosomes play a crucial role in protein synthesis, a fundamental process in biology. They are formed in the nucleus and subsequently enter the cell's communication network, working tirelessly to produce proteins, the building blocks of life.

The core manifestation of Samekh in biology is circularity and cycling, both within a biological system and in its relationship to the circularity outside the system. Think of an intricate "wheel within a wheel." This concept reinforces the idea that within each living entity exists a self-contained cycle, much like "each flea having its *own* flea," illustrating the nested nature of biological systems.

Ayin - ‬ע

The letter "Ayin (ע)" plays a pivotal role in biology by bestowing the ability to perceive both the "seen" and the "unseen"—and to recognize the distinction between them. Samuel Avital refers to *Ayin* as "vision." Ayin symbolizes the post-fertilization state of "seeing all," with its boundless capacity to explore both the vast world and the minutiae that it contains. Carlo Suares identifies Ayin as "the key to freedom," as it embodies the real probability of a measurable and logical state with definite alternatives.

In a biological context, consider the proteins synthesized by ribosomes under the influence of Samekh. The proteins, essential for Life's functions, eventually face destruction due to various agents, whether physical, chemical, or biological. This state of dissolution, known as "lysis," gives rise to amino acids and lysosomes, which may or may not be salvaged or altered by proteins.

Ayin, in essence, signifies the transition from purpose (functional role) to action, encompassing both time and space as pathways to manifesting a biological entity's function. Using the language of quantum mechanics, we can say that *Ayin* has the power to "break the wave function." In biological terms, it reveals the unity of the entire "object-process," offering insight into the intricate interplay between form and function within living organisms.

Pei - פ

"Pei (פ)" embodies the capacity to express biological functions through communication. This communication can occur at various levels, ranging from interactions *within* an organism's internal systems (e.g., communicating between different organs within the body) to occurring *between* biological systems (e.g., organisms communicating with their environments). Examples of such communication mechanisms include chemical signaling between cells, humoral communication between tissues, and verbal or olfactory communication between animals. Morphologically, Pei can be visualized as openings in cells, organisms, and other biological structures. These openings assist the entry and exchange of "biological signals" between organisms.

Pei distinguishes between biological and non-biological entities. Each biological unit's uniqueness is expressed through a multitude of features and components, many of which may appear contradictory. However, these contradictions can be resolved through communication facilitated by Pai. Circular communication pathways are vital for the survival of biological systems. In biosemiotics, the scientific discipline introduced in Chapter 17, the significance of "biological signs," takes precedence over separate chemical elements and molecules. It not only differentiates between biological entities, but it also distinguishes between investigators, as they employ distinct languages and frameworks to describe the natural world.

Tzadi - צ

"Tzadi (צ)" in biology ensures the harmonious organization of each biological system. This guarantees that every element within the system functions in coordination. Such coordination is observed as a progression that starts at the simplest and most primitive cellular level and culminates in the complete expression of the system's functionality.

Tzadi embodies continuity in the biological system. It harmonizes all the constituent parts to perform their roles seamlessly, like a beautifully choreographed "biodance," a term coined by Larry Dossey. This concept finds parallels in various biological examples, such as the organizing role of the nucleus in cellular activities, the guidance of embryonic development, the transmission of hereditary information through genes, the proper functioning of all organs within a healthy organism, and even the coordinated behavior of birds within a flock!

Tzadi emphasizes how important it is that each element within the system performs its designated role without encroaching on other elements' functions. A balanced system is achieved when all elements function optimally while respecting their counterparts' roles. Thus, Tzadi highlights the structuring energy inherent in the entire system. It ensures that individual elements' activities are subordinate to the overarching system's structure. For instance, if we consider the kidney's role in purifying blood by eliminating waste products, this function doesn't operate in isolation. It functions in full cooperation with other organs, like a violinist in an orchestra. While the kidneys' extensive activity to remove certain substances from the blood might seem beneficial, it could potentially harm the organism, as it might disrupt the delicate balance between the organs.

The function of Tzadi can be likened to the notion of the "holon," the concept discussed in Chapter 12. A holon represents a principle that not only organizes the internal components to perform their appropriate roles; it also coordinates the role of the individual within the larger hierarchy of an organization.

Quf - ק

The letter "Quf (ק)" consists of two separate parts. Each biological object contains its internal structure but also has some components that cannot be separated from its environment and history. It can be seen as analogy with the concept of "extended phenotype" that we discussed in Chapter 8. The biological role of an organism depends on where it is placed. As for the "history," the genome of each organism contains genetic information that is not functional for this organism. However, it does represent previous phylogenetic lineages.

The letter Quf itself demonstrates what proximity is. The curved part of the letter can come very close to the straight part, but cannot touch it. In Quf, straight and curved parts form the body. If they stay too far away from each other, they will be like two separate letters: either Vav and Kaf or Reish and Zayin, but it will not be Quf. At the same time, the curved part is always separate from the straight line; it doesn't touch it, as in Heit. The proximity creates relations.

Curved part
of Quf

Straight part
of Quf

Figure 65. The letter Quf is constructed of the straight and curved parts. The curved part is always separate from the straight line. The proximity creates relations.

Interestingly, Michael Epelbaum, a statistician from Nashville, Tennessee, developed an original theory and method that he designated "Quf." The proposed theory and method enable reproduction, prediction, description, and an explanation of correlated phenomena. It verifies that proximity leads to correlations and that correlated phenomena are nonlinear and multidimensional. This method provides procedures to transform a distance matrix into a proximity matrix designated as a "ק-matrix." There, the symbol "ק" stands for the letter Quf.[31]

This letter represents the duality of "circularity" (non-linear) and "linearity" for each biological object. An object consists of elements strictly dependent on the laws and elements that allow flexibility, circulation, adaptation, and development. Consider an example with a normal functioning range and the concept of a healthy response to the fluctuation from the lowest to the highest limits of its existence.

---

[31] Michael Epelbaum, "A new theory and a new method for temporal, spatial, and other correlated phenomena." Complexity, 2005, vol.11, No.1, pages 36-44.

Reish - ר

The letter "Reish (ר)" extends the concept of a container within the Hebrew alphabet. While the three letters "Beit (ב)," "Kaf (כ)," and "Reish (ר)" represent containers, each has distinct characteristics. Beit is the archetype of concrete containers and has its roots in resistance to life. Kaf, on the other hand, represents the containment of potential and the transformation of it into actuality. Reish, the universal container of all existence, has its origins in intense organic movement at a higher level.

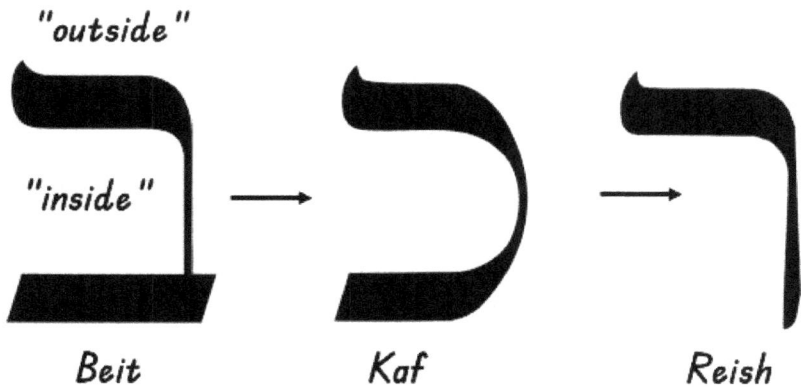

Figure 66. Within the Hebrew alphabet, the concept of a container progresses. Beit symbolizes the cardinal property of segregation between "inside" and "outside." Kaf represents the containment of potential and the transformation of it into actuality. Reish is called the universal container of all existence.

The letter, Reish, literally means "head" in Hebrew. In Biology, this letter indicates the initiation and start of any upcoming process with all advances and challenges. When representatives of an animal species venture into new territory, they enter unfamiliar land and encounter numerous challenges. They are not adapted to the environment, lack their own territory, and may be vulnerable to local predators and infections. These newcomers can be called "poor," the word associated with Reish. Later, some individuals can adapt to the new environment over time and thrive. To be "poor" in this context means experiencing fear, trembling, and being shaken, yet it also indicates openness to any opportunity and gratitude for any gift or sensitivity to any sign.

As Samuel Avital suggests, "...life is based on fruitfulness." Reish represents an exclusive form of container that allows leakage. This concept parallels the biological principle that organisms can do more than receive and contain waste—they can also excrete it. Excretion is the essential process of eliminating waste products and useless materials from an organism. Single-celled organisms discharge waste products directly through the cell's surface. Multicellular organisms employ more complex excretory methods. Animals excrete waste through organs like kidneys and skin, while plants eliminate gases through stomata on their leaves.

Reish signifies the perpetual flow between "yes" and "no," stimulating ongoing cycles and rejuvenation every moment. To initiate a new movement, something must end, like the life and death of cells within an organism or the birth and death of individuals in a population. In this sense, loss becomes the impetus for new life and movement.

The primary function of Reish is its ability to recognize the need for a change of direction. It can determine whether one is still on a straight path or whether the road is gradually turning. A profound illustration of this is the observance of Shabbat. Shabbat teaches us that after a period of continuous movement, a cycle is completed. It is essential to pause, recognize the end of this phase, and prepare for a new turn. The same is true of all natural processes, including biological cycles.

Shin - ש

The letter "Shin (ש)" symbolizes fire, which is essential for transformation and the existence of Life itself. Samuel Avital notes that Shin represents the fiery element required for transformation. It embodies the three pillars necessary for organic life manifesting as "The Triangle Principle." This principle is fundamental to biological life. Avital delves into Shin's fundamental question, which centers on the concept of "change." This raises questions about whether an entity should remain the same or undergo transformation. Shin

530

represents the entire Tree of Life within each object. Its interpretation is based on the central pillar of Shin, which guides the process of change.

Suares says Shin signifies a progressive enlargement of movement, beginning with the uncontrolled functional action represented by Gimel. It moves through the controlled connecting agent, Lamed, extending as far as the universal, Shin. This progression implies a transition from uncontrolled, instinctive action to a more controlled and connected state, eventually reaching a universal or all-encompassing understanding.

Suares also draws a parallel between Shin and the symbol, *psi* (Ψ), which looks very much like this letter. This symbol is used in quantum mechanics to represent the wave function or the *psi* field in extrasensory perception. Shin, in this context, signifies motion beyond the speed of light, suggesting a profound and rapid transformation. This letter also raises questions about the need for change and encapsulates the entire process from thought to action within each biological entity.

Tav - ת

The letter "Tav (ת)" represents the endpoint—the culmination of a cycle and the attainment of wholeness and fullness within a biological context. Samuel Avital characterizes this letter as the culmination of a journey, marking the end of the beginning and the beginning of the end. It signifies the final form of biological function, representing the completion of a process.

Avital underscores that dominion, or control, over biological functions, is a foundational aspect of life. Tav represents the essence of control and mastery over biological processes. In general terms, Tav introduces the concept of "to be AND not to be," highlighting the synthesis of opposing forces. This is in contrast to the classical dichotomy of "to be OR not to be."

Suares elaborates on the triad of Dalet-Mem-Tav, emphasizing their roles in physical resistance within biological structures. These three letters represent different aspects of this resistance, with Dalet symbolizing the maternal waters, Mem representing the origin of Life, and Tav exemplifying the pinnacle of world existence's capacity to resist Life and death.

Tav, in this interpretation, embodies an energy both equal and opposite to Alef, the first letter of the Hebrew alphabet. The interplay between Alef and Tav, with running and returning, is seen as the origin of Life. This cyclic movement reflects the continuous flow between the source (Alef) and its ultimate realization (Tav) in biology. In essence, Tav represents the culmination of complexity and simplicity in biological function. It also underscores the importance of balance in the journey from phenomenon to essence within life science.

Alef - א

"Alef (א)" is the first letter in the Hebrew alphabet. I am presenting it after introducing other letters because Alef unites all of them. It serves as the foundation upon which all other letters and aspects of biology rest. The Sefer Yetzirah describes a sign through which one knows that he/she has attained the status. One must go through the entire array, "Alef with them all, and all of them with Alef."[32]

Alef embodies the principles of unity, power, stability, and continuity. It encourages us to consider biological objects in their entirety, as holistic entities. Researchers often work with isolated molecules or tissues from specific organisms. Imagine a biologist studying a specific organism or delving into DNA or protein molecular details. While this approach is necessary for detailed analysis, Alef encourages us to acknowledge that these molecules and tissues are parts of the larger and integral whole – the organism itself.

This broader perspective is essential for a deeper understanding

---

[32] Arieh Kaplan, *Sefer Yetzirah.* 1997, page 135.

of biological life. It prompts us to view organisms not as isolated entities but as intricate systems with complex relationships to their environments, natural history, and all the interconnected components that define their existence. Alef signifies the essence of any biological object or process. It reminds us that before delving into biology specifics and minutiae, we must first recognize and acknowledge Life's unity and wholeness. While specialization and focused study are critical aspects of biology, Alef demands that we not lose sight of the organism as a whole.

Structurally, the letter, Alef, is a visual representation of profound meaning. Comprising three other letters – two Yuds and a Vav – its image encapsulates essential concepts of biology. The dual Yuds, positioned above and below, symbolize the two fundamental levels of existence inherent in any biological phenomenon.

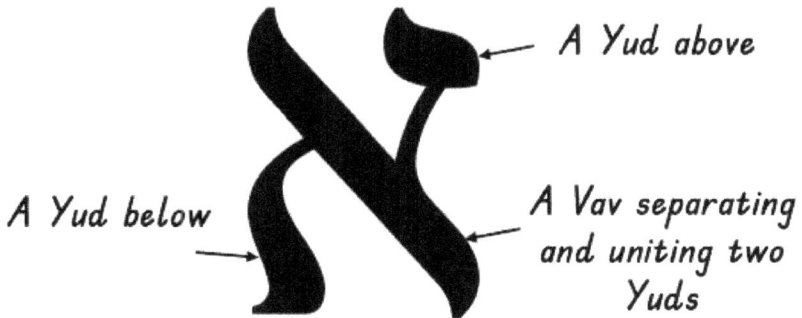

Figure 67. The letter, Alef, is comprised of three other letters – two Yuds and a Vav.

The lower Yud signifies the visible manifestation, the tangible aspects of the biological world. This encompasses physical traits, structures, and processes readily apparent to researchers. It represents biology's empirical, observable nature. Conversely, the higher Yud represents biology's unseen, intangible aspects. It delves into the intricate web of connections, meanings, and relationships that underlie every biological object or process. This level encompasses the hidden complexities, the underlying principles, and the interdependence of various elements within the biological realm. The inclined line, forming the Vav in Alef, serves as a bridge and unifying force between these two levels. It emphasizes that biology should not be fragmented or compartmentalized into visible and invisible aspects. Instead, Alef encourages us to integrate these dimensions, recognizing that the observable and the hidden are inextricably linked. It is at the convergence of these two levels that we gain a more comprehensive understanding of biology.

The role of Alef in the science of life is paramount. Samuel Avital calls Alef, "the foundation of life." It reminds us that a holistic perspective is essential for us to grasp the complete picture of any biological object or process. To truly understand biology, we must consider both surface-level observations and the underlying intricacies, seamlessly integrating the visible and the invisible. Alef serves as a guiding principle, urging us to explore the unity within diversity, the tangible and the intangible. It also recognizes that every biological phenomenon is a complex interplay of these two fundamental levels of existence.

Twenty-two Letters as Pathways Connecting Sefirot

As we discussed in Chapter 11, the number of Hebrew letters is related to the 22 paths, or channels, that connect ten sefirot. Aryeh Kaplan pointed out that the Hebrew word for "paths" in Sefer Yetzirah is *Netivot*. Its singular is *nativ*. However, this word occurs rarely in scripture. Much more common is the word *derekh*. The Zohar states that there is a significant difference between these two words. The word *derekh* is a public road used by many. In contrast, *nativ* is a personal route. It is a hidden path, without markers or signposts, which one must discover independently and then tread upon on its own devices.[33]

There is a correspondence between the type of letters and their position as channels connecting sefirot. There are three horizontal channels, seven vertical channels, and twelve diagonal channels. The three horizontal channels correspond to the mother letters. The seven vertical channels are related to seven double letters. The twelve

---

[33] Aryeh Kaplan, *Sefer Yetzirah*. 1997, page 10.

diagonal channels relate to twelve single letters. Each pathway between a pair of Sefirot has a specific quality.

The origin of the paths has been the subject of much speculation. Z'ev ben Shimon Halevi, the founder of the London-based Kabbalah Society, noticed that some authors put forward the concept that the Tree of Life is a solid whole with the paths between the Sefirot marking the mid-zone division of balanced function. According to such a view, Hesed and Gevurah are separated, but not parted. Others say that the Sefirot were first formed, perhaps like crystals emerging from a cosmic solution, and that the path's patterns were the connecting rays of relationship.[34]

To me, the main lesson provided by the Alef-Beit letters is the awareness of switching between different life stages, as symbolized by the Sefirot. In this regard, the letters can be seen as "doorways" leading from one "room" to another. Samuel Avital wrote in *The Invisible Stairways*:

"The art of passing through a doorway is a common, everyday occurrence. One doesn't give to it a very great significance ... Because of this, there is a reciprocal lack of awareness when shifting consciousness from one state to the next. In order to increase the general awareness of the student on the path of Light, specific attention is drawn to passing through the "door" into the "room," or in other words, crossing the "threshold.""[35]

In order to successfully apply the Alef-Beit letters to comprehend both the visible and invisible aspects of biological processes, one must recognize each letter's distinct functions and maintain a vigilant awareness of their roles.

---

[34] Z'ev ben Shimon Halevi, *Introduction to the Cabala*. 1991, page 95.
[35] Samuel Avital, *The Invisible Stairway*. 1982, page 45.

Chapter 21

Ninth Dimension: Individuality as the Foundation of Biological Organization

"What is a biological individual?" It seems like an easy question. Isn't it a biological object that is autonomous, distinguishable from its environment, capable of maintaining its integrity, and capable of feeding and reproducing? However, when the question is probed more deeply, it becomes clear that none of these criteria apply to all the entities referred to as "biological individuals."

The concept of biological individuality has long perplexed scientists, prompting a search for a more inclusive and nuanced understanding of this fundamental biological trait. Recent years have seen robust discussions in biology concerning how to conceptualize biological individuals. These discussions concern their roles in pivotal processes such as natural selection, adaptation to the environment, and organismic development. In the quest to define individuality in biology, traditional boundaries, and definitions have been challenged by complex, interconnected ecosystems and symbiotic relationships. This chapter delves into the multifaceted nature of biological individuality, drawing insights from Kabbalah to shed light on this intricate phenomenon.

An Individual is Not Always an Organism

It is common for biologists to make no distinction between organisms and individuals. Current thinking in the area assumes that questions about biological individuality, in general, are intimately related to questions about the nature of the organism. Pepper and Herron wrote that "among biologists, the question of what constitutes an individual is usually identical with the question of what constitutes an individual organism."[1] However, defining what counts as a "biological individual" is more complex than simply equating it with organisms.

While organisms have been the primary focus in discussions of biological individuality, it's now recognized that other categories also qualify as biological individuals.

From this perspective, biological individuals encompass:

- Organisms like wasps, whales, and potentially endosymbionts and slime molds;

- Certain parts of organisms, such as hearts, placentas, and plasmids; and

- Some groups formed by organisms, like a zebra dazzle and a colony of bacteria.

This view is often referred to as "organism-centered" because each of its three components relates to organisms in some way. It allows us to recognize a biological entity as an individual, even when we're unsure whether it's an organism, a part of an organism (like an endosymbiont), or a group of organisms (like a colony of eusocial insects). The initial formulation of the organism-centered view doesn't specify which parts of organisms or groups of organisms qualify as biological individuals, nor does it define what constitutes an organism. For example, genes and groups have also been proposed as types of biological individuals that act as selection units.

---

[1] John Pepper and Matthew, *Does Biology Need an Organism Concept?* Biological Reviews, 2008, vol. 83, no.4, page 622.

Additionally, some argue that genomes function as individual entities responsible for encoding organismal development.

Although concentrating on organisms as a foremost category of biological individuals provides valuable insights into the nature of these entities, we must extend our examination to encompass biological individuals more broadly. This enables us to address overarching inquiries regarding the structure of natural phenomena and their role in determining the identities of biological objects in all their diversity. While organisms are the most recognizable biological individuals, defining what truly constitutes an organism is complex.

The concept of a biological individual extends beyond the simple definition of an organism, and it brings with it two main challenges. First and foremost, it's not always *clear* what constitutes an organism. While common sense may identify certain organisms, biology reveals a more intricate picture. Some organisms, like a deer or a dove, are easily distinguishable. But this is not always the case. I will provide broader perspective on biological individuality. Second, some biological entities at levels below and above organisms can also be recognized as individuals. The concept of biological individuality doesn't exclusively apply to organisms. Rather than completely discarding the idea of organisms as biological individuals, let's consider an organism-centered view as a useful starting point to understand the problem of biological individuality.

Nearly everything in biology consists of smaller subunits in a series of hierarchical levels. It is tempting to explore the analogy between organisms and ecological entities. Philippe Huneman, professor at CNRS-Paris, examines what happens when we define an individual by isolating it from all the interactions with other basic entities. Since this still incorporates a level of fuzziness, Huneman calls this "a weak definition of individuality," unlike a strong definition. This requires strong cohesion and integration to allow individuals to become adaptation units in their own right.[2]

A simple look at the natural world reveals an astonishing variety of living entities. From ants and beetles to alligators and elephants, and even microscopic organisms like bacteria and protists, the living world teems with diverse individuals. Whether it's herds of zebras, expansive coral reefs, or vast fungus complexes, each entity can be seen as a biological individual. How do we distinguish natural individuals from the multitude of entities in our conceptual biological landscapes?

The Tripartite View of Organisms

Robert Wilson, professor of philosophy at the University of Western Australia, notes that "The living world is made up of living things, and living things are agents. We think about the dimensions of the living world, and the vague and contentious boundaries to it... When we think of the living world, we think of the individual agents in it, the properties those individuals have, and the relationships they enter into."[3]

Wilson proposed the so called "Tripartite View of Organisms." This perspective posits that organisms are a subset of living entities, central to the biological sciences. Furthermore, organisms are set apart from other living entities by two key characteristics: they belong to a specific reproductive lineage and they possess a distinct type of autonomy.

According to the Tripartite View, any organism is:

1) A living entity at some point in its life;

2) A part of a reproductive lineage where some members can engage in an intergenerational life cycle;

3) And has minimal functional autonomy.

---

[2] Philippe Huneman, *Individuality as a Theoretical Scheme*. Biological Theory, 2014, vol.9, pages 361-373.

[3] Robert Wilson, *Genes and the Agents of Life: The Individual in the Fragile Sciences Biology*. 2004, page 4.

Organisms possess life cycles that allow them to establish specific reproductive lineages. A life cycle comprises replicable stages or events through which a living agent progresses, involving processes defined as "development." Although life cycles follow standard sequences, they can vary considerably among different organisms, such as those undergoing metamorphosis or inhabiting multiple host organisms. Wilson emphasizes that organisms exhibit a minimal level of functional autonomy. They possess a degree of self-control and relative freedom concerning other entities, including fellow agents and their environment. Autonomy implies that organisms act as loci of control, contributing to their distinct identities and life experiences.

Significantly, Wilson's view differs from approaches that rely on single criteria to identify organisms. Instead, he recognizes the complexity of living things. Wilson acknowledges that no single property is universally essential to the category of "living things." Furthermore, his view accommodates heterogeneity within this category, allowing for various life forms, including non-reproducing or malfunctioning ones.

Wilson views organisms' properties as part of an interconnected cluster. It doesn't require every member of an organism's lineage to reproduce or even to have the capacity to reproduce. For example, some social insects have a caste structure, with certain individuals designated as sterile. Nevertheless, these sterile individuals still possess a life cycle and contribute to the lineage's existence. Wilson's main conclusion is that it's vital not to conflate organisms with all living things. Many biological entities, such as cells or organs, satisfy some Tripartite View criteria but do not qualify as organisms.

The "Man-of-War" Dilemma

The biological world's complexities remind us that defining organisms can be daunting. This requires a nuanced understanding of their unique characteristics and developmental processes. But first, let's have a look at some examples that can be used to illustrate the challenge of defining an organism, highlighting the need for a broader perspective on biological individuality.

Counting organisms in the biological world can be deceptively simple when dealing with familiar cases like cats and dogs. However, biology often presents perplexing challenges, as highlighted by Jack Wilson, Professor of Philosophy at Washington and Lee University: "...the same intuitions that allow us to confidently count puppies and tomato plants leave us confounded when attempting to tally colonial siphonophores such as the Portuguese man-of-war."[4]

The Portuguese man-of-war (the Latin name *Physalia physalis*) is a marine hydrozoan found in the Atlantic and Indian Oceans. It resembles a jellyfish, but it is a siphonophore, a colonial organism made up of many smaller units called "zooids."[5] This aquatic animal possesses many traits typical of individual organisms, including the ability to reproduce, a uniform genome, and a coordinated set of parts functioning as a single unit.

Nevertheless, biologists have proposed an alternative perspective due to its unique developmental process where the fertilized ovum buds into distinct structures. This peculiar development raises questions about whether or not it should be considered a colony composed of multiple distinct organisms. All zooids develop from the same single fertilized egg and are, therefore, genetically identical. They remain physiologically connected throughout life, and essentially function as organs in a shared body. Hence, a Portuguese man-of-war constitutes a single individual from an ecological perspective. However, it is made up of many individuals from an embryological perspective.

Coral reefs

Coral reefs serve as a compelling case study when grappling with biological individuality. These awe-inspiring natural wonders,

---

[4] Jack Wilson, *Biological Individuality*. 1999, page 1.
[5] Catriona Munroet al. *Morphology and development of the Portuguese man of war, Physalia physalis.* Scientific Reports. 2019, vol. 9, no.1, page 15522.

although threatened by climate change, hold a special place in our natural world. Comprising of two primary components, calcite deposits and tiny polyps; coral reefs invite us to ponder what constitutes a biological individual. Polyps, like sea anemones and jellyfish, are undoubtedly organisms. Conservation biologists categorize coral reefs, encompassing both polyps and deposits, as living entities that exhibit growth and eventual demise. This characterization raises the intriguing question of whether the coral reefs themselves could be considered organisms.

This perspective leads us to contemplate whether the dependency of coral reefs on polyps negates their status as organisms. Interdependence is common in biology. Polyps lean on single-celled algae called "zooxanthellae" to access glucose, essential for respiration and calcification. Additionally, these zooxanthellae contribute to the vivid pigments responsible for the vibrant hues of living corals. Any reduction in their presence or absence signals a potential threat to coral reef survival. However, even zooxanthellae do not operate in isolation. They secure a vital feeding den by infecting the polyps, showcasing a complex web of dependencies. Coral reefs, therefore, challenge our traditional understanding of biological individuality by highlighting the intricate interactions and interdependencies within the natural world.

Slime mold

Slime molds, the "myxomycetes," are a fascinating group of organisms that challenge the traditional notions of biological individuality due to their unique characteristics. They can exist as unicellular protists or form multicellular structures with many nuclei, blurring the lines between individuality and colony.

There are two primary types of slime molds, "plasmodial" and "cellular." Plasmodial slime molds are essentially giant single cells with thousands of nuclei. They start as individual flagellated cells that swarm together and fuse, creating a massive cytoplasmic structure with numerous nuclei. This mass is referred to as a "plasmodium" and is frequently observed as threads of "slime" on rotting wood. The plasmodium matures into a network of interconnected filaments, which slowly moves as a unit as its protoplasm streams along the network. These plasmodia can be quite large; some species have been recorded to be over thirty square meters in size.

Cellular slime molds, in contrast, spend most of their lives as separate single-celled amoeboid protists. When an individual cell encounters a food source, it sends out a chemical signal that attracts others of its kind, drawing them in until they form a mass capable of moving in an amoeba-like fashion, with each cell maintaining its integrity. The fruiting bodies of cellular slime molds release spores, each of which becomes a single amoeboid cell when it germinates. Cellular slime molds are rarely visible to the naked eye.

Despite their differences, both types of slime molds share a life cycle reminiscent of fungi. When conditions become unfavorable, they form sporangia, clusters of spores that are then dispersed to new habitats to restart the life cycle. This unique behavior makes slime molds valuable for studying how cells interact to create a multicellular organism.

As Raima Larter, a former professor of chemistry at Indiana University-Purdue University who became a writer, put it, "The slime mold cells seem perfectly comfortable with going back and forth between an individual existence and one as a member of a greater whole, but there is no guarantee when you band them together that you will ever be free to break … apart again."[6] Larter thinks that if individual amoebae become a multicellular organism where each cell loses its individuality, they might not survive alone like skin, heart, or bone cells. Those little cells become *dependent* on the life of the new organism in a way that they weren't before as isolated individuals. [7]

---

[6] Raima Larter, *Spiritual Insights from the New Science*. 2021, page 49.
[7] Ibid., page 50.

Giant Fungus

In Chapter 16, we looked at the fascinating case of the colossal giant fungus known as *Armillaris bulbosa,* which was discovered in Michigan's Upper Peninsula. Biologists were astounded to find remarkable genetic similarities among samples of this fungus. This hinted that this extensive fungal network might constitute a single organism spread over a vast area. In the professional and popular press, it has been widely claimed to be the "largest organism on earth." However, this notion raises a pertinent question: *Can we genuinely regard it as a single organism?*

To address this query, we must explore several key aspects. First, we need to ascertain whether this fungal entity forms a continuous biological structure. Is there a seamless connection between its various parts, or are they disjointed entities? Second, we must examine its growth pattern. Does it exhibit a coherent and consistent development pattern akin to individual organisms, or is its growth more sporadic and irregular? Third, reproduction comes into play. Can we detect a unified reproductive mechanism that produces new generations of this fungal entity? Does it follow a reproductive pattern that aligns with our conventional understanding of organisms?

However, it's essential to note that these considerations alone might not resolve the debate. To answer whether the giant fungus constitutes a single organism or not, we must also reflect on our understanding of what defines an organism. In addition, we must reflect on how these entities operate as biological individuals. Scientific investigations and rigorous analyses of the giant fungus' structure, growth, and reproductive processes can shed light on its true nature.

It is by carefully examining tangible evidence that we can refine, adapt, and even challenge our preconceived notions about biological individuality. Sometimes, empirical research provides a more accurate understanding than relying solely on intuitive or common-sense judgments. In this context, the case of the giant fungus is a captivating example of how the boundaries of biological individuality can be blurred and redefined when confronted with complex and extraordinary phenomena.

Aspen Trees: A Subterranean Perspective

The quaking aspen (the Latin name *Populus tremuloides*) is one of the most widely distributed trees in North America, from Canada to central Mexico. To the casual observer, aspen trees appear to be a collection of individual trees with their own trunks, branches, and leaves. It appears as if each tree lives its life independently, growing, photosynthesizing, and reproducing. With their distinct appearances, aspen trees naturally look like biological individuals.

However, the true complexity of aspen trees unveils itself beneath the forest floor. Aspen trees are not solitary entities; they are part of a vast, interconnected network known as a "clone." These clones arise from a single, ancient seedling and propagate through the growth of root suckers. As the root system extends horizontally, new shoots from the roots' nodes, ultimately developing into individual-looking trees. The crucial revelation is that every tree within a clone shares an identical genome, making the trees genetically identical. As all trees in a given clonal colony are considered part of the same organism, one clonal colony, named Pando, is considered the heaviest and oldest living organism on the planet. Pando spans 43 hectares, weighs six million kilograms, and is around 80,000 years old.[8]

The subterranean connection among aspen trees challenges our conventional understanding of individuality. This underground network facilitates nutrient and resource sharing, allows for coordinated responses to environmental stressors, and even enables collective reproduction through seed production. The aspen clone's root system functions as the circulatory and nervous system of the

---

[8] Michael Jenkins, *Fire History Determination in the Mixed Conifer/Aspen Community of Bryce Canyon National Park.* The UW National Parks Service Research Station Annual Reports. 1993, vol.17, pages 31–35.

huge organism, providing communication and resource allocation among its constituent trees. Just as organs work together to sustain an organism's life, individual aspen trees collaborate to ensure the survival and proliferation of the clone.

How many individuals are present when we see a pregnant individual?

This question was formulated by Anne Sophie Meincke, a philosopher at the University of Vienna. Are there one or two? The standard answer: two individuals. This is typically championed by scholars endorsing the predominant "containment view" of pregnancy, according to which the fetus resides in the gestating organism as if in a container. The alternative answer: one individual. This has found support in the "parthood view," according to which the fetus is a part of the gestating organism. Meincke, contemplating the question "one or two," proposes a third answer: "A pregnant individual is neither two individuals nor one individual but something in between one and two. This is because organisms are better understood as processes than as substances."[9]

William Morgan, research associate on the MetaScience project at the University of Bristol, calls this multifaceted challenge the "Fetus Problem."[10] It underscores the need for a more nuanced understanding of pregnancy. Elselijn Kingma, a Professor in Philosophy and Medicine at King's College, London, challenges the conventional view of pregnancy that regards females as mere "containers" for their offspring. Instead, she proposed that, in cases of mammalian placental pregnancy, the "foster" (the post-implantation embryo) is part of the "gravida" (the pregnant organism).[11]

Within the context of pregnancy, Laura Nuño de la Rosa at the University of Murcia in Spain and her colleagues introduced the

[9] Anne Sophie Meincke, *One or Two? A Process View of Pregnancy.* Philosophical Studies, 2021.

[10] William Morgan, *Biological Individuality and the Foetus Problem.* Erkenntnis, 2022.

[11] Elselijn Kingma, *Biological Individuality, Pregnancy, and (Mammalian) Reproduction.* Philosophy of Science, vol.87, no.5, pages 1037-1048.

concept of "historical individuality."[12] These authors argue that this perspective coexists with other views of biological individuality applied to pregnancy, including the physiological, evolutionary, and ecological viewpoints. Historical individuality offers fresh insights that can dispel misconceptions surrounding pregnancy and its complex nature.

Moira Howes, associate professor of philosophy at Trent University, approaches pregnancy from an immunological standpoint. She highlights that both the traditional container model and the notion of the embryo as part of "the mother's flesh" fail to acknowledge the dynamic material relationships between females and embryos. This perspective draws parallels with other intricate biological systems, such as insect colonies or symbiotic organisms. In these systems, clear boundaries between distinct, self-contained, and independent individuals become blurred.

The Concept of a Superorganism

Beyond single organisms, we encounter so-called "superorganisms.". In 1911, Harvard ecologist William Morton Wheeler gave a lecture entitled *The Ant-Colony as an Organism* at the Marine Biological Laboratory in Woods Hole, Massachusetts. In his lecture, he argued that ant colonies are not just analogous to organisms, but that they *are* organisms in their own right.[13] Essentially, a superorganism can be defined as a group of organisms of the same species working together as a cohesive unit to achieve collective goals. These goals can encompass various activities, such as ants collecting food or avoiding predators or bees selecting an appropriate nest site.[14] These are species where the division of labor is extremely specialized, and

---

[12] Laura Nuño de la Rosa et al. *Pregnant Females as Historical Individuals: An Insight from the Philosophy of Evo-Devo.* Frontiers in Psychology, 2021, vol.11, e572106.

[13] William Morton Wheeler, *The ant-colony as an organism.* Journal of Morphology, 1911, vol. 22, pages 307–325.

[14] N.F. Britton et al. *Deciding on a new home: how do honeybees agree?* Proceedings of the Royal Society B. 2002, vol. 269, pages 1383–1388.

individual members cannot survive independently for extended periods.

The concept of a superorganism raises the question, "What constitutes an individual?" Many social insect species, such as ants, bees, and termites, exhibit collective characteristics similar to those of biological individuals rather than mere groups. A colony of the most complex species is, therefore, often considered a unit of natural selection.

In their book, *The Superorganism – the Beauty, Elegance, and Strangeness of Insect Societies*, Bert Hölldobler and Edward Wilson revive the idea of insect societies as superorganisms. They describe these societies as "working individuals" who sacrifice their individual interests for the greater good of the colony.[15] Throughout their book, they provide four definitions of superorganisms:

1. **Reproductive Division of Labor** - This involves reproductive and sterile castes, overlapping generations, and non-reproductive members caring for the young.

2. **Efficiency-Driven** - This definition applies primarily to colonies with advanced social structures. Conflicts related to reproduction among colony members are reduced, and worker individuals are selected to maximize colony efficiency.

3. **Self-Organized Division of Labor** - Here, a superorganism is defined as a colony of individuals organized through a division of labor and united by a closed communication system.

4. **Analogous to Single Organisms** - In this view, a superorganism is a society that exhibits organizational features similar to the physiological properties of individual organisms.

---

[15] Bert Hölldobler and Edward Wilson, *The Superorganism – the Beauty, Elegance, and Strangeness of Insect Societies*. 2008, 544 pages.

In essence, a superorganism represents an organized colony composed of numerous specialized organisms. Individual organisms within the colony perform specific functions, and their survival depends on collective efforts.

An Ant Colony as a "Giant Ant."

The question of whether individual ants and ant colonies can be considered genuine biological individuals has intrigued biologists since Wheeler's report. It delves into the heart of individuality and challenges conventional notions of what constitutes an organism.

On the one hand, individual ants *are* organisms. Their hard chitin exoskeleton forms a physical barrier that separates each ant from the external environment, including other ants. Within this exoskeleton, each ant houses essential internal systems such as the digestive tract, respiratory system, circulatory system, and other organs necessary for its survival and proper functioning.

On the other hand, individual ants are part of a larger collective— the ant colony. Each ant has a specific role within the colony, contributing to tasks such as foraging, nursing, and defending the nest. In Wheeler's view, an ant colony possesses specific organismal characteristics. It is legitimate to view individual ants as components of the larger whole, much like cells in a multicellular organism. Just as cells in a body work together to maintain homeostasis and ensure survival, individual ants within a colony cooperate to achieve collective goals. Thus, separate ants lack autonomy, relying on the collective for their survival and function. This raises the question: *can we consider the ant colony as a superorganism, akin to a single organism, despite its being composed of individual ants?*

Indeed, some researchers argue that ant colonies meet the criteria of a superorganism. Notably, Hölldobler and Wilson in *The Superorganism* emphasize the unity and interdependence of individuals within an ant colony. This perspective further blurs the distinction between individual ants and the colony, supporting the notion of the ant colony as a superorganism. Ants are not passive automatons; they engage in intricate behaviors, communicate through chemical signals, and exhibit a division of labor that ensures colony survival and prosperity.

Ant colonies function as a tightly coordinated unit, responding collectively to environmental challenges such as locating food sources and defending against predators. Communication is a fundamental aspect of ant colonies' organization and effectiveness. Chemical signals, including pheromones and social fluids, are central to their communication modes. These signals convey essential information about the environment, danger, and the need for specific tasks to be carried out. In this context, the colony behaves like a single entity, with individual ants acting as its organs or cells.

However, the debate is not one-sided. Critics argue that the concept of a superorganism may oversimplify intricate relationships within ant colonies. They contend that while ant colonies display remarkable coordination, they are still collections of individual ants, each with its own genetic interests. From this viewpoint, ant colonies are highly organized social groups. The question of whether individual ants and ant colonies constitute real biological individuals depends on the context and level of analysis.

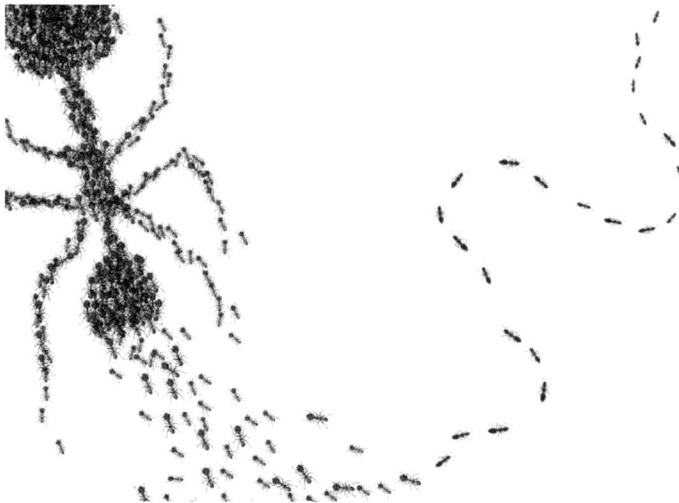

Figure 68. A classical illustration of a superorganism is a colony of individual ants forming "one giant ant." Modified from a Chris Madden cartoon (used under license).

The Many Faces of Biological Individuality

Biological individuality is not always easy to define, as we discussed. The problem of biological individuality remains elusive, and there is no one-size-fits-all definition. However, embracing co-creative thinking and network-focused value models through a multi-sided platform offers a promising path forward. By combining different views and harnessing collective intelligence, we can unravel biological individuality and gain deeper insights into the living world.

Suppose we consider biological individuality as a broad category that encompasses various types of biological individuals, (i.e., those related to evolution, development, living organisms, etc.). In that case, we can break down the issue of biological individuality into two main aspects. The first aspect deals with individuality in a general sense. It explores what defines something as an individual, regardless of its specific type or nature. The second aspect focuses on biology. It delves into what distinguishes an individual as biological instead of being classified as physical, sociological, or any other category.

Austin Booth, a Ph.D. student at Harvard University at that time, argued that there are two fundamental explanations of biological individuality. These reflect the distinction between "evolutionary" and "non-evolutionary" thinking.[16] Referring to the "non-evolutionary" branches of biology, Booth included fields like developmental biology, physiology, and immunology.

The development of evolution theory and the subsequent debate about the potentially hierarchical nature of the evolutionary process ensured that biological individuality would be central. This is so for two reasons. First, an evolutionary explanation requires the individuation of organisms that constitute evolving populations. Second, individuals are conceived as products of evolution and are found at many levels of the biological hierarchy. Fitness calculations require the ability to distinguish parents from their offspring. Counting individuals is thus essential to doing the kind of

---

[16] Austin Booth, *Essays on Biological Individuality*. Dissertation at Harvard University, 2014, page 10.

demographic work required for an evolutionary explanation.

According to Booth, we can distinguish between two key aspects when discerning "evolutionary individuals" from "organism individuals" -- reproduction and persistence. Evolutionary individuals are often seen as part of evolving lineages, primarily defined by their role in the process of reproduction. However, it's crucial to recognize that biology also explains how individual living things persist.

Living beings consist of intricate parts that interact in specific ways, granting them particular abilities. These parts engage in activities and configurations that give rise to, or are integral to, processes like development, physiology, and immunity within individuals. Explanations of how these parts function and contribute to the endurance of the whole entity are not necessarily tied to evolution.

For instance, investigating the mechanisms governing the physiological functions of a living system doesn't require knowledge of its reproductive outcomes. Organisms can be viewed as physiologically stable and self-sufficient entities. They intake nutrients and energy from their surroundings, maintain their metabolic stability despite external fluctuations, operate as a cohesive unit, and display a certain level of independence within the biological realm. Booth proposes that concepts such as homeostasis and autonomy clarify the distinctions between organisms, complex non-living natural systems, and components of living systems.[17]

It is common for biologists to count biological individuals at a time, but there are also questions about counting biological individuals *over* time. For example, consider the division of an amoeba into two distinct cells. Does this mean that two new offspring have been created and replaced the parent amoeba, or is it just that the original amoeba has grown larger and now has two cells?

Let's return to the process of metamorphosis, such as the

---

[17] Ibid., page 16.

transformation of a caterpillar into a butterfly. Does this event involve only one individual? The very same individual that is first a caterpillar and then a butterfly? Or does it involve two individuals? The caterpillar individual that ceases to exist and is replaced by a distinct butterfly individual? To answer these questions, we need to know what it takes for organisms to persist. Thomas Pradeu, a professor at CNRS and the University of Bordeaux, claims that one of the main attractions of the "physiological" approach is that it does a good job of explaining how organisms persist through time.[18]

William Morgan, the British philosopher we mentioned earlier regarding the Fetus Problem, called for "pluralism about biological individuality." According to Morgan, both the "physiological" approach and the "evolutionary" approach are correct! More specifically, pluralists hold that each approach selects a distinct kind of biological individual — the physiological approach is about biological individuals that are physiological wholes, while the evolutionary approach is about biological individuals that are units of selection. Viruses, for instance, are "evolutionary individuals" given their ability to reproduce, but they are not "physiological individuals" because they lack metabolic activity.[19]

It is imperative to note that there are more kinds of biological individuals than physiological and evolutionary individuals.[20] Immunology makes a major contribution to this question. The immune system plays a key role in monitoring every part of the organism and maintaining the cohesion between the components of that organism. This makes each organism unique and constantly re-establishes the boundaries between the organism and its environment.[21]

An essential aspect of biological individuality is the interaction between an organism and its immune system. This dynamic interplay

---

[18] Thomas Pradeu, *The Limits of the Self: Immunology and Biological Identity.* 2012, pages 237–238.

[19] Peter Godfrey-Smith, *Individuality and Life Cycles.* In *Individuals Across the Sciences,* 2016, pages 85–102.

[20] James DiFrisco, *Kinds of Biological Individuals: Sortals, Projectibility, and Selection.* The British Journal for the Philosophy of Science, 2019, vol.70, no.3, pages 845–875.

[21] Thomas Pradeu, Immunology and Individuality. 2019, vol.8, page 1.

allows an organism to maintain its distinct identity while defending itself against external threats. A major function of the immune system is to target and destroy foreign bodies. Interfaces, such as the gut lumen, belong to the "outside" of the organism, and boundaries are strict and fixed. The immune system can eliminate "non-self" elements and tolerate "self" elements, according to the "immunological individual" framework. Additionally, it strengthens the bond between the body's constituents. The immune system constantly redefines the boundaries of an individual within this framework.

Evolution, developmental biology, ecology, genetics, physiology, behavioral science, cognitive science, and various other disciplines provide distinct viewpoints on biological individuality. As a result, biologists consider distinct categories such as "evolutionary individual," "ontogenetic individual," "physiological individual," "immunological individual," and "ecological individual," among others.

Metaorganism: An Organism as an Ecosystem

In 1676, Antoni van Leeuwenhoek, a Dutch microscopist and amateur naturalist, peered through a microscope of his own design. He then proceeded to describe a world that would be misunderstood for the next 300 years. Microorganisms living in our bodies challenge the conventional definition of self and redefine our understanding of individuality.

In Chapter 8, we discussed the hidden world of "microbiota." Animal and plant bodies are not only composed of their "own" cells but are also host to a multitude of microorganisms. Scientists have calculated that the number of bacterial cells occupying a host body is much higher than the number of host cells.

I would like to address another aspect of hosting so many microbes in multicellular organisms. Multicellular organisms with a multitude of microbes extend the traditional limits of identity, like quaking aspen and ants. We can consider the totality of the microbiota as part of the host body. Such a perception reflects the fact that any animal body coexists with a multitude of microorganisms essential for its survival. Thus, it redefines our understanding of biological individuality. Knowledge of the human microbiota has significantly changed our perception of our health and promoted effective methodologies to treat and prevent many diseases. With such a perspective, our comprehensive understanding of human evolution and personalized characteristics is challenged.[22]

Such a view led to the concept of "metaorganism" as an integral combination of the host and its microbiome. The term was first used in 1998 by British writer and evolutionary biologist Graham Bell, in reference to organisms that are between two levels of organization. Later, the term was increasingly used to refer to any multicellular organism derived from coevolution with microbiota.[23] Metaorganism research aims to deepen our understanding of host-microbe interactions. As such, humans and all other organisms are truly metaorganisms composed of a host and a complex microbiome. There are no solitary organisms; instead, every plant and animal on earth is a metaorganism.

The idea of metaorganism is similar to the holobiont concept we discussed in Chapter 8. The latter concept considers the host and its microbial partners to be a single integrated unit, as a *holobiont*. This unit functions and evolves together. Although each holobiont could be a large organism, such as an animal, fungus, or plant, together with its microbiome, the holobiont itself cannot be physically seen. Importantly, holobionts as complex systems express "emergent properties"(see Chapter 3). This means that the behavior of the holobiont is unpredictable, as it is not the sum of the properties of the host and its microbial components.

An organism is not whole without its symbionts. For example, without its rumen microbiome, a cow is unable to digest plant

---

[22] Parag Kundu et al. *Our Gut Microbiome: The Evolving Inner Self.* Cell, 2017, vol., no.7, pages 1481-1493.

[23] Thomas Bosch and Margaret McFall-Ngai, *Metaorganisms as the New Frontier.* Zoology (Jena), 2011, vol. 114, no.4, pages 185-190.

material and, therefore, would not be a herbivore. Joan Roughgarden and her colleagues argue that holobionts are units of selection.[24] They provided multiple examples of how much organisms rely on their microbiome. These are essential physiological and metabolic processes, often required to provide nutrients. For example, corals rely on microalgae for photosynthesis and energy production, while legumes require rhizobial bacteria for nitrogen fixation. Endosymbionts are also a key part of development in many species. For example, *Wolbachia* bacteria are required for egg maturation in wasps of the genus *Asobara*.

Another example of endosymbionts is the light-producing organ of the Hawaiian bobtail squid that requires the symbiotic bacteria *Vibrio fischeri* to mature fully. After an adult squid releases these bacteria into the water, the bacteria can colonize the immature organs of newly hatched larvae. The squid's immune system specifically tolerates these bioluminescent bacteria, allowing a stable association between bacteria and the host by producing a functional light organ.

Jinru He and Thomas Bosch from the Zoological Institute at Christian-Albrechts-University in Kiel, applied the concept of "metaorganism" using, as a model, the hydra, a small freshwater member of the Cnidaria phylum. These organisms, less than 15 millimeters long, offer a simple biological model to study complex host-microbe interactions. One fascinating aspect of hydra is their apparent immortality, attributed to their unique stem cells capable of endless regeneration. He and Bosch's research focused on a gene, which is vital for this regenerative ability. It is also linked to the hydra's immune system and microbiome interactions. The study explored how the hydra's nervous system shapes its microbiome, influencing microbe distribution in its body. The research sheds light on the inseparable relationship between a host and its microbiome,

---

[24] Joan Roughgarden et al., *Holobionts as Units of Selection and a Model of Their Population Dynamics and Evolution.* Biological Theory, 2018, vol. 13, pages 44–65.

emphasizing complexity even in simple organisms like hydra.[25]

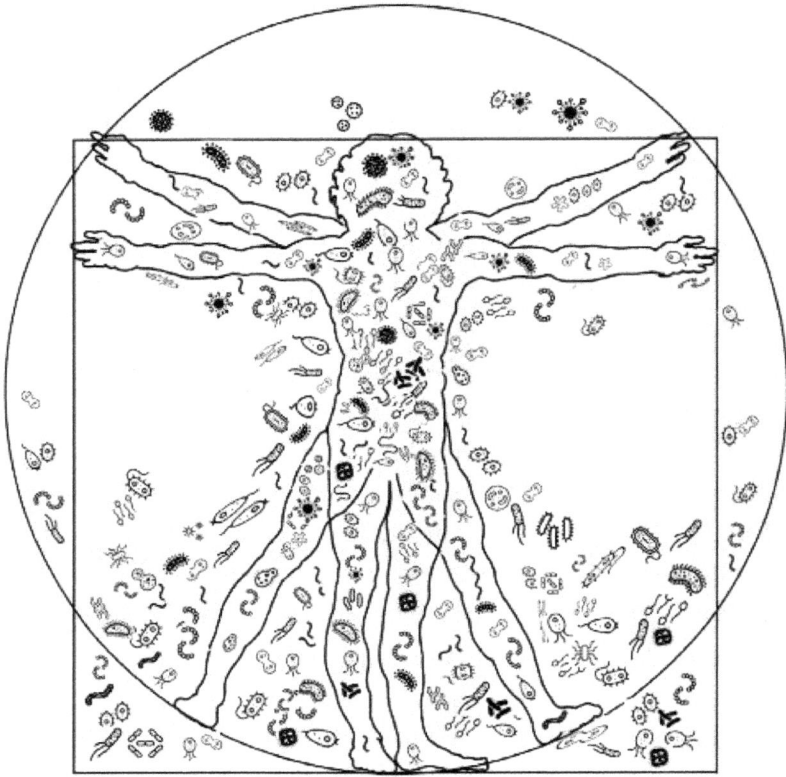

Figure 69. Human microbiomes are inextricably bound up with Man as a biological entity and the microbes that coexist with him.

In 2019, the European Research Council supported the interdisciplinary summer school on "Microbiota, Symbiosis and Individuality: Conceptual and Philosophical Issues" in Biarritz, France. Experts from biology, philosophy, and the history of science discussed cutting-edge empirical research, which provided insights into how microbiota research impacts our understanding of biological individuality. In light of the microbiome "revolution," there is increasing recognition that biological identity includes a

---

[25] Jinru He and Thomas Bosch. *Hydra's Lasting Partnership with Microbes: The Key for Escaping Senescence?* Microorganisms. 2022, vol.10, no.4, page 774.

dynamic "dialogue" with an organism's microbiota and is dependent on environmental factors.[26]

The concept of "metaorganism" as an "extended" personality transcends strictly biological disciplines. Amber Benezra, a sociocultural anthropologist at the Stevens Institute of Technology, develops the "anthropology of microbes" to address global health problems across biological and social disciplines. She said: "As an anthropologist, ... what I love about the microbiome is that it brings together social intimacies of life with our biological selves in ways that show us that those two things are inextricably entangled." [27]

Exogenetic Information as a Component of an Animal Population

A significant shift in our understanding of viruses and microorganisms has occurred. This shift is moving beyond viewing them solely as sources of infection. It recognizes them as carriers of essential genetic information for their host organisms. This transformation in perspective can be attributed to Howard Temin's discovery of the "protovirus," a free-living virus that was a hypothetical primitive or ancestral virus.[28] Temin proposed the concept of viruses' dual nature. According to this concept, "protovirus" (a DNA sequence capable of mutating into an oncogenic virus) plays a pivotal role in regulating the normal development and functioning of their host organisms. They also may break free to become infectious agents circulating within the host population. Soviet virologist Vadim Agol even referred to many viruses as "runaway protoviruses."[29]

Around three decades ago, before the concepts of microbiome,

---

[26] Isobel Ronai et al. *Microbiota, symbiosis and individuality.* Microbiome. 2020, vol. 8, No.1, 117.

[27] Suzanne Ishad et al. *Introducing the Microbes and Social Equity Working Group: Considering the Microbial Components of Social, Environmental, and Health Justice.* mSystems, 2021, vol.6, Issue 3, e00471-21

[28] H.M. Temin, *Viruses, protoviruses, development, and evolution.* Journal of Cellular Biochemistry, 1982, vol. 19, pages 105–118.

[29] Vadim Agol, *Protoviruses as carriers of genetic information.* Priroda ("Nature"), 1973, issue 11, pages 28–35 (Russian).

holobiont, and metaorganism were formulated, I proposed the concept of "exogenetic information."[30] I pointed out that every animal population interacts with a diverse but unique range of bacteria and viruses, each carrying specific genetic information. The circulation of this genetic information within a host population can lead to major changes in that population. Thus, the concept of "exogenetic information" was defined as the entirety of genetic information encoded in a microbial community linked to a specific animal population.

This concept recognizes two distinct dynamic processes regarding the circulation of genetic material within an animal population. The first process encompasses changes within communities of various microorganisms residing in animals of a particular local population. The second process involves the amalgamation of genetic information carried by microbes, which is foreign to animals. Nevertheless, the sum of this genetic information, shared by all internal microorganisms, is highly specific to a particular animal population. This consequence can profoundly affect animal hosts' ecology, behavior, physiology, and overall fitness.

Evidence suggests that animal hosts and microbes drive coordinated evolution. Animal-microbial partnerships are fundamental to animal development. Margaret McFall-Ngai, director of the Pacific Biosciences Research Program at the University of Hawaii at Mānoa, discerned two potential influences of bacteria on animal developmental programs: (1) the nonspecific influences of microbes as ubiquitous and critical components of the environment and (2) the specific influences of bacterial cells that have co-evolved with animals in tight, multi-generational associations.[31]

The intensity and nature of the circulation of exogenetic information by viruses and microorganisms are intricately regulated by the structure of animal populations. Factors such as population

[30] Michael Kosoy, *Conception of Exogenetic Biological Information in Animal Populations*. In Evolutional and Genetic Investigations in Mammals. 1990, pp. 201–214.

[31] Margaret McFall-Ngai, *Unseen forces: The influence of bacteria on animal development*. Developmental Biology, 2002, vol.242, pages 1–14.

density, social behavior, contact rates among individuals, and environmental conditions play pivotal roles. Animals residing in densely populated areas, large social groups, or with promiscuous mating systems create conditions conducive to the transfer of exogenetic information among individuals due to their close proximity and frequent interactions.[32]

Furthermore, the organization and behavior of host populations can influence not only the diversity and prevalence of microbes, they also impact the fitness advantages of various transmission strategies for carriers of exogenetic information. Consequently, exogenetic information is a meaningful characteristic when viewed from the perspective of a host population. While exogenetic information shares a physical structure (encoded in nucleotides) with hereditary (genetic) information, its mechanisms of exchange bear fundamental similarities to other forms of "signal information," such as visual, acoustic, or olfactory information transmitted through direct or indirect contact among animals.

We discussed the perception of the microbiome as part of an individual organism. Exogenetic information connects all animals belonging to the same population. The acceptance of the concept of exogenetic information leads to an understanding of a particular population as an individual entity. From this perspective, each animal population could be defined as an individual superorganism, not just ants or bees.

The Ocean's Microbiome: A Colossal Superorganism
Almost every biological system is home to "foreign passengers," in the words of Thomas Pradeu, some of which play important functional roles in the host. This raises critical issues regarding the boundaries of biologically individual entities. One major question that Pradeu calls the "holobiont debate" is about the degree of unity

---

[32] Sonia Altizer, *Social Organization and Parasite Risk in Mammals: Integrating Theory and Empirical Studies*. Annual Review of Ecology, Evolution, and Systematics, 2003, vol.34, pages 517-547.

and cohesion found in a holobiont.[33] At a particular level of unity and cohesion, each ecosystem, including the biggest one — the ocean, can be considered a holobiont or a superorganism.

Recent scientific insights lead to a fascinating perspective: the ocean's microbiome might function like a colossal individual superorganism. The ocean is home to about two million marine species ranging from a single-celled protozoan (one micrometer) to a blue whale (30 meters). All marine animals and plants share a symbiotic relationship with the vast diversity of microorganisms.[34] For example, there are approximately one million bacteria and ten million viruses in one milliliter of water, which is equivalent to three drops. These relationships, historically studied in single host-symbiont systems, range from simple sponges to complex vertebrates, all maintaining homeostasis in the diverse ocean systems.

This indicates a deeply integrated system, where multicellular and unicellular organisms contribute to the ocean's microbiome functioning as a larger whole. Host-microbe interactions play crucial roles in marine ecosystems. The ocean, driven by a complex network of microbes producing oxygen and regulating biogeochemical cycles, functions much like a single, vast organism.[35]

Given the connectivity and the unexplored biodiversity specific to the ocean's ecosystem, a deeper understanding of such complex systems requires further technological and conceptual advances. The ocean's microbiome has incredible diversity. Shinichi Sunagawa at the European Molecular Biology Laboratory in Heidelberg, Germany, and his colleagues created the Ocean Microbial Reference Gene Catalogue. It contains over 40 million genes from more than 35,000 species. The most striking result is a surprisingly high fraction

---

[33] Thomas Pradeu, *The many faces of biological individuality*. Biology and Philosophy, 2016, vol.31, page 767.

[34] Amy Apprill, *Marine Animal Microbiomes:*
Toward Understanding Host–Microbiome Interactions in a Changing Ocean. Frontiers in Marine Science, 2017, vol.4.

[35] Eric Karsenti et al. *A Holistic Approach to Marine Eco-Systems Biology.* PLoS Biology, 2011, vol.9, issue 10, e1001177.

of its abundance is shared with the human gut microbiome. An international team made this discovery of researchers during the Tara Oceans' scientific expedition around the world. The investigators identified ocean microbial core functionality and revealed that >73% of its abundance is shared with the human gut microbiome, despite the differences between these two ecosystems.[36]

Key environmental factors, particularly temperature, play a crucial role in determining the ocean microbiome composition. As global climate change continues to alter these environments, the diversity and functionality of these microorganisms are at risk, potentially impacting the entire ocean ecosystem. We still lack adequate baseline data with which to compare contemporary observations that will determine whether climate variability alters microbial metabolism and ocean ecosystem.[37] A microbial imbalance in a symbiotic community that affects the health of marine ecosystems is called "dysbiosis."[38]

This suggests that the ocean's microbiome, with its immense diversity and critical role in biogeochemical processes, could be considered a giant, interconnected superorganism crucial to the balance and health of our planet's ecosystems.

Leviathan Metaphor

The concept that the ocean's microbiome operates as an immense superorganism is awe-inspiring. But how can we grasp such a profound idea? We have data revealing the staggering abundance and diversity of microbes in our oceans. We also have information about the intricate interconnections within ocean ecosystems, providing essential building blocks for such knowledge. However, these facts alone may not be enough to form a vivid mental picture of ocean

---

[36] Shinichi Sunagawa et al., *Structure and function of the global ocean microbiome*. Science, 2015, vol. 348, Issue 6237.

[37] Chris Bowler et al. Microbial oceanography in a sea of opportunity." Nature, 2009, vol.459, no.7244, pages 180-184.

[38] Suhelen Egan and Melissa Gardiner, *Microbial dysbiosis: rethinking disease in marine ecosystems*. Frontiers in Microbiology, 2016, vol.7, e.991.

superorganisms.

As we've explored throughout this book, knowledge occurs within the realm of Briah. At this level of understanding, metaphorical thinking emerges as a vital tool in unraveling the essence of natural phenomena. Metaphors serve as bridges between the familiar and the abstract, enabling us to relate to complex concepts in more accessible ways.

Consider ocean superorganisms as a vast, interconnected web. Each microbe is a tiny thread woven into the fabric of this enormously vast living creature. This metaphorical perspective allows us to perceive the ocean's microbiome not merely as a collection of individual microbes but as a grand, collective entity. Moreover, likening it to a superorganism aligns with the notion that the whole is greater than the sum of its parts. Just as a human body consists of trillions of cells working in concert to sustain life, the ocean superorganism thrives through the coordinated efforts of myriad microorganisms. This metaphorical framework invites us to envision the ocean as a living, breathing being pulsating with vitality.

Finding a working metaphor to describe the existence of a colossal oceanic superorganism might ease public fears and expectations about the microbiome as an individual. To grasp such a concept, the Written and Oral Torah provides a powerful metaphor! I refer to the Leviathan, a vast aquatic creature of some kind. Rabbi Joel David Bakst called Leviathan "Judaism's most elusive crypto-zoological phenomenon." Bakst wrote: "...the secret of Leviathan has been kept well-hidden by the sages and mystics of Israel for thousands of years. It is only now in our generation that permission has been granted to begin to conceive of its deeper truth and imminent relevancy."[39]

The Hebrew word for "Leviathan" is derived from the root *Lamed-Vav-Hei*, which means "wreathed, twisted in fold."[40] Bakst

---

[39] Joel David Bakst, *The Secret Doctrine of the Gaon of Vilna.* Volume II. 2009, page 105.

[40] *Gesenius' Hebrew and Chaldee Lexicon to the Old Testament Scriptures: Numerically Coded to Strong's Exhaustive Concordance, with an English Index of More Than 12,000 Entries.* 1979.

indicated that the root of the Hebrew word *Levyatan* also means "accompany," "bring together," and "unite."[41]

There are a handful of references to Leviathan in the scripture. The Book of Job describes a creature beyond compare, something that defies human attempts to subdue or tame it. No weapons can overcome it. "No one is fierce enough to rouse it. Who then is able to stand against me?" (Job 41:10). The literary description of the Leviathan may lead to speculation that gigantic creatures inhabited seas and oceans, such as prehistoric whales, crocodiles, or dinosaurs. Any reading of the texts beyond a literary understanding requires a deeper understanding of allegorical meaning and interpretations. One of the interpretations is the identification of the Leviathan as the embodiment of chaos. Commonly, it is perceived as a metaphor for a powerful enemy.

The interpretations of Leviathan in the Zohar are not so negative and are very meaningful. In *Haqdamat Sefer ha-Zohar* (1:6a), we read:

> "He was a fish, circumnavigating the vast ocean from one end to the other. So grand and splendid, ancient of days, he would swallow all the other fish in the ocean, then spew them alive, thriving, filled with all goodness of the world. So strong, he could swim the ocean in one moment. He shot me out like an arrow from the hand of a mighty warrior, secreting me in that site I described. Then he returned to his site, disappearing into the ocean."[42]

The phrase "like an arrow," that the commentators use refers to *Hagigah*, one of the tractates comprising Mishnah (15a): "Any emission of semen that does not shoot forth like an arrow does not fructify."

Another important comment about the provided piece from the *Haqdamat Zohar* indicates that the "fish" (*nuna*) symbolizes the

---

[41] Joel David Bakst, *The Secret Doctrine of the Gaon of Vilna.* Volume II. 2009, page 109.

[42] *The Zohar.* Pritzker Edition, vol.1, page 37.

Sefirah Yesod. There are many indications in the *Zohar* on Leviathan as Yesod – 2:12a, 48b, 50b; 3:58a, 60a, all quoting Psalms (104:26): "So is this great and wide sea, wherein are things creeping innumerable, both small and great beasts ... There is that Leviathan, whom thou hast made to play therein." The connection of the concept of individuality to Yesod is discussed below.

Bakst taught that the three main themes of Leviathan in Scriptural terminology refer to "eating," "knowing," and "union."[43] He stressed that "eating, in this context, is also about taste, smell, texture, and even sound, along with the physiological process of transforming a substance from the external environment into an ever-refining internal environment."[44]

The Leviathan metaphor, evoking the image of a massive, singular entity, serves as a compelling conceptual tool to reevaluate how we define individual organisms. This is particularly true in the context of the vast, interconnected microbial networks of the ocean.

Figure 70. The Leviathan is described as a massive, powerful creature that rules over the aquatic realm. This aligns with the concept of the oceanic microbiome as a dominant force in the oceans, encompassing a vast network of microorganisms.

---

[43] Joel David Bakst, *The Secret Doctrine of the Gaon of Vilna.* Volume II, page 108.
[44] Ibid., page 109.

The oceanic microbiome, much like the mythological Leviathan, represents a collective entity that is more than the sum of its parts. It comprises countless interconnected microorganisms, each playing a role in the larger functioning of the ocean's ecosystems. This complex network challenges the simplistic view of individuality, suggesting a more composite, integrated understanding of life. The oceanic microbiome, like the metaphorical Leviathan, represents a collective entity that is more than the sum of its parts. It comprises countless interconnected microorganisms that interact in such a way that the complex ocean ecosystem becomes adaptive and exhibits emergent properties.

This perspective aligns with the concept of "holobionts," where a host and its associated microbiota are considered a unified living system. In the vast ocean, microbial communities interact and function in such a tight symbiosis that they could be perceived as one colossal organism. The metaphor speaks to biological individuality's fluid boundaries. Horizontal gene transfer, the exchange of physiological substances and signal information, and other interactions between countless microorganisms in the ocean blur the lines of individuality. This presents a more dynamic and interconnected view of life.

This perception also challenges anthropocentric views of life and individuality. It prompts a reevaluation of human relationships with the environment, emphasizing the interdependence of all life forms. Recognizing the oceanic microbiome as a Leviathan model underscores the importance of preserving these ecosystems upon which the planet's health depends.

"Anna Karenina Principle"

Biological individuality is a multifaceted and dynamic concept that challenges traditional definitions in contemporary science. David Krakauer, the President of Santa Fe Institute, and his colleagues proposed a robust model of individuality. It situates the hierarchical organization of life into nested trophic and functional levels. The team's insight highlights the existence of multiple parallel levels of individuality and allows measuring and quantifying each subsystem's degree of individuality.[45]

All biological systems, ranging from simple microorganisms to ocean ecosystems, have a hierarchical organization. Thus, it is necessary to select and measure each system's individuality in a specific situation. The situations are constantly changing because of internal and environmental perturbations. Each biological system studied to date is associated with symbiotic communities of microorganisms which itself leads to the question of whether they are a part of this system or aliens.

The question is not purely theoretical. The changes in hierarchical structure induced by many perturbations are stochastic and, therefore, lead to transitions from stable to unstable states of each system. However, despite their apparent ubiquity, these patterns are easily missed or discarded by some common workflows.

In Kabbalistic biology, there is no long-lasting balance between stable ("healthy") and unstable ("unhealthy") states. However, there is a central point in the biological system's dynamics when internal and external factors jointly lead the system to balance. We discussed this principle in Chapter 15 while defining "health" and other biological phenomena by using the "Third dimension" of Kabbalistic biology.

Considering the relationship between stress and stability in animal microbiomes, Jesse Zaneveld of the University of Washington Bothell and his colleagues formulated the so-called "Anna Karenina principle." This paralleled Leo Tolstoy's dictum that "all happy families look alike; each unhappy family is unhappy in its own way."[46] They argue that "Anna Karenina principle" effects are a common and meaningful response of microbiomes to stressors that reduce the

---

[45] David Krakauer et al. *The information theory of individuality.* Theory in Biosciences, 2020, vol.139, no.2, pages 209-223.

[46] Jesse Zaneveld et al. *Stress and stability: applying the Anna Karenina principle to animal microbiomes.* Nature Microbiology, 2017, vol.2.

ability of the host or its microbiome to regulate community composition.

According to their study, patterns consistent with "Anna Karenina" effects have been found in systems ranging from the surface of threatened corals exposed to above-average temperatures to the lungs of patients suffering from HIV/AIDs. The associations between microbiome instability and many stressors and diseases suggest that microbiome resistance and resilience are hallmarks of healthy physiology. This is consistent with animals' evolution in a "sea" of microorganisms and viruses.

Yesod: The Nexus of Biological Individuality

In biology, individuality touches on questions of existence, unity, and the interconnectedness of life. A confrontation and combination of perspectives from different biological disciplines can enrich the concept of a biological individual. To bridge the gap between these diverse, sometimes conflicting viewpoints, Kabbalistic biology's approach aims to harmonize and synthesize the various facets of biological individuality.

Kabbalistic biology proposes a novel perspective on biological individuality, offering a more holistic understanding that transcends traditional disciplinary boundaries. Instead of relying on a single, static definition, Kabbalah introduces the concept of a "hardness-softness" scale, allowing for a more dynamic understanding of individuality. This scale accommodates the diversity of life forms, from seemingly isolated organisms to deeply interconnected ecosystems. Such a perspective introduces an innovative lens through which we can explore this concept. This lens allows us to combine various aspects. Yesod, the ninth of the ten sefirot in the Kabbalistic Tree of Life, emerges as a symbol of individuality in biology.

In the Zohar, Yesod is called the "Vitality of the Worlds that animates all of existence."[47] This Sefirah forms the essential foundation upon which individuality is manifested in the organic

---

[47] *The Zohar*, Daniel Matt comment, vol. 1, page 176.

world. The Yesod's significance lies in its role as a transmitter, facilitating communication between the six Sefirot of Zeir Anpin and the realm of Malkhut, which represents tangible reality.

The word Yesod means "foundation" in Hebrew. The idea of a foundation suggests that there is a substance that lies behind physical matter and "informs it" or "holds it together," something less structured, more plastic, more refined, and rarified. Yesod interfaces the rest of the Tree of Life to Malkhut. In the Kabbalistic tradition, Yesod is often pictured as the procreative masculine organ that brings together productive energies from parents to offspring. Thus, Yesod is considered a "narrow" channel from the infinite potential of procreation to the actual manifestation in the progeny of all living creatures.

Yesod plays a pivotal role in shaping biological individuality. It acts as a reservoir that combines the various sources contributing to this individuality. These facets are enriched through the confrontation and combination of the six Sefirot forces that lead to different aspects of biological individuality.

Hesed — Ecological Individuality
Hesed signifies individuality that reflects environmental diversification. Ecological niches form distinct individuals, contributing to life's rich tapestry. When it is difficult to decide whether a given entity is cohesive enough to be considered an individual, a group of slightly distinguishable but closely interacted biological elements can serve the role of an "Ecological individual."

Gevurah — Genetic Individuality
Genetic individuality is defined by a common genotype formed from the information received from one's ancestors. In sexual organisms, the fusion of a seed and an egg creates a unique combination of genetic material, leading to the development of a new individual. In asexual organisms, each individual is composed of multiple genetic units that share a common genetic makeup through clonal growth. Genetic determinism contributes to the forming of canalizing

trajectories in biological development. Like valleys guiding water flow, developmental canalization directs and selects certain gene manifestations.

Tiferet — Evolutionary Individuality

Tiferet represents the evolutionary aspect of individuality. It reflects the potentially hierarchical nature of evolution, where individuality is central to the process. The "Tiferet individual" constitutes an identifiable, accountable, and cohesive unit shaped by evolution. Evolutionary individuality ensures unified reproductive lineages, their fitness success, and their capacity to adjust adaptation to surrounding conditions.

Netzah — Physiological Individuality

If a group of biological entities engage in a significant amount of physiological, especially metabolic, interactions with each other, such an aggregation can form a "physiological individual." The exchange of biologically significant signals, such as olfactory signals, also contributes to this formation. Netzah embodies the ability of individuals to persist. This persistence is manifested in the capacity of individuals to respond as a whole to changes in their environment. Netzah also encompasses transformations through life stages while maintaining individuality. An example is the "caterpillar-cocoon butterfly" model, which encompasses the stages of the same "evolutionary individual."

Hod — Immunological Individuality

For a group of biological elements to be considered an immunological individual, there needs to be a system that determines which elements are part of that entity and which elements do not belong. Hod involves discrimination between the Self and the Non-Self through the immune system. The immune system acts as a "barrier" that separates possible intruders from the elements that are parts of the Self.

Yesod – Functional Individuality

For a biological entity to be a functional individual, the components from which it is constructed must be integrated into an organic whole. Jack Wilson suggested a general principle about functional individuals, where the current causal relations between parts, not just their natural history or genetic composition, matter.[48] Yesod serves as the foundation upon which these diverse forces converge. It forms a coherent basis for the manifestation of biological individuals in the physical world, represented by the Malkhut (Figure 71).

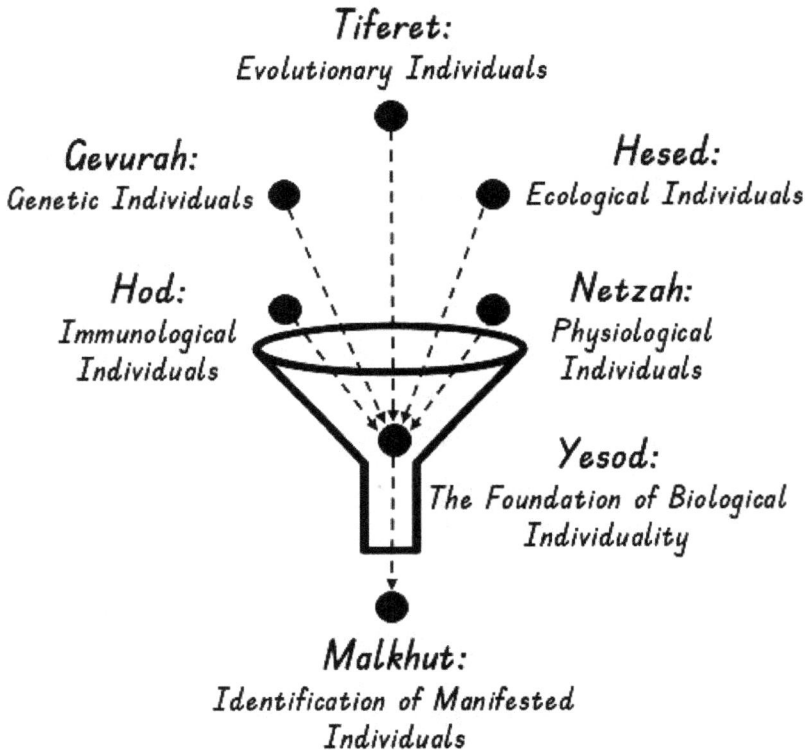

Tiferet:
Evolutionary Individuals

Gevurah:
Genetic Individuals

Hesed:
Ecological Individuals

Hod:
Immunological
Individuals

Netzah:
Physiological
Individuals

Yesod:
The Foundation of Biological
Individuality

Malkhut:
Identification of Manifested
Individuals

Figure 71. Yesod serves as the foundation that forms a coherent basis for the manifestation of biological individuals in the physical world, represented by the Malkhut.

---

[48] Jack Wilson, *Biological Individuality*. 1999, page 89.

The terms that are used to describe different aspects of biological individuality are simply labels intended to highlight these aspects. An individuality can be defined by recognizing the forces that are driving manifestations of the properties that define how it manifests itself in a particular way. The six Sefirot of Zeir Anpin are driving forces, not specific manifestations. Individual manifestations are expressed in *Malkhut*. Within *Malkhut*, biologists can discern criteria to define an individual at various levels, from cells to ecosystems.

Biological individuality is shaped by the harmonious interplay of forces formed by the six sefirot of Zeir Anpin: Tiferet, Hesed, Gevurah, Netzah, Hod, and Yesod. The interplay of these forces creates a distinctive space of variants for estimating a position of any definition of biological individuality under certain times and circumstances.

The Ninth Dimension empowers us to select specific criteria to understand individuality, recognizing that the definition may vary across different biological contexts. This approach opens an opportunity for measuring multiple degrees of individuality at all levels of biological organization.

Chapter 22

Tenth Dimension: The Journey to Knowledge

During our journey through the realms of biology, we've explored numerous facets of Life, from the microscopic to the macroscopic. We've delved into the intricacies of ecosystems, the fascinating world of genetics, and the awe-inspiring beauty of evolution. In the first chapter of this book, we defined science as "the process of obtaining knowledge."

Isn't this statement clear? Yet, as we've ventured deeper into the subject, a recurring theme has emerged that transcends the boundaries of scientific inquiry. It is the journey from "information" to "knowledge." As we embark on this chapter, we will be introducing the final dimension of our understanding of Kabbalistic biology. This dimension is directed toward our acquisition of knowledge.

Knowledge is Always Personal

There are numerous definitions and qualities of "knowledge." These include: "accurate information reflecting the actual situation," "useful information for decision-making," or "a blend of contextual information, values, experience, and rules." Often, knowledge is characterized by its role of assigning "meaning" to information.

In Chapter 5, we delved into the statement that "information is NOT knowledge." This distinction is far from a mere matter of semantics; it holds substantial significance. Whether conveyed verbally, through written texts, or in today's digital age, information can be transmitted from one individual to another. Knowledge, on the other hand, is distinct. It is intrinsically linked to the individuals or groups who possess it. It is these individuals or groups who determine how knowledge is applied and utilized. This inherent connection between knowledge and its bearers underscores the critical importance of recognizing the ethical considerations and repercussions associated with how knowledge is applied in various contexts.

As the English writer and philosopher Aldous Huxley remarked: "There are things known and there are things unknown and in between are the doors of perception."[1] The phrase "doors of perception" was originally a metaphor written by Blake in a 1790 book. This expression refers to the ways in which we perceive and interpret the world around us. Our perception limits our understanding of reality, and there could be more to the world than we currently know or understand.

In essence, knowledge is not an abstract entity; it is deeply personal, shaped by the experiences, values, and intentions of those who wield it. Knowledge not only brings an accumulation of information but, more importantly, it actively transforms information based on preexisting knowledge. "To know" means building relationships with an object from different sides and levels. The Kabbalistic interpretation of the Biblical text equates the word "know" (*yada,*) to the utmost intimate relationship: "Adam *knew* his wife, Eve" (Genesis 4:1).

There was no appropriate word in English until 1961 when the American writer Robert Heinlein coined the word "grok," a neologism in his science fiction novel, *Stranger in a Strange Land.* The Oxford English Dictionary summarizes the meaning of "grok" as: "to understand intuitively or by empathy, to establish rapport with, to empathize or communicate sympathetically (with), or to experience enjoyment. "Grok" also means "to understand so

---

[1] Cited from Danny Altmann and Rosemary Boyton, *Replace pathogens with 'perceptogens'.* Nature, 2015, vol. 518, no.7537, pages 35-35.

thoroughly that the observer becomes part of the observed – to merge, blend, intermarry, or lose one's identity in the experience."[2]

Samuel Avital beautifully stated in his book, *From Ecstasy to Lunch:* "For information to become knowledge, it must be tested, verified, and realized with one's own knowing. Otherwise, this is just pretending to know. Once we learn to distinguish between information and knowledge, we can prevent accidents and avoid confusion … Knowledge requires deep exploration."[3]

The Subjectivity-Objectivity Axis

Understanding the human dimension of knowledge is essential when navigating the intricate research terrain. Thanks to quantum physics, an astounding phenomenon known as the "observer effect" suggests that the mere act of observing can profoundly alter subatomic particles' behavior. According to the Copenhagen interpretation of quantum physics, as presented by Niels Bohr in 1927, the so-called "objective" world, at least at the subatomic level, depends on how we choose to see it.

Max Planck wrote: "The beginning of every act of knowing, and therefore the starting-point of every science, must be in our personal experience."[4] Such a view challenges the conventional concept of objectivity and forces us to acknowledge the inseparable link between the observer and the observed. However, when we apply this concept to biology, it may be more apt to call it the "participant effect." This adaptation arises from the intricate connection between the biologist and the biological entities being studied. Investigating a biological phenomenon now becomes a participatory engagement. Phillip Berg noted that the great Ari (Isaac Luria, 1534-1572), went even further when he stated that the terms "observer" and "participator" should be replaced by the term "determinator."[5]

---

[2] Rafeeq McGiveron, *From Free Love to the Free-Fire Zone: Heinlein's Mars, 1939–1987.* Extrapolation, 2001, vol. 42, no. 2.

[3] Samuel Avital. *From Ecstasy to Lunch.* 2020, page 272.

[4] *Einstein and Buddha,* 2002, p.35.

[5] *Gate of Divine Inspiration, Writings of the Ari.* vol. 12, page 10 (Cited from Phillip Berg).

At the same time, the notion of personalized knowledge introduces a delicate nuance. Objective information remains the cornerstone of our scientific exploration, providing the strongest and most reliable foundation. It's crucial to emphasize that Kabbalistic biology is very cognizant of the significance of objective information; however, it proposes a spectrum that stretches between the poles of "objective information" and "personalized knowledge." At one end of this spectrum lies the pursuit of reliable data, untainted by individual perception biases, offering a clear and unbiased view of the subject. At the other end, we encounter deeply personal experiential knowledge. This is where the biologist's unique perspective and interactions with the subject shape understanding.

The spectrum, encapsulated in the "subjectivity-objectivity axis," serves as a valuable conceptual tool. It grants biologists the freedom to base their selection of research approaches on their individual priorities, their unique circumstances, and their personal values. When we acknowledge the existence of this axis, we recognize that aspects of both objectivity and subjectivity have their place in scientific inquiry. Such a position allows researchers to strike a balance that aligns with their objectives as well as the nature of the phenomena under investigation. This newfound perspective encourages a holistic exploration of the biological world, fostering a deeper understanding of its intricacies and ethical implications.

When delving into the subjective aspects of knowledge, it's crucial to remember that Kabbalah acknowledges the presence of objective laws in nature that are independent of human influence. Understanding these fundamental laws of nature is intricately linked to our connection with the higher sefirot of Hokhmah ("wisdom") and Binah ("understanding").

"We don't see things as they are; we see them as we are."

The saying above is often attributed to Anais Nin. She indeed used these words in the "Seduction of the Minotaur" in 1961. Since then, many other authors have voiced their agreement with her statement. The idea dates back to the Talmudic tractate Berakhot (55b) when Rabbi Samuel ben Nahman, who lived in the Land of Israel from the beginning of the 3rd century until the beginning of the 4th century, said, "A man is shown in a dream only what is suggested by his own thoughts."

Solomon ben Jacob Almoli, a rabbi and physician who lived in the Ottoman Empire from the end of the 15th century to the mid-16th century, said, "Every interpretation should follow the dreamer's work and interests," and that "dream interpretation must be based on a prior detailed knowledge of the dreamer's private life."[6]

My teacher, Joel Bakst, often repeated during his lectures that it is not so much what you are looking at but *from where* you are looking. A Kabbalistic motto dictates: "Choose your point of view!" Bakst also wrote: "Remember, … it is not about what and how much you know, but rather it is about how you know what you know, no matter how little you know."[7]

This thought relates to the notice that the accommodation of new information is based on preexisting knowledge. When studying biological life, the ability to acquire knowledge comes from the dynamic interplay between what we already understand and the influx of relevant information. This profound interaction is central to our quest to comprehend the natural world. Before delving into how novel insights align with or challenge existing understanding, we must first appreciate the foundational role of preexisting knowledge. This knowledge, a product of years of learning, observation, and scientific inquiry, acts as the scaffold upon which we construct our understanding of the complex, dynamic biological world.

Our existing knowledge serves as a filter through which we perceive biological phenomena. When confronted with an unexpected observation or data point, we rarely approach it with a blank slate. Instead, we subconsciously apply our existing knowledge

---

[6] Joel Covitz, *Visions of the Night*. Shambhala, 1990, 149 pages.
[7] Joel Bakst, *Beyond Kabbalah: The Teaching That Cannot Be Taught*. 2012, page 36.

and experience that significantly influenced our initial interpretation. It's akin to viewing a painting through colored glasses; our prior experience inevitably tints the hues and shades we perceive.

Crucially, in biology, the lens of preexisting knowledge is not static. As we encounter new discoveries and insights, our developed knowledge framework can expand, refine, or even undergo a paradigm shift. However, this accommodation process is not always straightforward. In some instances, updated information may challenge existing views and paradigms, prompting resistance, cognitive dissonance, or skepticism about our ability to gain objective knowledge.

To illustrate this process, let's consider an ornithologist with years of expertise in studying bird species in Colorado. Her knowledge encompasses behavior, habitat preferences, and migratory patterns, along with other aspects of avian ecology. Now, imagine this ornithologist embarking on a research expedition to a remote tropical rainforest. Upon encountering the bird community, she discovers a strikingly complex ecological structure. Initially, she interprets these observations through the lens of what she already knows about avian ecology.

However, as she observes and collects data, a puzzling pattern emerges. It becomes increasingly evident that the bird community's functionality does not align with her current understanding of avian biology. At this juncture, she faces a choice: dismiss these observations as anomalies or consider that the existing knowledge may need to be expanded or adapted to accurately incorporate the relevant information.

Accommodating new information can be challenging. It may even require that long-held perceptions of bird behavior and ecology may need to be revised. Yet, the ornithologist was encouraged to critically examine the evidence and, if necessary, adjust the preexisting knowledge to better reflect the newly discovered biological reality. This example is intentionally simple. Accepting novel concepts about complex biological phenomena, such as birds' evolution or their role as reservoirs of flu viruses, can be far more challenging. Herein lies a critical aspect of knowledge acquisition: the role of personal perspective.

Biologists must decide which sources of information to prioritize. While objective information is vital, a personal perspective allows them to select and evaluate sources that align with the existing knowledge framework and emerging requirements. This choice does not reject objectivity, but it does recognize that not all information holds equal strength to shape human understanding.

The dynamic interplay between cherishing existing knowledge and embracing new information is familiar to many biologists. However, the more intricate challenge lies in navigating our own personal orientation along the "subjectivity-objectivity" axis. As scientists, we are influenced by a multitude of factors that shape our perspectives and guide our choices. These influences are as diverse and unique as our upbringing and cultural background, mentors, peers, the scientific literature we engage with, and the broader media landscape.

As discussed in Chapter 1, we have stringent criteria for testing and falsifying hypotheses. As long as we maintain the stance that knowledge should be devoid of personal biases and adhere to rigorous criteria of repeatability, we appear to avoid complications. However, we encounter the limitations of this approach when delving into biology's complex and enigmatic realm, particularly its imperceptible aspects. David Hume, the Scottish philosopher of the 18th century, wrote in his book, "Treatise of Human Nature," that no logical argument can prove that "the incidents about which we have no experience are similar to those for which we have experience."[8]

Such a statement should not be perceived as discouraging. It warns us about using the process of interpreting information about visible aspects of life to gain knowledge about invisible aspects. The danger here lies in the potential rejection of scientific truths

---

[8] David Hume, *Treatise of Human Nature.* 1986, 688 pages.

altogether. The prevailing notion is that data must be utterly reliable, information should aspire to completeness, and knowledge is simply the product of information-gained prominence during the post-modern era. This trend has led to a growing decline in public confidence in science's capacity to elucidate and predict biological catastrophes. This is in a world that encompasses not only the external environment but also our very selves.

Navigating the Complex Web of Values
Navigating this invisible web of influences requires self-awareness and a conscious effort to strike a balance between subjectivity and objectivity. It calls for humility in acknowledging the potential biases that may arise from these influences. It also demands being committed to critical thinking, evidence evaluation, and the ability to consider alternative perspectives. When we embrace the necessity of incorporating subjective elements into our acquisition of new knowledge, we enter uncharted territory.

Science doesn't exist in isolation but is deeply entwined with the contexts within which it operates. These contexts are both external and internal. One cannot divorce science from its broader context, which encompasses social issues and culture. Scientists are influenced by prevailing social norms, political ideologies, and cultural beliefs of the societies in which they live and work. These external forces can shape scientific inquiry, impact research funding, and even influence scientific results interpretation. In contemporary times, for example, debates over issues like climate change, stem cell research, and genetically modified organisms are often fueled by clashes between different ideologically affiliated groups.

Even more critical is the scientist's internal context. Personal values, ethical norms, education, and the willingness to embrace alternative thinking all play a significant role in shaping the scientific process. While science aspires to objectivity, it is conducted by humans, who bring their biases, perspectives, and personal values to the laboratory bench or fieldwork. Value-free science, as an ideal, exists only under unnatural and short-lived conditions. Scientists are not automatons but thinking and feeling individuals. They make decisions about what to study, how to design experiments, and how to interpret data. These decisions are influenced by their own moral compasses, worldviews, and ethical principles.

The question, "Do external and internal factors impact research activity positively or negatively?" is complex. Some may argue that the influence of the external context is negative, while personal values are positive. However, Kabbalistic biology suggests that it's not a matter of evaluating these influences as favorable or undesirable. Rather, the issue lies in the conjunction, "or."

When the conjunction "or" becomes an "and," — we can integrate both external and internal influences to enrich our research endeavors. In essence, the impact of these factors on research activity is not a matter of judgment but of choice. Researchers can harness the synergy of both external and internal influences, shaping their work in both innovative and meaningful ways. This perspective aligns with the idea that acquiring knowledge is a dynamic, evolving process. This is where the interplay of diverse influences contributes to a richer understanding of the world around us, including the biological world.

In science, we, as researchers, can choose how much influence both external and internal factors will exert on our work. It's not a binary choice between positive and negative. Instead, it's about recognizing that we have the autonomy to navigate the intricate interplay between these influences. This perspective encourages us to embrace the complexity of research. It acknowledges that external factors can offer valuable insights and challenges, just as personal values can provide motivation and ethical considerations.

Ethical issues permeate science, from society's norms to individual scientists' moral codes. At the institutional level, ethical committees, such as IRB (Institutional Review Board that reviews research studies involving human subjects) and IACUC (Institutional Animal Care and Use Committee), play a pivotal role in safeguarding the integrity of research. However, they also pose challenges. Committees, even if well-intentioned, can sometimes distort a scientist's morality by imposing external standards that may not align with the individual researcher's values.

Ethics are particularly noticeable in biology. Biologists grapple with such ethical issues as the use of animals in research, patient privacy protection in medical studies, the responsible use of genetic manipulations, genetic screening, experimentation with embryos, and many other moral dilemmas. On the other hand, biology explores issues directly related to ethics, such as the origin and evolution of cooperation, competition, and altruism.

Recognizing the role of values in science does not diminish science's credibility; rather, it enriches it. It invites a broader conversation about scientific inquiry and its relationship to the surrounding world. Science's traditional linear models, where objective observations lead to objective conclusions, are no longer tenable at the level of nuanced understanding.

Da'at, the Nexus of Knowledge

Fortunately, engaging in Kabbalistic practice offers a distinct advantage in honing one's ability to shift perspectives by raising the level of discernment required when acquiring new knowledge. Doubts about the existence of true knowledge are, in fact, foreign to Kabbalistic thinking.

Kabbalah holds that it is only the knowledge of "particulars" that fails to place things in a wider context of life. Sanford Drob wrote: "When knowledge is separated from life, it becomes purely technical... For the Kabbalists, such knowledge is an act of contemplation that does not embrace the sefirot in their totality but that isolates an individual *Sefirah*, particularly the final one, *Malkhut* or Kingship... When knowledge is disconnected from life, ... the tree of knowledge becomes a tree of death."[9]

Chapter 11 described the alternative sefirah Da'at, literally

---

[9] Sanford Drob, *Symbols of the Kabbalah.* 2000, page 356.

meaning "knowledge" in Hebrew. It functions as a nexus of knowledge that integrates all sefirot except Keter. As illustrated in Figure 32, Keter and Da'at have an alternative relationship. Th ere are ten sefirot in the Tree of Life, and Keter is one of them, while Da'at is hidden. We also have ten sefirot in the Tree of Knowledge, including Da'at, while Keter is hidden. We cannot know anything about Keter's essence. In the Tree of Knowledge, Da'at temporarily replaces Keter during a learning process.

In the Tree of Knowledge, Da'at occupies a pivotal position, serving as the bridge that connects the higher and lower realms of understanding. Da'at serves as the connection between the higher and lower sefirot and oversees the ongoing maintenance of this profound connection. This sefirah represents the transformative moment of an unprecedented fusion of the subjective and the objective. It is here that the "participant dimension" finds its symbolic anchor.

In the words of Berke and Schneider, Da'at denotes a moment when "the knower and the known merge." They compare this moment to the recognition that the archer, the arrow-in-flight, and the target are one.[10] Kabbalistic sources often used the erotic imagery of Da'at — "Knowledge" — as a sexual union between a man and woman with an intertwining of selves. They also revealed the essence of Da'at by rearranging the Hebrew letters of this word to get *aydoot*, meaning "witness."

This contemplation was connected to the tradition that requires two witnesses to make a testimony. According to this view, Hokhmah ("Wisdom") and Binah ("Understanding") are witnesses that should be consistent with each other. The act of unification demands prior differences – Adam and Eve, arrow and target, Hokhmah, and Binah.[11]

---

[10] Joseph Berke and Stanley Schneider, *Centers of Power*. 2008, page 101.
[11] Ibid., page 102.

A Narrow Bridge above the Abyss

Because of the almost impassable barrier between the upper and lower sefirot, this gap is called the "abyss." The lifeforce flows along the path structure of the Tree of Life except when it jumps between Binah and Hesed. The notion of a gap can be found in ancient Kabbalistic literature as the "Great Abyss" – *tehom* in Hebrew (*tov-hey-vav-mem*). Understanding the symbolic meaning of this barrier is crucial for us to comprehend the difference between "information" circulating at the lower level and "understanding" and "wisdom" at the upper level. Daʿat — "knowledge" is only one way to bridge the "abyss" (Figure 72).

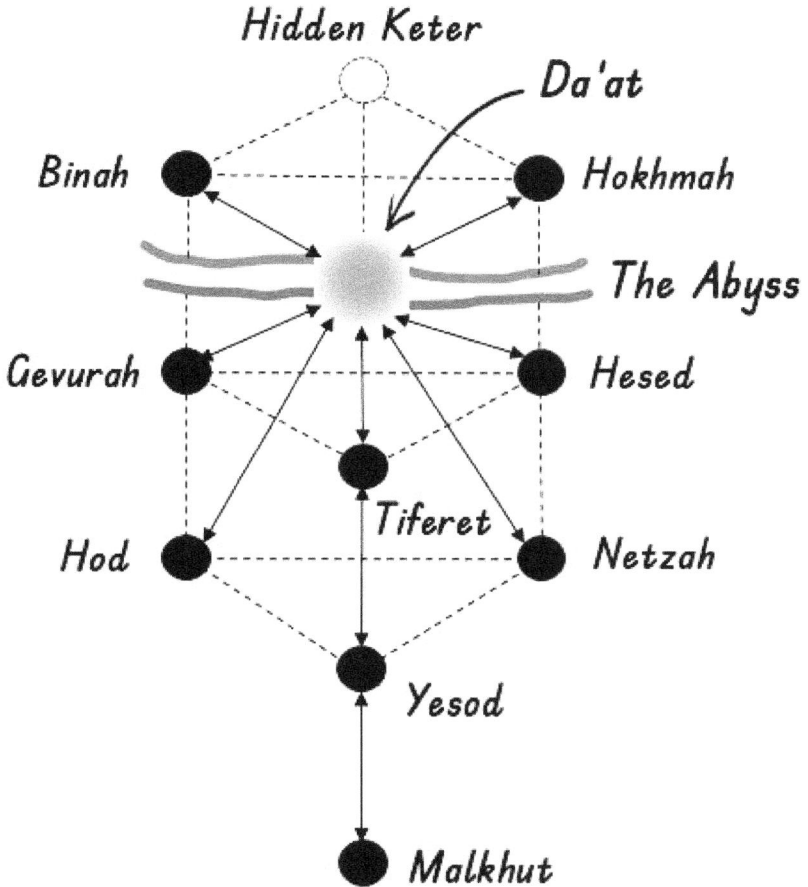

Figure 72. Da'at ("Knowledge") is only one way to bridge the "abyss," the gap between the upper and lower Sefirot.

Imagine a narrow, tightrope-type bridge above a deep abyss. Each step shakes the bridge. Walking such a narrow and shaking bridge requires courage, along with an overwhelming feeling of fear and awe. Both "fear" and "awe" have the same word in Hebrew – *ira* (*yud-reish-alef*). Searching for true knowledge is like walking a tightrope across a bottomless gulf. How many scientists experience fear and awe while searching for knowledge? Should they? If they gather accurate data, perhaps not. If they precisely analyze information, it is

587

not necessary. However, we must avoid calling this "knowledge" in terms of the meaning of Daʿat as Kabbalah teaches it. Sir Karl Popper, one of the 20th century's most influential philosophers of science, wrote: "The phenomenon of human knowledge is no doubt the greatest miracle in our universe."[12] It is time to stop taking for granted that knowledge is a direct result of information analysis!

How often do people misuse the words "understanding" and "wisdom?" The precarious path we must tread and often overlook, is the bridge called "knowledge." "Understanding" is a term commonly used to signify a grasp of facts, concepts, or information. We can encounter phrases like "I understand the text" or "I understand the evolution process." In these instances, understanding is often reduced to mere comprehension—a superficial awareness of the subject matter. However, true understanding reaches deeper, delving into profound knowledge. Wisdom, on the other hand, exists on an even more elusive plane. It transcends understanding itself, encompassing the discernment of deeper truths in life. It encompasses the intricacies of biological and human nature and the profound interconnectedness of all things.

Kabbalah, with its symbolism embodied in the sefirot Binah and Hokhmah, provides well-defined meanings for understanding and wisdom. To grasp the authentic essence of understanding and wisdom, we must embark on a treacherous journey through knowledge. Knowledge serves as the bridge leading us to understanding and wisdom.

There exists a well-known Hebrew song called *"Kol Ha'Olam Kulo,"* originating from the wisdom of the revered Hasidic Rabbi Nachman of Breslov (1772–1810). The lyrics of this song resound with the words, "The entire world is a very narrow bridge, and the main thing is to have no fear at all." Yet, a more precise translation of the second part of this verse reveals a profound truth: "The essential thing is not to become paralyzed by your fear." This distinction

---

[12] Karl Popper, *Objective Knowledge: An Evolutionary Approach.* 1972, page VII.

carries a valuable lesson. Fear is an inherent part of human knowledge and experience. Instead of allowing fear to immobilize our will and hinder our search for knowledge, we can confront it with courage, resilience, and humility.

Nilton Bonder retells a Hasidic story about a conversation between a rabbi and a trapeze artist on the secret of not losing balance. The tightrope walker inquired, "Where do you think you should look to keep your balance?" The rabbi replied, "Certainly neither at the ground nor at the rope." "Correct," the performer said. "You have to keep looking at the poles at the end of the rope. And when is the most dangerous moment?" The rabbi answered, "The moment when you have to turn around, and you lose your reference point for a second." "Precisely." The performer agreed. Bonder commented on this story, "Any researcher will tell you that the moments when you turn around – when you are left without a pole to guide you – are key to keeping your balance and therefore to successfully completing your process of inquiry."[13]

Human knowledge is more than a linear path; it's a nuanced dance on a tightrope. At times, this dance leads us to a precipice, where familiar ground leads to the unknown. It's in these moments when we are left without a pole to guide us that we discover a profound truth: balance is the key to the process of acquiring knowledge. Knowledge, like the Tree of Life, is not static. It's a dynamic, ever-changing stream we navigate as participants. To truly grasp the depths of this knowledge, we must learn to balance ourselves on the narrow bridge between Da'at and the other sefirot that encompasses our understanding of the biological world.

Hesed encourages us to be open-minded, to embrace new ideas, and to approach the unknown with curiosity. Gevurah, on the other hand, compels us to be critical thinkers, to assess information rigorously, and to discern fact from conjecture. *It is in the balance between the Sefirot that we find the essence of Da'at – true knowledge.* Knowledge requires both the willingness to explore new frontiers and the discipline to evaluate their worth. It thrives when we

---

[13] Nilton Bonder, *Yiddishe Kop.* 1999, page 36.

dare to question our preconceptions and acknowledge the limitations of our current understanding.

This harmonious interplay of sefirot is not one-size-fits-all. During a specific period and under specific circumstances, the unique configuration of the relationship between sefirot enables us to acquire knowledge. As we navigate this intricate dance, let us remember that it is the moment when we turn around, when we relinquish the safety of the known, that we will find ourselves propelled toward a new awareness of the invisible aspects of biological life.

Concluding Remarks from This Journey:

"The Map is *Not* the Territory!"

We are about to conclude our journey through the Ten Dimensions that await exploration. However, as you prepare to investigate the invisible realm of biological life, there is one final caution you must heed: "The map is *not* the territory." This profound phrase, simple but powerful, is attributed to the Polish-American philosopher Alfred Korzybski. Its message is as relevant in biology as it is in our daily lives. Korzybski used it to warn us not to mistake models and representations of reality for reality itself!

In this book, we've explored how these Ten Dimensions can help us orient ourselves during our searches, similar to how we would use topographic symbols on a map. However, while these dimensions are invaluable guides to gaining insight into the complexities of biological life, they are *not* a substitute for the actual journey. *Maps, no matter how detailed or accurate, can never replace the experience of traversing the terrain they represent.*

In the realm of biology, as in geography, having the right "map" is only one aspect of the journey. It can be compared to possessing accurate maps and the skills to read them to prepare for a journey through
uncharted wilderness. Yet, even with the most comprehensive maps

and navigational skills, the journey through biology's wild, unexplored territories requires more than theoretical information. The real test will be successfully making the journey and gaining new information about Life itself.

To truly explore and understand biological life, one must become a "biologist" in the fullest sense of the word. This means acquiring a solid education, engaging in meaningful interactions with colleagues, immersing oneself in specialized literature, and, most importantly, actively participating in biological research. Such preliminary training is akin to becoming a seasoned traveler, one who not only possesses maps and navigational skills but also knows how to navigate the unpredictable challenges of the terrain.

The "map-territory" relationship relates to the connection between an object and how it appears to others. This reminds us of the potential pitfalls of confusing terminology with what the words are meant to signify. Just as mistaking the map for actual territory is a logical fallacy, melding theoretical knowledge with biological practices can also lead to misconceptions and limitations. In the words of Alan Watts, an English writer and philosopher, "The menu is not the meal." Just as studying a menu is not equal to the experience of savoring the actual food, acquiring knowledge about biology is not the same as being a biologist in the truest sense.

The practice of the Ten Dimensions proposed in this exploration extends far beyond scientific research. It beckons us to apply these insights and orientations to our everyday lives. To truly embody Kabbalistic biology is to seamlessly integrate these dimensions into our existence outside the laboratory or field.

As we navigate the complex terrain of our personal and professional lives, these Ten Dimensions serve as guiding stars. They illuminate our paths and enable us to make choices that are not only informed by knowledge—they are also enriched by wisdom and ethical discernment. In this way, Kabbalistic biology actually becomes a way of life, bridging the gap between theory and practice. It harmonizes our scientific pursuits with the broader tapestry of existence itself. As we engage with this holistic practice, we become not only biologists but also stewards of life's profound interconnections, both within and beyond the laboratory walls.

The central conclusion of our exploration is clear: a biologist is not merely someone who talks about biology or possesses theoretical knowledge. A biologist dares to venture into the complex terrain of biological life, armed not just with maps but with the skills, experience, and commitment required to navigate this fascinating realm.

In the end, it is both the journey and one's immediate experience that truly define a biologist. Therefore, let us embark on this journey with curiosity, humility, and a sense of gratitude for the opportunity to enter and explore the secret realms of biology.

Acknowledgments

It would be impossible to list those to whom I feel grateful for their impact on my development as a person, a biologist, and a Kabbalah student. All of them have contributed to this book, whether willingly or unwittingly. This is especially true of my numerous biology colleagues. How many teachers, students, laboratory co-workers, and field expedition fellows have I encountered during my almost half-century professional journey? Hence, I acknowledge their unique contributions without mentioning their names.

There are a few teachers whom I particularly wish to thank for introducing me to Kabbalah. My first instructor in climbing the Tree of Life was Chirley Chambers, the founder of the Kabalah Karin Center in Atlanta; she created a unique blend of Kabbalistic knowledge and Eastern mysticism. After moving to Colorado, I became engaged in a more traditional approach to Kabbalah. During my first decade there, I was strongly influenced by the lectures of Rabbi Zalman Schachter-Shalomi, who lived and taught in neighboring Boulder.

It took years of learning from many fine teachers before I found a teacher who spoke to my soul. Samuel Ben-Or Avital was born in the Atlas Mountains of Morocco. He was educated in the home of a remarkable family that traced its lineage to 15th-century Spain, passing the Kabbalistic tradition from father to son in an unbroken line. I joined a small group of his students that gathered weekly for an introduction to the Authentic Kabbalah. Only after 12 years of studying the hidden meaning of Hebrew letters, I was allowed to explore the Tikuney haZohar and Sefer haZohar with my personal teacher.

The impact of my Kabbalistic education led to my decision to postpone writing the book. I had intended to share my experience as a scientist investigating invisible realms of biological life, using the tools provided by the Kabbalistic tradition. However, the more I learned, the more aware I became of how deeply I needed to reach before sharing it with others. Eventually, it became unbearable to keep the knowledge I had acquired "in the closet."

After deciding to write the book, I had to consider the question: "Who is the target audience?" Would they be fellow biologists? Explorers of the authentic Kabbalah? Metaphysics seekers? Nature lovers? To gain clarity on this question, I checked the names of editors and literary agents who had worked on books related to Kabbalah.

My first choice was Claire Gerus, and I emailed her to introduce the book idea on the coming Rosh Hodesh Nisan. In a few minutes, she responded, inviting me to call her. She was fascinated by the topic! For the next two years, Claire became my mentor as I wrote the book. Her guidance as I entered the world of book publishing was as valuable as her editorial corrections and suggestions. I found her to be very gentle and delicate as I walked on the narrow bridge above the abyss between modern biology and ancient knowledge. Claire also encouraged me to reach a wider circle of readers than I had initially anticipated.

It was not an easy decision to write for a wider audience. I had never planned to write popular books. At the same time, I realized that an expanded readership would allow me to meet those outsides of my usual circle of contacts. I intuitively felt the need for a visual representation of complex and difficult-to-understand issues in theoretical biology and Kabbalistic sources. My darling daughter, Jenia, was a professional artist and designer, although she had never worked on book illustrations. To my delight, Jenia enthusiastically agreed to join me on this adventure.

By then, Jenia had become an accomplished designer who designed offices for many Manhattan companies, including the luxury book publishing house Assouline, Neoscape agency, Square Mile Capital Management, Cache headquarters in New York's Garment District, Martha Stewart Living Omnimedia, Martin Scorsese's office, and others. When the COVID pandemic badly damaged New York's office design business, Jenia found more time for her new creative activity: book illustration. As the work on the illustrations was about to come to an end, Jenia suddenly passed away from this world to a Higher one.

THIS BOOK IS DEDICATED TO HER MEMORY.

I was blessed not only by working with my daughter but also with my son, Roman, one of the finest biologists I know. Roman provided some very helpful insights into the draft version.

Finally, this book is possible because of two women, a mother and a wife who share the name, Olya. One gave birth to me; another made me a *mensch*. I owe this book to both of them with the deepest love.

Bibliography

Abbott, Edwin A. *Flatland: A Romance in Many Dimensions*. New York, Dover, 1992.

Afilalo, Raphael. *Kabbalah Concepts*. Kabbalah Editions, 2006.

Afterman, Allen. *Kabbalah and Consciousness*. Sheep Meadow Press, 2005.

Ahl, Valerie and T.F.H. Allen. *Hierarchy Theory: A Vision, Vocabulary, and Epistemology*. New York, Columbia University Press, 1996.

Alexander, Christopher. *The Nature of Order*. Book One: *The Phenomenon of Life*, 2002.

Alexander, Victoria. *The Biologist's Mistress: Rethinking Self-Organization in Art, Literature, and Nature*. Litchfield Park, Arizona, Emergent Publications, 2011.

Alexenberg, Mel. *The Future of Art in a Digital Age: From Hellenistic to Hebraic Consciousness*. Bristol, UK, Intellect, 2006.

Arzy, Shahar and Moshe Idel. *Kabbalah: A Neurocognitive Approach*. Yale University Press, 2015.

Ashlag, Yehudah L. *In the Shadow of the Ladder*. Safed, Israel, Nehora Press, 2002.

Aviezer, Nathan. *Modern Science and Ancient Faith*. KTAV Publishing House, 2013.

Avise, John. *The Genetic Gods*. Harvard University Press, 1998.

Avital, Samuel. *The Invisible Stairway*. Boulder, Kol-Emeth Publishers, 1982.

Avital, Samuel. *Mime and Beyond: The Silent Outcry*. Prescott Valley, Hohm Press, 1985.

Avital, Samuel. *From Ecstasy to Lunch*. Boulder, Colorado, Kol-Emeth Publishers, 2020.

Bakst, Joel D. *The Secret Doctrine of the Gaon of Vilna*. Manitou Springs, Colorado, City of Luz Publications, 2009.

Bakst, Joel D. *Beyond Kabbalah: The Teaching That Cannot Be Taught*. Manitou Springs, Colorado, City of Luz Publications, 2012.

Bakst, Joel D. *Kabbalah of the Adamic Messiah.* Manitou Springs, Colorado, City of Luz Publications, 2020.

Barash, David P. *Buddhist Biology.* Oxford University Press, 2014.

Barbieri, Marcello. *The Organic Codes: An Introduction to Semantic Biology.* Cambridge University Press, 2002.

Bateson, Gregory. *Steps to an Ecology of Mind: Collected Essays in Anthropology, Psychiatry, Evolution, and Epistemology.* San Francisco, Chandler Pub, 1972.

Berke, Joseph and Stanley Schneider. *Centers of Power: Convergence of Psychoanalysis and Kabbalah.* Lanham, Maryland, Jason Aronson, 2008

Berthoz, Alain. *Simplexity.* New Haven, Yale University Press, 2012.

Bird, Richard J. *Chaos and Life.* New York, Columbia University Press, 2003.

Bláha, Josef. *Azriel of Gerona: Commentary on the Ten Sephiroth.* The Karolinum Press, 2015.

Bloom, Harold. *Kabbalah and Criticism,* New York, Continuum, 2005.

Bohm, David. *Wholeness and the Implicate Order.* Routledge, 2002.

Bonder, Nilton. *Yiddishe Kop.* Boston & London, Shambala, 1999.

Bonder, Nilton. *The Kabbalah of Time.* Bloomington, Indiana, Trofford Publishing, 2009.

Bouchard, Frederic and Philippe Huneman (eds). *From Groups to Individuals.* Cambridge, Massachusetts, The MIT Press, 2013.

Burger, Dionys. *Sphereland: A Fantasy About Curved Spaces and an Expanding Universe.* Harper and Row Publishers, 1983.

Capra, Fritjof. *The Web of Life.* New York, Anchor Books, 1996.

Capra, Fritjof and Pier L. Luisi. *The Systems View of Life.* Cambridge University Press, 2014.

Carey, Nessa. *Junk DNA: A Journey Through the Dark Matter of the Genome.* Columbia University Press, 2017.

Chambers, Shirley. *Kabalistic Healing.* Los Angeles, Keats Publishing, 2000.

Chase, Jonathan and Mathew Leibold. *Ecological Niches: Linking Classical and Contemporary Approaches.* University of Chicago Press, 2003.

Chilton, Bruce. *Redeeming Time*. Peabody, Massachusetts, Hendrickson Publishers, 2002.

Cohen, Irun. *Rain and Resurrection*. Landes Bioscience, Austin, Texas, 2010.

Cohen, Jack and Ian Stewart. *The Collapse of Chaos: Discovering Simplicity in a Complex World*. Viking, 1994.

Cole, Peter. *The Poetry of Kabbalah: Mystical Verse from the Jewish Tradition*. The Margellos World Republic of Letters, Yale University Press, 2014.

Cordovero, Moshe. *Pardes Rimonim, Orchard of Pomegranates*. Providence University, Multilingual edition, 2007.

Darwin, Charles. *The Origin of Species*. New York, Gramercy Books, 1979.

Dawkins, Richard. *The Extended Phenotype*. Oxford University Press, 1999.

Dolgopolski, Sergey. *The Open Past: Subjectivity and Remembering in the Talmud*. Fordham University Press, New York, 2013.

Dossey, Larry. *Space, Time & Medicine*. Boulder, Colorado, Shambhala Publications, 1982.

Drob, Sanford. *Symbols of the Kabbalah*. Northvale, New Jersey, Jason Aronson Inc., 2000.

Drob, Sanford. *Kabbalistic Metaphors*. Northvale, New Jersey, Jason Aronson Inc., 2001.

Drob, Sanford. *Kabbalistic Vision*. New Orleans, Louisiana, Spring Journal Books, 2009.

Dupre, John. *Process of Life*. Oxford University Press, 2012.

Eldredge, Niles. *Why We Do It: Rethinking Sex and the Selfish Gene*. W. W. Norton & Company, 2005.

Eldredge, Niles and Steven J. Gould. *The Pattern of Evolution*. New York, W.H. Freeman and Company, 1999.

Elior, Rachel. *The Paradoxical Ascent to God: The Kabbalistic Theosophy of Habad Hasidism*. State University of New York Press, 1992.

Emmeche, Claus and Kalevi Kull. *Towards a Semiotic Biology: Life is the Action of Signs*. Imperial College Press, 2011.

Fain, Benjamine. *Law and Providence*. Israel, Urim Publications, 2011.

Firestein, Stuart. *Ignorance: How It Drives Science.* Oxford University Press, 2012.

Frankel, Estelle. *Sacred Therapy.* Boston & London, Shambala, 2005.

Frankel, Estelle. *The Wisdom of Not Knowing.* Boulder, Shambala, 2017.

Frankl, Viktor. *Man's Search for Meaning.* Beacon Press, 2006.

Friedman, Bruce. *Mystery of Black Fire, White Fire.* iUniverse, 2016.

Friedman, Norman. *Bridging Science and Spirit.* St. Louis, Living Lake Books, 1994.

Ginsburgh, Yitzchak. *The Hebrew Letters.* Jerusalem, Gal Einai Publications, 1990.

Ginsburgh, Yitzchak. *What You Need to Know about Kabbalah.* Jerusalem, Gal Einai, 2006.

Ginsburgh, Yitzchak. *137: The Riddle of Creation.* Kfar Chabad, Gal Einai, 2018.

Ginsburgh, Yitzchak. The Breath of Life. Jerusalem, Gal Einai, 2018.

Ginsburgh, Yitzchak and Moshe Genuth. *Wisdom: Integrating Torah & Science.* Kfar Chabad, Israel, Gal Einai, 2018.

Gissis, Snait et al. (eds). *Landscapes of Collectivity in the Life Sciences.* Cambridge, Massachusetts, The MIT Press, 2017.

Gleick, James. *Chaos: Making a New Science.* Penguin Books, 1987.

Goldschlager, Ron and Adin Steinsaltz. *The Mystery of You: A Journey through the Paradoxes of Life.* Hybrid Publishers, 2011.

Goldstein, Jeffrey. *The Unshackled Organization: Facing the Challenge of Unpredictability Through Spontaneous Reorganization.* Portland, Oregon, Productivity Press, 1994.

Gould, Stephen J. *The Structure of Evolutionary Theory.* Harvard University Press, 2002.

Griffiths, Paul and Karola Stotz. *Genetics and Philosophy: An Introduction.* Cambridge University Press, 2013.

Grobstein, Clifford. *The Strategy of Life.* San Francisco, W.H. Freeman and Company, 1974.

Halevi, Z'ev. *The Way of Kabbalah*. Samuel Weiser, York Beach, Maine, 1991.

Haralick, Robert. *God Consciousness, The Exercises: Working the Sefirot and Netivot*. Pomona, New York, Torah Books, 2014.

Henderson, Linda D. *The Fourth Dimension and Non-Euclidean Geometry in Modern Art*. The MIT Press, 2018.

Hoffman, Edward. *The Way of Splendor: Jewish Mysticism and Modern Psychology*, Jason Aronson, Inc. 1993.

Hoffman, Edward. *The Hebrew Alphabet: A Mystical Journey*. Chronicle Books, 1998.

Hoffmeyer, Jesper. *Biosemiotics: An Examination into the Signs of Life and the Life of Signs*. University of Scranton Press, 2008.

Hölldobler, Bert and Edward Wilson. *The Superorganism – the Beauty, Elegance, and Strangeness of Insect Societies*. London and New York, W.W. Norton, 2008.

Jablonka, Eva and Marion Lamb. *Evolution in Four Dimensions*. Cambridge, Massachusetts, The MIT Press, 2005.

Kaku, Michio. *Hyperspace*. New York, Anchor Books, 1994.

Kaku, Michio and Jennifer Thompson. *Beyond Einstein: The Cosmic Quest for the Theory of the Universe*. New York, Anchor Books, 1995.

Kaplan, Aryeh. *The Bahir Illuminated*. Boston, Weiser Books, 1979.

Kaplan, Aryeh. *Inner Space*. Jerusalem, Moznaim Publishing Corporation, 1990.

Kaplan, Aryeh. *Sefer Yetzirah - The Book of Creation: in Theory and Practice*. San Francisco, Weiser Books, 1997.

Kauffman, Stuart. *The Origins of Order: Self Organization and Selection in Evolution*. Oxford University Press, 1993.

Klein, Gary. *Streetlights and Shadows*. Cambridge, Massachusetts, The MIT Press, 2011.

Klein, Eliahu. *Kabbalah of Creation: The Mysticism of Isaac Luria, Founder of Modern Kabbalah*. North Atlantic Books, 2005.

Kosko, Bart. *Fuzzy Thinking*. New York, Hyperion, 1993.

Kull, Kalevi et al. *An Introduction to Our View on the Biology of Life Itself.* In *Towards a Semiotic Biology.* Imperial College Press, 2011.

Kushner, Lawrence. *The River of Light.* Woodstock, Vermont, Jewish Lights Publishing, 1993.

Labowitz, Shoni. *Miraculous Living: A Guided Journey in Kabbalah through the Ten Gates of the Tree of Life.* New York, Fireside, 1996.

Laitman, Michael. *Kabbalah Science and the Meaning of Life.* Toronto, Canada, Kabbalah Publishers, 2006.

Lamm, Norman. *The Shema: Spirituality and Law in Judaism as Exemplified in the Shema, the Most Important Passage in the Torah.* Philadelphia, The Jewish Publication Society, 2000.

Larter, Raima. *Spiritual Insights from the New Science.* World Scientific, 2021.

Lewin, Roger. *Complexity: Life at the Edge of Chaos.* Chicago, The University of Chicago Press, 1999.

Lima, Manuel. *The Book of Trees: Visualizing Branches of Knowledge.* New York, Princeton Architectural Press, 2008.

Lipton, Bruce and Steve Bhaerman. *Spontaneous Evolution.* Hay House, 2009.

Luzzatto, Moshe Ch. *138 Openings of Wisdom.* Translation by Avraham Greenbaum, Jerusalem, Azamra Institute, 2005.

MacArthur, Robert. *Geographical Ecology.* Princeton, Princeton University Press, 1984.

Mansfield, Victor. *Synchronicity, Science, and Soul-Making.* Open Court, 1998.

Margulis, Lynn and Dorion Sagan. *What is life?* New York, Simon & Schuster, 1995.

Margulis, Lynn and Dorion Sagan. *Acquiring Genomes: A Theory of the Origins of Species.* Basic Books, 2002.

Maturana, Humberto and Francisco Varela. *The Tree of Knowledge.* Shambhala, 1998.

Matt, Daniel. *God & the Big Bang.* Woodstock, Vermont, Jewish Lights Publishing, 1996.

Menzi, Donald and Zwe Padeh. *The Tree of Life. The Palace of Adam Kadmon*. New York, Arizal Publications Inc. 2008.

Moore, John A. *Science as a Way of Knowing: The Foundations of Modern Biology*. Harvard University Press, 1993.

Munk, Michael. *The Wisdom of the Hebrew Alphabet*. Mesorah Publications, Brooklyn, 1986.

Negoita, Constantin. *Expert Systems and Fuzzy Systems*. Benjamin-Cummings Pub Co., 1985.

Nelson-Isaacs, Sky. *Living in Flow: The Science of Synchronicity and How Your Choices Shape Your World*. North Atlantic Books, 2019.

Neuman, Yair. *Reviving the Living: Meaning Making in Living Systems*. Elsevier, 2008.

Newberg, Andrew and David Halpern. *The Rabbi's Brain*. Nashville, Tennessee, Turner Publishing Company, 2018.

Nicolescu, Basarab. *The Hidden Third*. New York, Quantum Prose, Inc., 2016.

Ouspensky, Piotr. *The Fourth Dimension*. Kessinger Publishing, 2010.

Parrington, John. *The Deeper Genome: Why There Is More to the Human Genome Than Meets the Eye*. Oxford University Press, 2017.

Parrish, David. *Nothing I See Means Anything*. Boulder, First Sentient Publications, 2006.

Peat F. David. *Infinite Potential: The Life and Times of David Bohm*. Helix Books, 1997.

Pietsch, Theodore. *Trees of Life: A Visual History of Evolution*. The John Hopkins University Press, Baltimore, 2012.

Pigliucci, Massimo and Gerd Muller. *Evolution – The Extended Synthesis*. Cambridge, Massachusetts, The MIT Press, 2010.

Pinson, DovBer. *Thirty-Two Gates of Wisdom*. BenYehuda Press, 2008.

Pocheville, Arnaud. *The Ecological Niche: History and Recent Controversies*. In *Handbook of Evolutionary Thinking in the Sciences*. Dordrecht, Springer, 2015.

Pollack, Rachel. *The Kabbalah Tree*. St. Paul, Minnesota, Llewellyn Publications, 2005.

Popper, Karl. *The Logic of Scientific Discovery*. Harper & Row Publishers, 1968.

Pradeu, Thomas. *The Limits of the Self: Immunology and Biological Identity*. Oxford University Press, 2012.

Prigogine, Ilya and Isabelle Stengers. *Orders Out of Chaos: Man's New Dialogue with Nature*. Bantam Books, 1984.

Primack, Joel and Nancy Ellen Abrams. *The View from the Center of the Universe*. Riverhead Books, 2006.

Raskin, Aaron. *Letters of Light*. Brooklyn, N.Y., Sichos in English, 2003.

Reid, Robert. *Biological Emergence*. Cambridge, Massachusetts, The MIT Press, 2007.

Rosen, Robert. *Life Itself*. New York, Columbia University Press, 1991.

Rosenak, Michael. *Tree of Life, Tree of Knowledge*. Westview Press, 2001.

Rosenberg, Eugene and Ilana Zilber-Rosenberg. *The Hologenome Concept: Human, Animal and Plant Microbiota*. Springer, 2016.

Salthe, Stanley. *Development and Evolution*. Cambridge. Massachusetts, The MIT Press, 1993.

Sapp, Jan. *Genesis: The Evolution of Biology*. Oxford University Press, 2003.

Schipper, Hyman. *Kabbalistic Panpsychism: The Enigma of Consciousness in Jewish Mystical Thought*. Iff Books, 2021.

Schachter-Shalomi, Zalman. *Credo of a Modern Kabbalist*. Victoria Canada, Trafford Publishing, 2005.

Shneur Zalman of Liadi. *Tanya - Likutei Amarim*. Kehot Publication Society, 2014.

Schochet, Jacob. *The Mystical Tradition: Insights into the Nature of the Mystical Tradition in Judaism (The Mystical Dimension)*. Kehot Pubns Society, 1990.

Scholem, Gershom. *On the Kabbalah and Its Symbolism*. New York, Schochen Books, 1969.

Schmalhausen, Ivan. *Factors of evolution: The theory of stabilizing selection*. Blakiston Co., 1949.

Schneider, Sarah Y. *Evolutionary Creationism: Kabbala Solves the Riddle of the Missing Links.* Jerusalem, Israel, A Still Small Voice, 2005.

Schneider, Sarah. *Kabbalistic Writings on the Nature of Masculine & Feminine.* Jerusalem, A Still Small Voice, Israel, A Still Small Voice, 2007.

Schrodinger, Erwin. *What is life?* Cambridge, UK, Cambridge University, 1944.

Schwartz, Arturo. *Kabbalah and Alchemy.* Northvale, N.J., Jason Aronson, Inc., 2001.

Schwartz, Gary and Linda Russek. *Living Energy Universe: A Fundamental Discovery that Transforms Science and Medicine.* Hampton Roads Publishing Company, 1999.

Seidenberg, David M. *Kabbalah and Ecology.* Cambridge University Press, 2015

Sheinkin, David. *Path of the Kabbalah.* Paragon House, 1998.

Sheldrake, Merlin. *Entangled Life.* New York, Random House, 2020.

Sheldrake, Rupert. *A New Science of Life.* Park Street Press, Rochester, 1995.

Sheldrake, Rupert. *Morphic Resonance: The Nature of Formative Causation.* Park Street Press, Rochester, 2009.

Shokek, Shimon. *From Kabbalah and the Art of Being.* London, Routledge, 2001.

Shyfrin, Eduard. *From Infinity to Man.* White Raven Publishing, 2019.

Shulman, Y. David. *The Sefirot.* Northvale, New Jersey, Jason Aronson Inc. 1996,

Shulman, Jason. *Kabbalistic Healing.* Rochester, Vermont, Inner Traditions, 2004.

Solomonick, Abraham. *A Theory of General Semiotics: The Science of Signs, Sign-Systems, and Semiotic Reality.* Cambridge Scholars Publishing, 2015.

Steinsaltz, Adin. *The Thirteen Petalled Rose.* Basic Books, 2006.

Sterling, Peter. *What is Health?* Cambridge, Massachusetts, The MIT Press, 2020.

Stewart, Ian. *Flatterland: Like Flatland, Only More So.* New York, Basic Books, 2002.

Suares, Carlo. *Cipher of Genesis.* Weiser Books, 2005.

Tauber, Alfred. *The Immune Self: Theory or Metaphor?* Cambridge University Press, 1997.

The Zohar, Pritzker Edition, Translation and Commentary by Daniel Matt, Stanford, California Stanford University Press, 2004-2016.

Theise, Neil. *Notes on Complexity: A Scientific Theory of Connection, Consciousness, and Being.* Spiegel & Grau, 2023.

Tirosh-Samuelson, Hava (ed.). *Judaism and Ecology.* Cambridge, Massachusetts, Harvard University Press, 2002.

Tishby, Isaiah and David Goldstein. *The Wisdom of the Zohar.* The Littman Library, 1991.

Turner, J. Scott. *The Extended Organism: The Physiology of Animal-Built Structures.* Cambridge, Massachusetts, Harvard University Press, 2000.

Uexkull, Jacob von. *A Foray into the Worlds of Animals and Humans.* Minneapolis, University of Minnesota Press, 2010.

Waddington, C. H. *Organizers & Genes.* Cambridge University Press, 1940.

Waddington, C.H. *The Scientific Attitude.* New York, Pelican Books, 1941.

Waddington, C.H. *Behind Appearance.* Cambridge, Massachusetts, The MIT Press, 1970.

Waddington, C. H. *Tools for Thought.* New York, Basic Books Publishers, 1977.

Wagner, Gunter (ed.). *The Character Concept in Evolutionary Biology.* Academic Press, 2000.

Waldrop, M. Mitchell. *Complexity: The Emerging Science at the Edge of Order and Chaos.* New York Simon and Schuster, 1993.

Webster, Gerry and Brian Goodwin. *Form and Transformation: Generative and Relational Principles in Biology.* Cambridge, U.K., Cambridge University Press, 1996.

Wilber, Ken. *Sex, Ecology, Spirituality: The Spirit of Evolution.* Shambhala, 2001.

Wilson, Jack. *Biological Individuality.* Cambridge, U.K., Cambridge University Press, 1999.

Wilson, Robert. *Genes and the Agents of Life: The Individual in the Fragile Sciences Biology.* Cambridge, UK, Cambridge University Press, 2004.

Winkler, Gershon. *Kabbalah 365: Daily Fruit from the Tree of Life.* Andrews McMeel Publishing, 2004.

Wolf, Fred A. *Mind into Matter: A New Alchemy of Science and Spirit,* Moment Point Press, 2000.

Wolf, Laibl. *Practical Kabbalah.* Harmony, 1999.

Wolfson, Elliot. *Alef, Mem, Tau: Kabbalistic Musing on Time, Truth, and Death.* University of California Press, 2006.

Zukav, Gary. *The Dancing Wu Li Masters: An Overview if the New Physics.* Bantam New Books, 1979.

Index